Structure and Flow
in Surfactant Solutions

ACS SYMPOSIUM SERIES **578**

Structure and Flow in Surfactant Solutions

Craig A. Herb, EDITOR
Helene Curtis, Inc.

Robert K. Prud'homme, EDITOR
Princeton University

Developed from a symposium sponsored
by the Division of Colloid and Surface Chemistry
at the 206th National Meeting
of the American Chemical Society,
Chicago, Illinois,
August 22–27, 1993

American Chemical Society, Washington, DC 1994

Library of Congress Cataloging-in-Publication Data

Structure and flow in surfactant solutions / Craig A. Herb, editor, Robert K. Prud'homme, editor.

    p.   cm.—(ACS symposium series, ISSN 0097–6156; 578)

    "Developed from a symposium sponsored by the Division of Colloid and Surface Chemistry at the 206th National Meeting of the American Chemical Society, Chicago, Illinois, August 22–27, 1993."

    Includes bibliographical references and indexes.

    ISBN 0–8412–3054–4

    1. Surface active agents—Congresses. 2. Rheology—Congresses.

    I. Herb, Craig A., 1951–   . II. Prud'homme, Robert K., 1948–
III. American Chemical Society. Division of Colloid and Surface Chemistry. IV. American Chemical Society. Meeting (206th: 1993: Chicago, Ill.) V. Series.

TP994.S76   1994
660'.2945—dc20   94–38931
                                                          CIP

The paper used in this publication meets the minimum requirements of American National Standard for Information Sciences—Permanence of Paper for Printed Library Materials, ANSI Z39.48–1984.

Foreword

THE ACS SYMPOSIUM SERIES was first published in 1974 to provide a mechanism for publishing symposia quickly in book form. The purpose of this series is to publish comprehensive books developed from symposia, which are usually "snapshots in time" of the current research being done on a topic, plus some review material on the topic. For this reason, it is necessary that the papers be published as quickly as possible.

Before a symposium-based book is put under contract, the proposed table of contents is reviewed for appropriateness to the topic and for comprehensiveness of the collection. Some papers are excluded at this point, and others are added to round out the scope of the volume. In addition, a draft of each paper is peer-reviewed prior to final acceptance or rejection. This anonymous review process is supervised by the organizer(s) of the symposium, who become the editor(s) of the book. The authors then revise their papers according to the recommendations of both the reviewers and the editors, prepare camera-ready copy, and submit the final papers to the editors, who check that all necessary revisions have been made.

As a rule, only original research papers and original review papers are included in the volumes. Verbatim reproductions of previously published papers are not accepted.

M. Joan Comstock
Series Editor

Contents

Preface xi

BACKGROUND AND THEORY

1. Viscoelastic Surfactant Solutions 2
   H. Hoffmann

2. Theoretical Modeling of Viscoelastic Phases............................... 32
   M. E. Cates

3. Dynamical Properties of Wormlike Micelles: Deviations
   from the "Classical" Picture.. 51
   François Lequeux and Sauveur J. Candau

4. Interesting Correlations Between the Rheological Properties
   of Rod-Shaped Micelles and Dye Assemblies 63
   H. Rehage

RHEOLOGY AND STRUCTURE

5. Microstructure of Complex Fluids by Electron Microscopy 86
   S. Chiruvolu, E. Naranjo, and J. A. Zasadzinski

6. Cryo-Transmission Electron Microscopy Investigations
   of Unusual Amphiphilic Systems in Relation to Their
   Rheological Properties 105
   D. Danino, A. Kaplun, Y. Talmon, and R. Zana

7. Structure of Sheared Cetyltrimethylammonium
   Tosylate–Sodium Dodecylbenzenesulfonate Rodlike
   Micellar Solutions.. 120
   Richard D. Koehler and Eric W. Kaler

8. Rheo-optical Behavior of Wormlike Micellar Systems 129
   Toshiyuki Shikata and Dale S. Pearson

9. **Spinnability of Viscoelastic Surfactant Solutions and Molecular Assembly Formation** 140
    Toyoko Imae

10. **Correlation of Viscoelastic Properties with Critical Packing Parameter for Mixed Surfactant Solutions in the L$_1$ Region** 153
    Craig A. Herb, Liang Bin Chen, and Wei Mei Sun

11. **Effect of Alcohol Chain Length on Viscosity of Sodium Dodecyl Sulfate Micellar Solutions** 167
    Eric Y. Sheu, Michael B. Shields, and David A. Storm

12. **Control of Flow Properties in Surfactant Solutions via Photoreactions of Solubilizates** 181
    Thomas Wolff

13. **Microstructure and Rheology of Nonionic Trisiloxane Surfactant Solutions** 192
    M. He, R. M. Hill, H. A. Doumaux, F. S. Bates,
    H. T. Davis, D. F. Evans, and L. E. Scriven

14. **Rheology of Sucrose Ester Aqueous Systems** 217
    C. Gallegos, J. Muñoz, A. Guerrero, M. Berjano

15. **Rheological Study of Polycrystalline Lyotropic Mesophases in the Cesium n-Tetradecanoate–Water System** 229
    Peter K. Kilpatrick, Saad A. Khan, Akash Tayal,
    and John C. Blackburn

16. **Optical Probe Studies of Surfactant Solutions** 239
    G. D. J. Phillies, J. Stott, K. Streletzky, S. Z. Ren,
    N. Sushkin, and C. Richardson

17. **Flexibility of Cetyltrimethylammonium 3,5-Dichlorobenzoate Micelles** 250
    P. D. Butler, L. J. Magid, and J. B. Hayter

SHEAR EFFECTS

18. **Shear Effects in Surfactant Solutions** 260
    Rhyta S. Rounds

19. **Formation of Nonequilibrium Micelles in Shear and Elongational Flow**.......................... 278
     Shi-Quing Wang, Y. Hu, and A. M. Jamieson

20. **Structure of Complex Fluids under Flow and Confinement: X-ray Couette Shear Cell and the X-ray Surface Forces Apparatus**..................... 288
     Stefan H. J. Idziak, Cyrus R. Safinya, Eric B. Sirota,
     Robijn F. Bruinsma, Keng S. Liang,
     and Jacob N. Israelachvili

21. **Relation Between Rheology and Microstructure of Lyotropic Lamellar Phases**............... 300
     Didier Roux, Frederic Nallet, and Olivier Diat

22. **Steady Shear Behavior of Ternary Bicontinuous Cubic Phases** 306
     Gregory G. Warr and Chih-Ming Chen

POLYMER–MICELLE INTERACTIONS

23. **Polymer–Surfactant Complexes** 320
     Yingjie Li and Paul L. Dubin

24. **Interactions Between Water-Soluble Nonionic Polymers and Surfactant Aggregates** 337
     Josephine C. Brackman and Jan B. F. N. Engberts

APPLICATIONS

25. **Viscoelastic Surfactants: Rheology Control Without Polymers or Particulates** 352
     Gene D. Rose and Arthur S. Teot

26. **Effect of Counterion Structure on Flow Birefringence and Drag Reduction Behavior of Quaternary Ammonium Salt Cationic Surfactants**.................... 370
     Bryan C. Smith, Lu-Chien Chou, Bin Lu, and
     Jacques L. Zakin

27. **Interfacial Rheology of β-Casein Solutions** 380
     Diane J. Burgess and N. Ozlen Sahin

INDEXES

Author Index **397**

Affiliation Index **398**

Subject Index **398**

Preface

ALTHOUGH THE RHEOLOGY OF SURFACTANT SOLUTIONS has been the focus of industrial and academic research for much of this century, only during the past decade has a quantitative understanding of the underlying phenomena begun to emerge. This understanding of the causes of flow behavior and the development of reliable models takes us beyond empirical characterization of individual samples and provides the ability to predict the behavior of whole classes of systems. In addition, the rheological behavior of these solutions provides a tool for gaining insight into the dynamic behavior of surfactants and the various self-assembling structures that they can form. Although further advances in our understanding of the rheology of surfactant solutions will continue, we have reached a point at which predictive tools can be successfully applied to the solution of many relevant problems.

Many papers have been published on this topic, and a number of excellent review articles have appeared in journals and as chapters in more general books on surfactants. To date, however, there has not been a comprehensive volume that deals exclusively with the subject. This book assembles for the first time a collection of papers on the rheology of surfactant solutions that range from discussions of theoretical models to a review of some of the practical industrial applications of these systems. Topics covered include the rheology of solutions containing novel surfactants, mixed surfactant systems, and solutions containing polymers and/or other additives. The modeling of viscoelastic surfactant phases in both the linear and nonlinear regimes is reported. The underlying structures responsible for the rheology are covered in many of the chapters, and include discussions of cryo-transmission electron microscopy, small-angle neutron scattering, small-angle X-ray scattering, light scattering, time-resolved fluorescence quenching, NMR spectroscopy, rheo-optical techniques, and optical probe diffusion as methods for studying the link between structure and rheology. The effect of shear fields on solution structure and phase behavior is also considered.

The multidisciplinary nature of the field is reflected in the range of backgrounds represented by the authors of this book. For example, solutions of cylindrical surfactant micelles provide an opportunity for the study of "living polymer" systems. Other surfactant systems form liquid-crystalline phases and "self-assemble" in ways that often mimic biological membranes. The wealth of phenomena and the analogies with other fields of condensed matter physics, physical chemistry, and biophysics makes the research reported here of interest to physicists, pharmaceutical scientists, chemical engineers, polymer scientists, mechanical engineers, and chemists. Almost one-fourth of the chapters include industrial authors representing the surfactant, pharmaceutical, petroleum, personal care, and chemical industries. Contributing authors represent the United States, France, Germany, United Kingdom, Japan, Netherlands, Israel, Spain, and Australia.

The book is arranged in five sections. The first section (Chapters 1–4) as well as the first chapter of each subsequent section (Chapters 5, 18, 23, and 25) provides sufficient review material to serve as an introduction to the topic for people just entering the field. At the same time, these chapters and the remainder of the book contain enough new material to provide a comprehensive update of the international research in progress in this important area of colloid and surface science.

Acknowledgments

The symposium on which this book is based was made possible by the very generous support of Helene Curtis, Inc., Henkel Corporation, Stepan Company, and Witco Corporation. Additional support was provided by Bohlin Instruments, Inc. and Physica USA, Inc. The funding provided by these companies made it possible for us to bring together a large group of researchers including a very significant percentage of the internationally recognized leaders in this field. We thank these donors for their important contribution to the success of the symposium. We also thank Liang Bin Chen, Wei Mei Sun, Michelle A. Long, and Trefor A. Evans for their assistance during the symposium and during the editing of this book.

CRAIG A. HERB
Helene Curtis, Inc.
4401 West North Avenue
Chicago, IL 60639–4769

ROBERT K. PRUD'HOMME
Department of Chemical Engineering
Olden Street
Princeton University
Princeton, NJ 08544–5263

August 31, 1994

Background and Theory

Chapter 1

Viscoelastic Surfactant Solutions

H. Hoffmann

Physical Chemistry, University of Bayreuth, 95440 Bayreuth, Germany

A review is given on rheological data of viscoelastic surfactant solutions. The viscoelastic properties in the discussed systems are due to entangled threadlike micelles. In such solutions that have this microstructure the zero shear viscosities depend strongly on conditions like the charge density of the micelles, the salt and cosurfactant concentration, and the chainlength of the surfactant. For a 1% solution the viscosity can vary between one and 10^6 mPas. Over extended concentration ranges the viscosities show simple power law behaviour ($\eta^0 = (c/c^*)^x$ with x = 1.5 - 8.5). The largest values for x are observed for systems with charged unshielded micelles and the smallest values for shielded or neutral systems which are close to the L_1/L_α-phase boundary. The various exponents are explained on the basis of different scission mechanisms as proposed by M. Cates. Many viscoelastic surfactant solutions show simple Maxwell behaviour.

In various areas of detergency, in particular in cosmetics, it is of importance to control the flow behaviour and the viscoelastic properties of fluids (*1*). Furthermore it is often necessary to prevent particles or oil droplets that have been dispersed in the aqueous phase from sedimenting or from creaming. All these objectives can be achieved with surfactants, which organize themselves into supramolecular structures. Such systems have viscoelastic properties. They are optically transparent, look alike and they contain only a few percent of surfactants or sometimes other additives. The networks which can exist in these solutions can, however, have a different origin and morphology. In this symposium we are mainly concerned with networks from long cylindrical micelles. These micelles have been described as worm-, thread- or rodlike in the literature. These terms were chosen in order to express that the micelles have some intrinsic flexibility and that the micelles cannot be visualized as being stiff if their contour length is longer than their persistence length. The persistence length depends, however, on many different parameters like charge density of the micelles, excess salt concentration, chainlength of the surfactant and so on and can vary a lot. Values for the persistence length have been determined which range from 100Å to a few 1000Å. So, to describe cylindrical micelles as stiff or flexible depends very much on the conditions and one's own liking. In this article the terms worm-, thread- or rodlike will be used interchangeably. Viscoelastic systems from entangled rodlike micelles can be made from practically every surfactant. They can be prepared from ionic surfactants in combination with strongly binding counter-ions by mixing cationic and anionic surfactants or by mixing surfactants with cosurfactants. The controlling parameter for the sphere rod transition is the packing parameter for a surfactant. In the article the rheological properties of some of these systems will be described. As will be seen, their rheological behavior depends very much on the conditions and can vary from system to system even though all of them are in the entangled region. Nevertheless, it will be possible to draw some general conclusions. At the end of the article a viscoelastic system will be discussed with a completely different behavior. It has a yield value while all the systems that are formed from entangled cylindrical micelles have a finite structural relaxation time and a well defined zero-shear viscosity.

Surfactant solutions with globular micelles always have a low viscosity. The theoretical basis for the viscosity η of solutions with globular particles is Einstein's law according to which the viscosity is linearly increasing with the volume fraction Φ of the particles.

$$\eta = \eta_s(1 + 2.5\Phi) \tag{1}$$

This volume fraction is an effective volume fraction and takes into account hydration of the molecules. In extreme situations like for nonionic surfactants or block-copolymers the effective volume fraction can be two to three times larger than the real volume fraction. But even under such situations the viscosity of a 10% solution is only about twice as high as the viscosity of the solvent. This is still true if nonspherical aggregates are present as long as they do not overlap which means as long as their largest dimension is shorter than the mean distance between the aggregates.

On the other hand many surfactant solutions can be very viscous at low concentrations of about 1%. From this result alone, i.e. that a dilute surfactant solution is much more viscous than the aqueous solvent, one can already conclude that the surfactant molecules in this solution must have organized themselves into some kind of a supramolecular three-dimensional network.

The viscosity of such a system depends very much on parameters like the salinity, the temperature and the cosurfactant concentration (2). When any one of these parameters is increased one often observes a maximum in the viscosity. In detergency, the parameter salinity is usually used to adjust the viscosity for a particular application. At the maximum the viscosity of the solution can be many orders of magnitude higher than the viscosity of the aqueous solvent. Solutions that are that viscous have usually also elastic properties because the zero shear viscosity η° is the result of a transient network, which can be characterized by a shear modulus G° and a structural relaxation time

$$\eta^\circ = G \cdot \tau_s \tag{2}$$

In a single phase micellar solution the network is normally composed of cylindrical micelles and the shear modulus is determined by the number density of the entanglement points

$$G^\circ = \nu \cdot kT \tag{3}$$

The networks of entangled cylindrical micelles can be made visible by cryo-electron microscopy. Beautiful micrographs of such networks have been published by Talmon et al. (4). While such pictures show us clearly the shape of the micelles and even their persistence lengths, they do not reveal their dynamic behavior. The viscosity is however always the result of structure and dynamic behavior, in this particular case of the structural relaxation time. This time constant depends strongly on many parameters and can vary many orders of magnitude for the same surfactant when, for instance, the counterion concentration is changed (5). This is demonstrated in a plot of the viscosity of cetylpyridiniumchloride (CPyCl) against the concentration of sodiumsalicylate (NaSal) in Figure 1. With increasing NaSal the viscosity passes over a maximum, then through a minimum and finally over a second maximum. Results like this have been obtained on many systems by various groups. The data clearly express that the viscosities of micellar solutions can behave in a seemingly complicated way. The complicated pattern is the result of different existing mechanisms for the relaxation of stress.

It is thus not possible to predict the viscosity for an entanglement network of cylindrical micelles. The viscosity of a 1% solution can be anywhere between the solvent viscosity and 10^6 times the solvent viscosity.

A schematic sketch of an entanglement network is shown in Figure 2. It is generally assumed that the effective network points which are responsible for the elastic behavior are due to entanglements (6). This might, however, not always be the case in all systems. The network points could also have their origin in adhesive contacts between the micelles or in a transient branching point which could be a many armed disklike micelle (7,8). Some experimental evidence for both possibilities have been observed recently. The entangled thread- or wormlike micelles have a typical persistence length which can range from a few hundred to a few thousand Å, and they may or may not be fused together at the entanglement points.

The cylindrical micelles have an equilibrium conformation in the networks. The micelles constantly undergo translational diffusion processes, and they also break and reform. If the network is deformed or the equilibrium conditions are suddenly changed, it will take some time to reach equilibrium again. If the network is deformed by a shear stress p_{21} in a shorter time than it can reach equilibrium, it behaves like any solid material with a Hookean constant G°, which is called the shear modulus and we obtain the simple relation

$$p_{21} = G° \cdot \gamma \qquad (4)$$

(γ: deformation). If on the other hand the network is deformed slowly it behaves like a viscous fluid with a zero-shear viscosity $\eta°$ and the shear stress is given by

$$p_{21} = \eta° \cdot \dot{\gamma} \qquad (5)$$

where $\dot{\gamma}$ is the shear rate.

A mechanical model for a viscoelastic fluid consists of an elastic spring with the Hookean constant G° and a dashpot with the viscosity $\eta°$ (Maxwell-Model). If this combination is abruptly extended the resulting stress relaxes exponentially with a time constant which is given by $\tau_s = \eta°/G°$.

The zero-shear viscosity is thus a parameter that depends on both the structure of the system, because G° is determined by the entanglement points (G° =vkT), and the dynamic behavior of the system through τ_s. Both quantities are usually determined by oscillating rheological measurements (9). From the differential equations for the viscoelastic fluid, the equations for the storage G' and the loss modulus G" can be derived.

$$G' = \frac{\hat{p}_{21}}{\hat{\gamma}} \cdot \cos\delta = G° \cdot \frac{\omega^2\tau^2}{1+\omega^2\tau^2} \qquad (6)$$

$$G'' = \frac{\hat{p}_{21}}{\hat{\gamma}} \cdot \sin\delta = G° \cdot \frac{\omega\tau}{1+\omega^2\tau^2} \qquad (7)$$

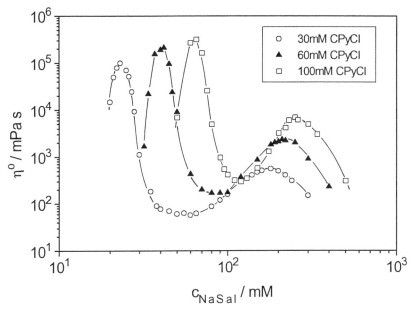

Figure 1. Double log plot of the zero shear viscosity η^0 for three CPyCl concentrations against the NaSal concentration.

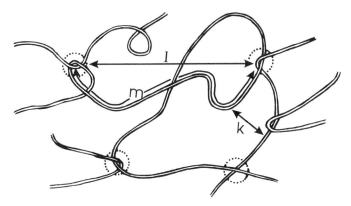

Figure 2. A schematic drawing of an entanglement network from long cylindrical micelles. Note the different length scales: k is the meshsize and l the mean distance between knots.

$$|\eta *| = \frac{\sqrt{G'^2 + G''^2}}{\omega} \qquad (8)$$

$$\lim_{\omega \to 0} |\eta^*(\omega)| = \lim_{\dot{\gamma} \to 0} \eta(\dot{\gamma}) = G^\circ \cdot \tau \qquad (9)$$

Here $\hat{p}_{21}$ is the amplitude of the shear stress, $\hat{\gamma}$ the amplitude of the deformation, δ the phase angle between the sinusoidal shear stress and the deformation and $|\eta^*|$ is the magnitude of the complex viscosity.

Many rheological results of viscoelastic surfactant solutions can be represented over a large frequency range with a single structural relaxation time and a single shear modulus G° as is demonstrated in Figure 3a. Fluids with such a simple rheological behavior are called Maxwell fluids (*10*). However there are other situations in which the rheograms are very different. This is shown in Figure 3b. In this situation there is no frequency independent modulus and the results cannot be fitted with one G° and one τ_s. In such situations the shear stress after a rapid deformation can usually be fitted with a stretched exponential function (*11*).

$$p_{21} = p_{21}^o \cdot e^{-(t/\tau)^\alpha} \qquad (10)$$

We note from Figure 3b that G'' is increasing again with ω. This increase can be related to a Rouse mode of the cylindrical micelles. On the basis of a recent theoretical model, the minimum of G'' can be related to the ratio of the entanglement length l_e and the contour lengths l_k of the cylindrical micelles in the semidilute region (*12*).

$$\frac{G'}{G''_{min}} = \frac{l_e}{l_k} \qquad (11)$$

Different Viscoelastic Solutions

All solutions with entangled threadlike aggregates are generally more viscous than the solvent. In this article the increase of the viscosity with the surfactant concentration is of general interest. To put this problem into a wider perspective it is interesting to look at the behavior of other non surfactant systems. Entangled rodlike micelles are complicated objects to deal with. They break and reform and the length of the aggregate and hence their molecular weight is certainly going to change with the total concentration of the surfactant. However in the entangled region there is no easy method from which we can determine the mean length of the rods. It would be ideal if we would have a method from which the number density of end caps could be determined. It is absolutely certain that the size distribution of the rods is going to effect the time constants for the disentanglement processes. It is thus a good basis for the evaluation of the data on the surfactants to compare their rheological data with the rheological results on other systems. Of interest in this context are systems which are used as thickeners in formulations, like water soluble polymers and polyelectrolytes. Such molecules do not change their molecular weight with concentration and they might thus serve as a reference for our surfactant results. In Figure 4 zero-shear viscosities of hydroxyethyl cellulose (HEC) and sodium-poly-styrene sulfonate (Na-pSS) and two surfactants are shown in a double log plot. For all systems we observe in this representation straight lines for the viscosities. The viscosities follow a power law behavior of the type $\eta \propto (c/c^*)^x$ where x has different values for the different

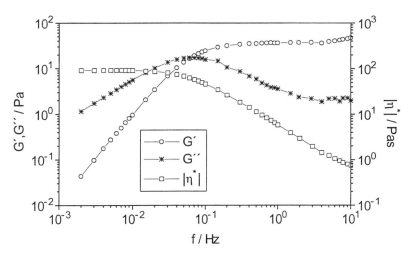

Figure 3a. Rheological data for a solution of 100 mM CPyCl and 60 mM NaSal. The solution behaves like a Maxwell fluid with a single modulus G^0 and a single structural relaxation time.

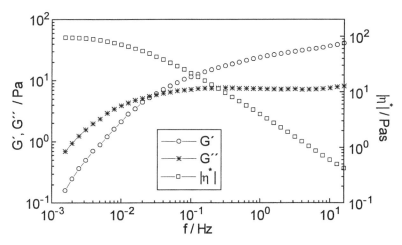

Figure 3b. Oscillating rheogram for a solution of 100 mM tetradecyl-, dimethyl-amineoxide ($C_{14}DMAO$) and sodium dodecylsulfate (SDS) 8:2 + 55 mM hexanol (C_6OH). Note the difference to the rheogram in Figure 3a. There is no plateau value for G' and G" does not pass over a maximum as it does for a Maxwell fluid. G' rises after the crossover with G" with a constant slope of about 0.25 for two decades in frequency.

systems. The slope for the poly-electrolyte is very small even though they are charged and have about the same molecular weight as the hydroxyethyl cellulose. We find that for small concentrations ($c < 0.5\%$) the viscosity of the polyelectrolyte is higher than for the uncharged polymer. The exponents for the surfactants are somewhere in between the exponents for the polymers and the polyelectrolyte. We also note that the viscosities of the surfactant and HEC-solutions rise rather abruptly at a characteristic concentration c^*. This concentration c^* marks the cross-over concentration from the dilute to the semidilute solution. For HEC we have nonoverlapping coils for $c < c^*$ while for $c > c^*$ the coils do overlap and form an entangled network.

Ionic Surfactants

Most single chain ionic surfactants form globular micelles in aqueous solutions. This is the result of the large repulsive interaction of the charged headgroups (*13*). As a consequence of this mutual repulsion an ionic surfactant occupies a large area at an interface whether this is a micellar interface or a bulk interface. Solutions with such surfactants have, therefore, low viscosities. Rodlike micelles in such surfactant solutions are formed when the charge density in the micellar interface is shielded. This can be accomplished by adding excess salt. In such conditions high salt concentrations are usually required for the sphere rod transition to occur (*14*). However, when salts are used with a counterion that can bind strongly to the ionic surfactant on the micellar interface, then only little salt is needed for the formation of rodlike micelles. Such counterions are usually somewhat hydrophobic and they act as small surfactant molecules or cosurfactants. Typical representatives are substituted benzoates, or benzensulfonates or trihaloacetates (*15*). Micelles of such surfactants are weakly charged because of the strong binding. In many combinations the micelles have a dissociation degree of less than 10 percent while ionic surfactants with hydrophilic counterions have usually a dissociation degree of 30 percent and more.

Such surfactants do form rodlike micelles at low surfactant concentration of less than 1% of surfactant. While the charge density on the micellar interface is low in these systems the intermicellar interaction in these systems is still large enough for the system to show a prominent scattering peak in scattering experiments (*16*). Rodlike micelles are also formed with surfactants that have two chains and hydrophilic counterions. For such combinations the correct spontaneous curvature at the micellar interface is determined by the large required area of the two chains. All these surfactant solutions become rather viscous and viscoelastic with increasing concentration. Some results of such systems are plotted in Figure 5. We note the remarkable result that all systems show a concentration region in which the slope of the viscosity against the concentration in the double log plot is the same and about 8, which is very high by any comparison. We also note that the viscosities rise rather abruptly from the viscosity of the solvent and within a small transition concentration follow the scaling law $\eta \propto (c/c^*)^x$. The exponent for all shown systems is the same even though the chemistry of the systems is very different. We have a double chain surfactant (*17*), a perfluorochain surfactant (*18*), two single chain surfactants (*15*) and one system that consists of a zwitterionic surfactant which was charged by adding an ionic surfactant (*19*).

Obviously the detailed chemistry of the systems is of no importance. The common feature of all the systems is that they form rodlike micelles and the rods are charged. Obviously the power law exponent must therefore be controlled by the electrostatic interaction of the systems. The slope of the systems with 8.5 ± 0.5 is much larger than the slope of 4.5 ± 0.5 which is expected for a large polymer molecule, which does not change its size with concentration. Qualitatively we can, therefore, understand the data by assuming that the rodlike micelles continue to grow for $c > c^*$. Indeed, it was shown by MacKintosh et al. that charged rodlike micelles beyond their overlap concentration should show an accelerated growth (*20*). The details for this

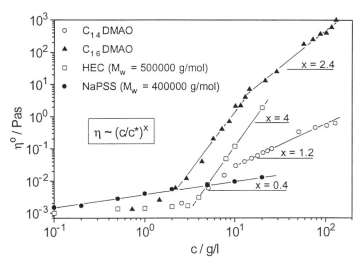

Figure 4. Rheological data for different systems. Double log plot of η^0 against concentration. All systems show power law behavior with different exponents.

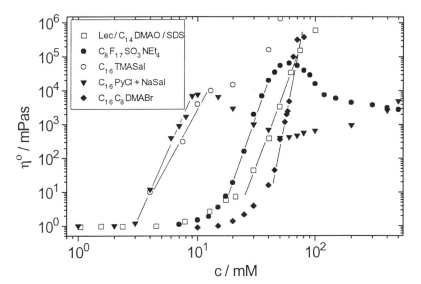

Figure 5. Double log plot of the zero shear viscosity against the concentration for several surfactant systems which are charged: mixture of lecithin, $C_{14}DMAO$ and SDS (Lec/C_{14}DMAO/SDS); tetraethyl-ammonium perfluoro-octanesulfonate ($C_8F_{17}SO_3NEt_4$), cetyl-, trimethyl-ammonium salicylate (C_{16}TMASal); equimolar mixture of cetyl-pyridinium chloride and sodium salicylate (C_{16}PyCl+NaSal); hexadecyl-, octyl-, dimethyl-ammonium bromide ($C_{16}C_8$DMABr).

Note that all the systems show the same power law exponent in a limited concentration region above c*.

situation, where the viscosities are controlled by the behavior of charged micelles, are given in a contribution from Leguex and Candau in this volume (21).

Charged Systems with Excess Salt

The results in Figure 6 for CPyCl + NaSal show that the high power law exponent is sometimes even observed in the presence of excess salt if the total concentration of surfactant and salt is low enough. When CPyCl is combined with NaSal in a stoichiometric ratio the rodlike entangled micelles are formed from CPySal, and excess NaCl is in the bulk solution. The high slope for the viscosity on this system is only observed in the relatively low concentration region between 3 and 10 mM. For higher total concentrations the viscosity decreases and finally increases again. Historically, this observation proved to be of special importance for the theoretical understanding of the processes in entangled rodlike systems and was the starting point for more studies (3). It could be imagined that the maximum and the minimum of the viscosities could be the result of a change in the micellar structure in the system. Detailed rheological oscillating measurements on this system revealed, however, that for concentrations that are above the first maximum the frequency dependent rheological data are determined by a single shear modulus and a single relaxation time, and the zero-shear viscosity is simply the product of these two parameters (equation 2). Furthermore, it was found that the shear modulus increases smoothly with concentration in the different concentration regions and follows the same scaling behavior. It turned out that the seemingly complicated dependence of the viscosity on the concentration - a behavior which is not found for polymer systems - is a result of a complicated behavior of the structural relaxation time. The relaxation time - concentration curves also show breaks in the slope at the same concentration as the viscosity. From the two observations it was concluded that the mechanism for the structural relaxation time can be controlled by different processes. It was assumed that in the concentration range with the high slope the relaxation time is as for polymers entanglement controlled, while after the first break it is kinetically controlled. It was also obvious from the data that there are even different mechanisms in the kinetically controlled region. It was then when the term kinetically controlled viscosities was coined (3, 22). This term was meant to illustrate a situation that occurs when the entangled network, upon deformation by shear, can relax faster to an undeformed state by breaking and reformation processes than by reptation processes as in polymers. That this was indeed the case could be proven by temperature jump measurements which showed that the micellar systems relax with about the same time constant when the system is forced out of equilibrium by either a change of temperature or deformation (23). Somewhat later M. Cates introduced the term living polymers to describe the situation in which the structural relaxation time is completely or partly controlled by kinetic processes (24). In retrospect it is also worth noting that in the first temperature jump experiments it was found that the structural relaxation time and the chemical relaxation time were practically identical. This is not necessarily the case as it was later shown by J. Candau (25). It is possible that the structural time constant is affected by the kinetics but not identical to the kinetic time constant. The resulting relaxation process depends very much on the conditions of the system, which will be expressed in the different power law exponent for the viscosity.

The different relaxation mechanisms for different concentration regions are also very obvious in experimental results in which the surfactant concentration for CPyCl is kept constant and the NaSal concentration is varied (Figure 1). With increasing NaSal concentration one observes for the zero-shear viscosity first a maximum which is followed by a minimum and then a second maximum. Again, detailed oscillating measurements showed that the structural relaxation time follows the viscosity, and the shear modulus, which is a measure of the density of the entanglement points, is the same in the different regions. The different viscosities are thus a result of a different dynamic behavior and not of the structure. These conclusions which were made on the basis of rheological measurements were recently confirmed by cryo-TEM measurements on such systems. Beautiful micrographs of the entangled threadlike micelles were obtained for all four concentration regions and it turned out that no difference in the appearance of the micellar structure could be detected (*4*). It is actually very remarkable that no large differences are noticeable on first view on the micrographs. At least the persistence length of the micelles in the different concentration regions should be somewhat different. For NaSal concentrations before the first viscosity maximum the micelles are obviously highly charged because the Sal/CPy ratio is less than one, while at the minimum of the viscosity the micelles are completely neutral, and at the second maximum they carry a negative charge. The persistence length should have a contribution from the charge density. The reverse of charge of the micelles with increasing NaSal can be proven by mobility measurements of emulsion droplets which are made from the surfactant solutions.

The power law behavior in the three regions is very different as is demonstrated in a plot of the viscosity against the concentration for the situations in which the viscosity is at the first and second maximum and at the minimum. The result show the three exponents 8, 1.3 and 2.5 (Figure 6). The exponent 1.3 is extremely low even for a surfactant system. No theoretical explanation is available at present for this low exponent even though it has been observed for several different systems, which vary completely in chemistry and conditions. The low exponent, therefore, again seems to represent a general behavior of a situation that is controlled by fundamental physics. Similar situations are encountered with zwitterionic and nonionic systems and a qualitative explanation will be given when the different mechanism for the relaxation behavior is discussed.

Zwitterionic Systems

Viscosities for several alkyldimethylamineoxides are given in Figure 7. The data again show power law behavior over extended concentration regions (*7*). Some of the results show breaks in the double log plots of the viscosity concentration curves. This is an indication that even uncharged systems can undergo a switch of the relaxation mechanism when the concentration is varied. At the lowest concentration region in which a power law behavior is observed, the slope is the highest and close to the slope that is observed for polymers. This might however be pure coincidence because we should expect to observe a higher slope with surfactants than for polymers if τ_s is entanglement controlled because it is likely that the rodlike micelles become larger with increasing concentration. It is thus likely that even under the conditions where the slope is in the range of the polymers, the dynamics of the system are already influenced by kinetic processes.

For higher concentrations we observe a lower exponent. Obviously for these conditions a new mechanism comes into play that is more efficient in reducing the stress than the mechanism that is operating in the low concentration region. This mechanism might actually still occur but it is now too slow to compete with the new mechanism. A mechanism that comes in with increasing concentration can only become determining when it is faster than the previous one. It is for this reason that the slope of the η^0-c plots can only decrease but never increase. The moduli, which have been determined by oscillating measurements, are given in Figure 8. Again they increase with the same exponent in the different concentration regions. Figure 7 shows that the absolute value of η^0 for $C_{16}DMAO$ and ODMAO differ by an order of magnitude even though the slope is the same in the high concentration region. This is the result of the fact that in a kinetically controlled region the breaking of micells depends very much on the chainlength of the surfactant.

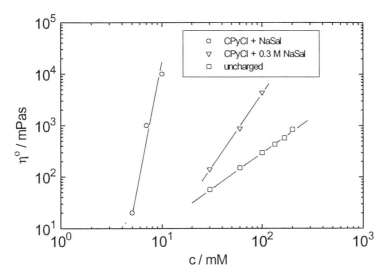

Figure 6. Double log plot of the zero shear viscosity η^o against the concentration of CPyCl for solutions that are at the first maximum (o), the second maximum (∇) and at the minimum ($\square$). For comparison see Figure 1.

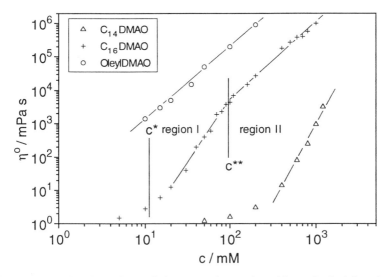

Figure 7. Double log plots of the zero shear viscosities of alkyldimethyl-aminoxides with different chain lengths against the concentration.

Alkyldimethylamineoxides and Cosurfactants

Cosurfactants can be looked at as surfactants with an extremely small headgroup. When they are added to normal surfactants and incorporated into micelles, the average area per headgroup of surfactant and cosurfactant is lower than an area that is occupied by the surfactant on its own. It is for this reason that one usually observes a sphere to rod transition when cosurfactants like n-alcohols with more than five carbon atoms are added to micellar solutions with spherical micelles. If rodlike micelles are already present in the micellar solution, the rods will become larger (*26*). As a consequence of this, the viscosity of the solutions will increase with the cosurfactant concentration. Such a situation is shown in Figure 9 where the viscosity of 100 mM $C_{14}DMAO$ is plotted against the cosurfactant concentration of several cosurfactants. The viscosities are increasing first and then pass over a maximum. The situation is thus similar as in Figure 1 where NaSal was added to CPyCl. The reason for the maximum is the same as in Figure 1. It is likely that the threadlike micelles for $c > c_{max}$ are still increasing in size with the cosurfactant concentration but the system switches from one mechanism on the left side of the η-maximum to a faster mechanism on the right side to reduce stress in the entangled threadlike system. The reason for the switch has probably to do with the rods becoming more flexible and deformable with increasing cosurfactant/surfactant ratio.

The different mechanisms, indeed, become obvious in double log plots of the viscosity against the surfactant concentration as shown in Figure 10 for mixtures of $C_{14}DMAO$ and decanol. The cosurfactant decanol has such a small solubility in water that we can safely assume that all of the added decanol is present in the mixed micelles and the ratio of cosurfactant /surfactant in the micelles remains constant. The plot shows that the viscosities of mixtures which are on the left of the maximum increase with the same slope while the mixture with the highest cosurfactant/surfactant ratio has the lowest slope. Higher mixtures could not be prepared because then the systems are no longer in the single phase region. It turns out now that the slope for the systems with highest cosurfactant/surfactant ratio is the same as for the CPySal system at the viscosity minimum in Figure 1. Even though the viscosities for the CPySal system are higher for the same concentration both systems seem to be controlled by the same mechanism. Actually there are more rheological data on other systems in the literature for which such power law exponents have been observed (*27, 28*). A summary of several systems for which the slope is the same is given in Figure 11. We note the chemistry of these systems is very different, their absolute values are different, yet the slope is the same. We determined the structural relaxation time from electric birefringence measurements and calculated a shear modulus. The results are given in Figure 12. They show that the modulus for a given surfactant concentration is very similar as for other systems. It furthermore scales with the same exponent. The low exponent for the viscosity comes about by the structural relaxation times becoming shorter with increasing concentration $\tau \propto (c/c^*)^{-1}$. The solution to the low viscosity exponent lies, thus, in finding an explanation for the exponent -1 for the structural relaxation time.

Influence of Charge Density on Surfactant Systems

The data in Figure 5 contained one system that was composed of a zwitterionic surfactant, which was mixed with an ionic surfactant. For such systems it is of interest to vary the charge density on the rodlike micelles and observe the influence of this variation on the viscosity. Such data for the ODMAO is given in Figure 13. We note that the sign of the charge plays a role in the amineoxide system. We observe a maximum in the viscosity with increasing charge density. The maximum is, however, much higher for the negatively charged system. We also observe that with SDS as an

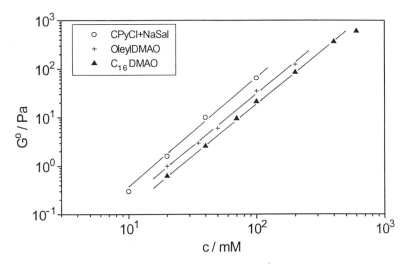

Figure 8. Double log plot of the shear modulus G^o against the concentration for different viscoelastic systems.

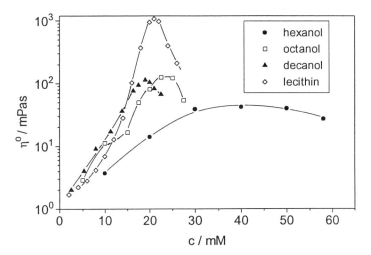

Figure 9. Semilog plot of the zero shear viscosity of a 100 mM solution of $C_{14}DMAO$ against the concentration of cosurfactant. Note that all curves pass over a maximum. The additive lecithin shows qualitatively the same behavior as the n-alcohols.

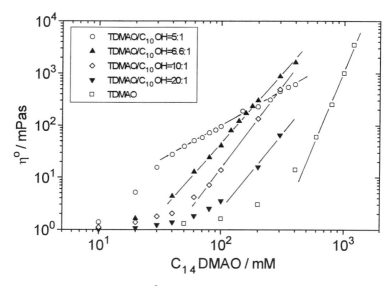

Figure 10. Double log plot of η^0 against the concentration of $C_{14}DMAO$ with different mixing ratios of decanol.

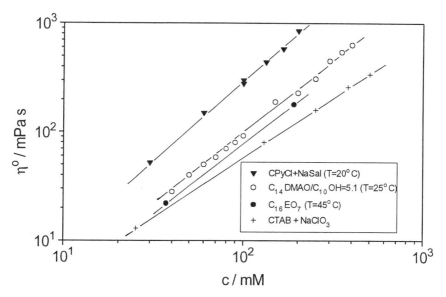

Figure 11. Double log plots of η^0 against the total surfactant concentrations for systems with the same power law exponent of 1.3.

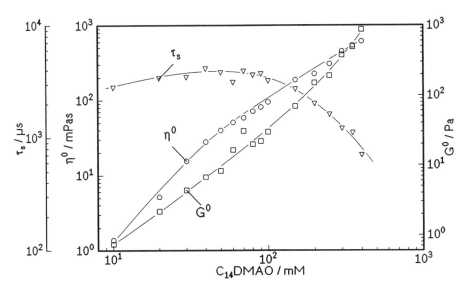

Figure 12. Rheological data for the system $C_{14}DMAO/C_{10}OH = 5:1$. Note that for high concentrations of $C_{14}DMAO$ the structural relaxation time τ_s is decreasing with c^o according to $\tau \propto (c/c^*)^{-1}$.

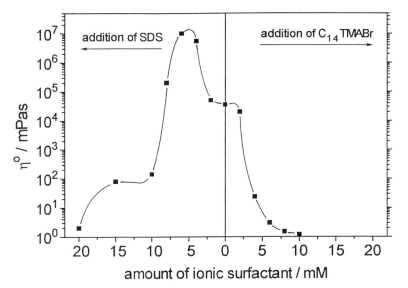

Figure 13. Semilog plot of η^o for a 50 mM solution of ODMAO with increasing amounts of cationic and anionic surfactant.

ionic surfactant the maximum occurs at a mole fraction of around 25% while with the cationic surfactant the maximum occurs already at about 5%. This system has been studied in detail and it was concluded that the maximum was caused by adhesion contacts between the rods (7). These adhesion contacts could slow down reptation processes. With an increase of the charge density, the adhesion contacts finally break and, therefore, the viscosity could become much lower. However, it is noteworthy to mention that even with higher charge densities, the system is still in the semidilute region.

The fact that the alkyldimethylamineoxide micelles can be charged more with anionic than with cationic surfactants before the viscosity decreases strongly has to do with the synergism between the amineoxides and the anionics. Mixtures of these two surfactants are more surface active than the two components by themselves. It can be argued that the synergism is a result from the cationic group of the zwitterionic head being closer to the hydrophobic chain than the negative charge. The system therefore has some cationic character. In combinations of uncharged surfactants that do not have this polarity, the dissymmetry of the sensitivity to charged surfactants disappears (29). This is demonstrated on results of mixtures of alkyldimethylphosphineoxides with SDS and CPyCl in Figure 14. For both combinations the maximum appears at the same mole fraction of charge and the maximum has about the same height. The growth of micellar rods can generally be achieved by the addition of cosurfactants to surfactant solutions (see Figure 9). This is also possible in mixtures of zwitterionic and ionic surfactants as is demonstrated in Figure 15 where the viscosities of mixtures of C_{14}DMAO and C_{14}TMABr are plotted against the added hexanol concentration (26). When the mole fraction of the cationic surfactant X_c is larger than 0.1 we observe viscosities which continously increase up to the phase boundary. The viscosity at the phase boundary is, however, the smaller the higher the charge density is. For the mixtures with the highest charge density the viscosity does not increase at all, and no rodlike micelles seem to be formed in these mixtures. In the presence of excess salt the viscosity is found to increase again as has been shown by several groups on solutions of C_{14}TMABr with different salinities (30).

The results in Figure 13 demonstrate that the sign of the charge in mixed ionic nonionic surfactant systems is of importance for the absolute value of the viscosity. Actually, the viscosities are even different for mixed systems with the same sign of the charge but where the headgroup that carries the charge is different (31). This is shown in Figure 16 where viscosity data of mixed alkyldimethylamineoxides with different positively charged headgroups are shown. In one system the positive charge was introduced by the surfactant C_{14}TMABr while in the other system it was introduced by adding HCl which transforms the amineoxide to the ammonium salt. Both cationic systems are thus very similar. They have the same chainlength and in the headgroup one CH_3-group is replaced by a OH-group. The two surfactant solutions differ already by a factor of 10. With increasing decanol concentrations the viscosities pass over a maximum and then decrease again. At the maximum the η-values are three orders of magnitude different. Surprisingly, for decanol concentrations above the maximum the viscosities are the same. It is unclear why the two systems have such a different behavior even though they are so similar in their chemical composition. It is conceivable that the differences are related to the distribution of the positions of the charges on the cylindrical micelles. At least two different microscopic distributions are possible. The charges could be randomly distributed or they could be concentrated at the end caps of the rods where a strong curvature would favour their location. In principle the distribution of the C_{14}TMABr in the micelles can be determined from SANS contrast variation measurements. Such measurements have been carried out for mixtures of SDS with C_{14}DMAO (32). Within experimental error random distribution in the mixed surfactants was observed. Since these mixtures behave in many ways very similar to the mixtures of C_{14}DMAO/HCl, it is likely that the last mentioned system also has a random distribution while the mixed C_{14}TMABr system

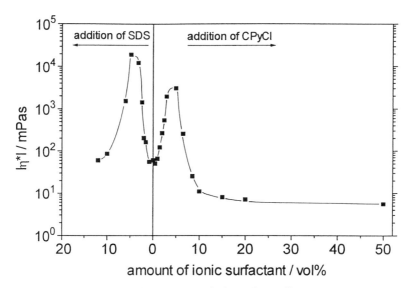

Figure 14. Semilog plot of a 50 mM solution of C_{14}-dimethylphosphineoxide (C_{14}DMPO) with increasing concentrations of cationic and anionic surfactant. Note the difference in the behavior of C_{14}DMPO with respect to ODMAO in Figure 13.

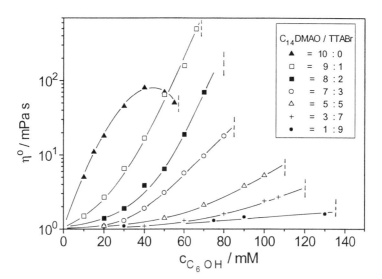

Figure 15. Semilog plot of η^0 for mixed surfactant solutions with a total concentration of 100 mM and various mixing ratios against the cosurfactant concentration. (Reproduced with permission from ref. *26*. Copyright 1993 Academic Press, Inc.)

would then have a nonrandom distribution. A nonrandom charge distribution would result in a different intermicellar interaction than for the random distribution.

The different intermicellar interaction energy between the rods in the two systems is reflected in the results of electric birefringence measurements. The system with C_{14}TMABr shows a single relaxation process, the time constant of which increases continuously with the hexanol concentration. This time constant is identical with the structural relaxation time which determines the zero-shear viscosity. This is evident in the plot of the time constant against η° in Figure 17. The system with HCl on the other hand shows four distinct relaxation processes with increasing hexanol concentration (*31*).

The disappearance of the difference for high decanol concentrations obviously means that the intermicellar interaction for the two systems now becomes the same. If this difference is, indeed, associated with non random charge distribution between the end caps and the cylindrical part of the micelle, the transition in the η°-c curve could indicate the formation of a real crosslinked network in which end caps no longer exist.

Viscosities and other Parameters

The results in the previous sections show that the scaling exponent for the viscosities can vary between 1.3 and 8 and depends on specific conditions of the system. The question, therefore, arises whether it is possible to predict the correct exponent from other experimental results. In order to be able to compare the rheological results with some other results, we show here some results of other techniques on the discussed systems. In Figure 18 light scattering data of some of the investigated systems are shown. The scattering data for CPySal at the composition of the minimum of the viscosities are compared with those of C_{14}DMAO with different molar ratios of decanol. For all systems we observe in the overlap regions a linear decrease of log R_θ with log(c). The slope of these curves is determined by the scaling theory of polymers. For good solvents we should expect a value of -0.3, while for θ-solvents we should observe a value of -1 (*33*). One of the observed slopes is close to -1 while the CPySal system has a slope of -0.4. We thus find no strict correlation between the exponents of the scattering data with the exponents of the viscosity data. The light scattering data with a different mixing ratio of C_{14}DMAO and C_{10}OH have, however, different scaling exponents in the semidilute region. The different scaling exponents for the viscosities of those systems are, thus, reflected in the scaling exponents of the scattering data. The scattering data reflect the structure and we may ask whether the differences of the systems are also reflected in the translational diffusion coefficients in the semidilute region. Some data are given in Figure 19. With the exeption of CPySal indeed we observe the largest slope in the semidilute region for the systems that have the lowest exponent for the viscosities. The dynamic light scattering data are, thus, consistent with the static scattering data. We find that the semidilute C_{14}DMAO/decane surfactant systems behave under these conditions like polymers in θ-solvents. Similar results as those described were found by Kato et. al. for the nonionic systems $C_{12}E_5$ and $C_{16}E_7$ (*34*) and (*35*). It should be noted that as was observed by Kato for $C_{16}E_7$ the maximum of the scattering intensity and the minimum of the collective diffusion constant does not occur at the overlap concentration c*which divides the dilute from the semidilute range and from which the viscosity begins to increases but at a higher characteristic concentration c** at which the scaling exponent changes from a value of 3.5 to a value of 1.3. It is interesting to note in this connection that Kato et. al. determined also self diffusion constants for the surfactants in their systems. For c < c** they observed results which were similar to the results that have been found by other groups and other systems. For giant cylindrical micelles one usually observes a strong decrease of D_s with concentration (*35*). Usually the data can be fitted with power laws with an exponent of 1.5. For c > c** Kato observed, how-

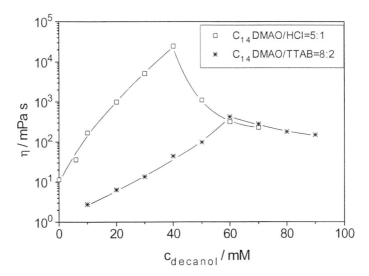

Figure 16. Semilog plot of η^0 against the concentration of the cosurfactant $C_{10}OH$ for 200 mM solutions of $C_{14}DMAO$ which were charged by adding 40 mM HCl or by replacing 40 mM of $C_{14}DMAO$ by $C_{14}TMABr$.

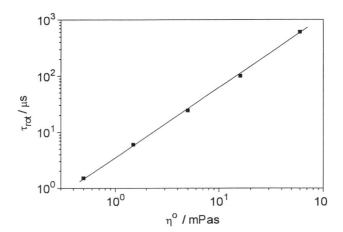

Figure 17. Double log plot of the rotational time constant τ_{rot} as measured by the electric birefringence method against η^0. (Reproduced with permission from ref. 26. Copyright 1993 Academic Press, Inc.)

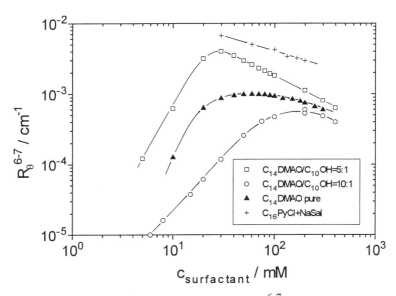

Figure 18. Double log plot of the Rayleigh factor R_θ^{6-7} for forward scattering (6°-7°) against the surfactant concentration.

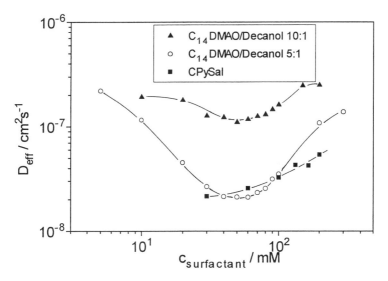

Figure 19. Double log plot of the collective diffusion constant D_{eff} in C_{14}DMAO/decanol mixtures and CPySal against the concentration.

ever, a change in the general behavior and observed an increase of D_s with the surfactant concentration. The relevance of this behavior for the micellar structures and the structural relaxation time will be discussed in the section on theoretical models. Interesting data on the various diffusion processes in CPySal systems and the underlying laws have also been published by Nemoto (*36*).

The existence of rodlike micelles is determined by the packing parameter of surfactants in a micellar aggregate. This parameter is reflected in the value of the interfacial tension of a micellar solution against a hydrocarbon (*37*). High interfacial tensions ($\sigma > 0.2$ mN/m) mean strongly curved interfaces and low interfacial tensions mean bilayer type structures ($\sigma < 0.1$ mN/m). Interfacial tension results of some of the systems for which viscosity data were discussed are given in Figure 20. The result for the CPyCl + NaSal system shows that the minimum of the viscosity coincides with the minimum of the interfacial tension. We also find a continuous decrease of the interfacial tension in the alkyldimethylamineoxide system with the decanol concentration. The smallest slope in the exponent corresponds with the lowest value in the interfacial tension for which single phase solutions could be prepared. It thus seems clear that the interfacial tensions can be related to the exponents of the viscosity. Systems with a low exponent are close to the phase boundary L_1/L_α while high exponents are found for systems that are far from this phase boundary. At present there is no good quantitative theory available to interrelate interfacial tensions with scaling exponents of viscosities. The presented data suggest, however, that such a relation exists. Obviously when the systems approach the phase boundary the systems can undergo fluctuation between rod and disklike micelles. It is then conceivable that entanglement networks are transformed to networks with branching points which consist of disklike micelles.

Viscoelastic Systems with a Yield Value

The different viscoelastic solutions that have been described so far have one feature in common: They all have a finite structural relaxation time. This means that a network that is deformed by stress will always relax within a finite time to an undeformed state. This means that the solutions have a well defined zero-shear viscosity and that for very small shear rates or oscillating frequencies the viscosity reaches a constant value. Particles or bubbles that are dispersed in the solution will therefore always sediment or rise and after a day or so we will see again a clear solution that is free of bubbles or dispersed particles. On this criteria the situation can be quite different for another type of viscoelastic solution that will be described now. It can also look perfectly clear and may have about the same elasticity as the normal viscoelastic solution (*38*). When we handle containers with the two fluids we might actually not notice a difference. However, when small bubbles are dispersed in this solution they will not rise to the top, not in a day, not in a month, and not in a year (Figure 21). In oscillating measurements we find that the storage modulus is independent of frequency and is always larger than the loss modulus (Figure 22).

If we slowly increase the shear stress in a stress controlled rheometer we find that the solution in the cell does not begin to flow until we reach the yield stress value (Figure 23). For small shear stress values the new system behaves like a real crosslinked network, while the previous viscoelastic solution behaved like a temporary network. Obviously the microstructures which exist in both cases must be very different. The shear modulus in the two types of networks has also a different origin. In the normal viscoelastic network the modulus is given by entropic forces, while in the real network it is determined by electrostatic interactions. This is demonstrated in Figure 24 where the shear modulus is plotted against the square root of the added salt concentration for two viscoelastic systems. In the first situation it is independent of salinity while in the second situation the modulus breaks down with the salinity. The difference in the behavior is also obvious in strain sweep measurements. While the

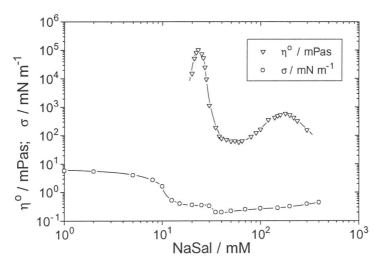

Figure 20. Comparison of the zero shear viscosity η^0 for the system CPyCl + NaSal with the interfacial tension measured against decane.

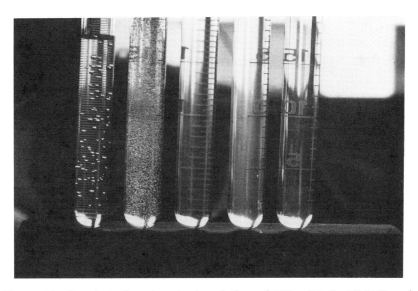

Figure 21. Samples of a viscoelastic solution of 100 mM $C_{14}DMAO$ and $C_{14}TMABr$ with a mixing ratio of 9:1 and 200 mM hexanol with various amounts of KCl. Note that the solutions with the trapped bubbles have a yield value. (Reproduced with permission from ref. 44. Copyright 1994 VCH Verlagsgesellschaft mbH.)

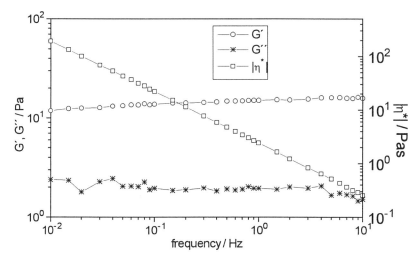

Figure 22. Rheogram of $|\eta^*|$, G' and G" against ω for a viscoelastic solution with a yield value. (Reproduced with permission from ref. 39. Copyright 1994 Academic Press, Inc.)

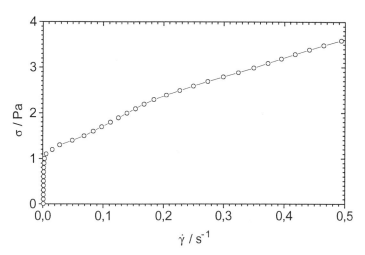

Figure 23. Plot of the shear stress σ against the shear rate $\dot{\gamma}$ to demonstrate the yield value.

entropy elastic solutions can be stretched by 100% before the modulus decreases, the energy elastic solutions can only be deformed by about 10% before the modulus decreases. Above this strain the system begins to flow. The yield value is therefore approximately one tenth of the value of the shear modulus.

The energy elastic system had a composition of 90 mM tetradecyldimethylamine oxide, 10 mM tetradecyltrimethylammoniumbromide and 200 mM hexanol in water. As mentioned it looks perfectly transparent and is a single phase. However, the system is not in the L_1-phase. The surfactant solution without the hexanol is a normal micellar solution of globular micelles. With increasing cosurfactant concentration the globules grow into cylindrical micelles and the solution becomes viscous. At about a concentration of 100 mM hexanol the solution becomes turbid and a two phase system. At a somewhat higher hexanol concentration the system is clear again and a single phase. At this state the surfactants are assembled in single and multilamellar vesicles as is shown in a freeze fracture diagram in Figure 25. The electron-micrograph shows that at the concentration of the sample the vesicles are more or less densely packed. Because of their charge, the bilayers furthermore repel each other and the vesicles can not pass each other in shear flow without being deformed. It is for this reason that the system has a yield value. When necessary, the charges on the vesicles can also be reversed by replacing the cationic surfactant by an anionic system (*39*).

Theoretical Models

All studied surfactant solutions follow the same qualitative behavior. The viscosity begins to rise abruptly at a specific concentration which is characteristic for the surfactant and which is the lower the longer is the chainlength of the surfactant. This specific concentration c* can be associated with the transition from the dilute to the semidilute regime. For c < c* we have isolated nonoverlapping cylindrical micelles while for c > c* the micelles do overlap and a network is formed. In principle this network can be an entanglement network like it is assumed in polymer solutions, or it can be a connected network in which the micelles are fused together or are held together by adhesive contracts. All types of network have been proposed and it is conceivable that all three different types can indeed exist (*40*). In the theoretical treatment of such networks it is usually assumed that the cylindrical micelles are "wormlike", i. e. their contour length l_k is larger than their persistence length l_p and the micelles are, therefore, considered flexible. While this can be the case for some systems - some micelles in this investigation might actually belong to this category -, it is certainly not the situation for the binary surfactant systems of this investigation. This can be unambiguously concluded from several experimental results. The OD-MAO and the HDMAO have very low cross-over concentrations. If we would have a similar situation as for polymers, we would expect to find for c < c*coiled micelles in the dilute state. This is, however, not the case. By electric birefringence measurements and by dynamic light scattering measurements we find rodlike micelles that have around c* the same length as the mean distance between the micelles. This shows that the cylindrical micelles for ODMAO are fairly stiff and have persistence lengths of several 1000 Å. In this context it should also be mentioned that the electronmicrographs which have been obtained from viscoelastic surfactants show cylindrical micelles with $l_p \propto 1000$ Å. Experimentally we find that the rodlike micelles are rather stiff, and at the same time we find that the viscosity is increasing abruptly when c > c*. For stiff rods this should not be the case even when we allow a further growth of the rods for c > c*. For many systems with rodlike particles it has been shown that the rotational time constant for the rods is very little effected at c*, and the solutions do not become viscoelastic for c > c*. We have to assume therefore that other interactions than just hard core interactions between the rods must exist and become effective when c > c* and must be responsible for the formation of the network. It is conceivable that the rods form adhesive bonds when they form contacts or that they ac-

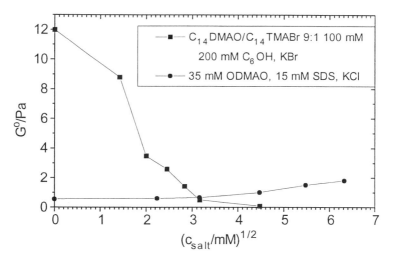

Figure 24. Comparison of the behavior of the shear modulus in two viscoelastic systems against salinity. The modulus in the systems with entangled rodlike micelles is insensitive to salt concentration while in the system with charged bilayers the modulus breaks down with increasing salt concentration. (Reproduced with permission from ref. 44. Copyright 1994 VCH Verlagsgesellschaft mbH.)

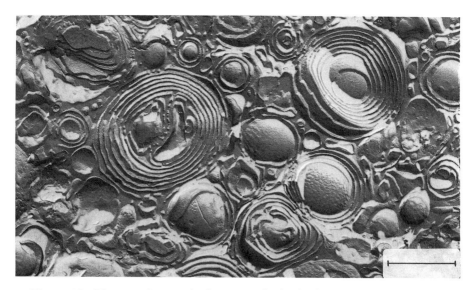

Figure 25. Electronmicrograph that was obtained with the freeze fracture method. The bar represents 1 μm.

tually form a connected network from fused rods as has been proposed by M. Cates (*41*). In such situations different types of networks have to be differentiated, namely saturated and interpenetrating networks. In the first case the mean distance between the knots or entanglement points is equivalent to the meshsize. In the second case the distance between knots can be much larger than the mean meshsize between neighbouring rods.

The viscosities for c > c* rise abruptly and can be represented by a power law behavior of $\eta \propto (c/c^*)^x$ with x > 3.5. The power law exponent for the shear modulus is about 2.3. The behavior in this concentration region is the same as the one that has been treated in detail by M. Cates et al. (*42*). For this situation the structural relaxation times which are measured in this range are affected both by reptation and by bond breaking processes and can be given by $\tau_s = (\tau_b \cdot \tau_r)^{1/2}$. M. Cates treats three different kinetic mechanisms, a scission mechanism in which rods break and two new end caps are formed. In the reverse recombination step the two rods have to collide end on in order to fuse into a new rod. In the second mechanism end caps collide with a rod and in a three armed transition state form a new rod and a new end cap. In the third mechanism, which Cates calls a bond intercharge mechanism, two rods can collide and can form, through an intermediate four armed transition state, two new rods. It is obvious that all three mechanisms can lead to a release of strain. At first glance it seems difficult to decide which of the mechanisms is the more effective one even though the microscopic situation is quite different in the solution. Mechanism three, for instance, does not assume the presence of end caps in the solution. For this mechanism we could then have a fused connected network. Even for such an assumption different possibilities exist. It is conceivable that the network points are locally fixed and the points cannot slide or, as has been assumed recently, that they can slide (*40*). The three different mechanisms lead to somewhat different power laws for the kinetic time constant on the concentration. But for all of the mechanisms the result is $\tau_n \propto (c/c^*)^x$ with x between 1 and 2. Mechanism 3 is probably less likely in systems where c* is low and the rods are rather stiff. This would make it difficult to the linear regions to diffuse and to collide. For this argument it is, therefore, likely that mechanism 2 or 3 is effective in the more concentrated region I of the pure alkyldimethylaminoxide solutions (Figure 7).

For both the ODMAO and $C_{16}DMAO$ we note that the slope in the double log plot changes suddenly at a particular concentration which we call c**. For both linear regions we find the same scaling law for the shear modulus while the power exponent for the relaxation times changes from one to zero. Since we find that the modulus increases with the same law on the concentration, it can be concluded that the structure of the viscoelastic solution is the same in both concentration region. The change in the slope must then be caused by a new mechanism that becomes effective for c > c**. The independence of the structural relaxation time on the concentration makes it likely that dynamics in region II (Figure 7) is controlled by a pure kinetically controlled mechanism and that a reptation process after a break of the coils or a contact between rods is no longer possible. This situation has not yet been treated theoretically. M. Cates mentions, however, that there might be situations where the reptation loses on importance. It is conceivable that we found experimental evidence for such a situation. The more effective mechanism in region II with respect to region I could be the bond interchange mechanism. It could also be that reptation effects in region II are still present and that the independence of τ_s on the concentration is a result of a compensation of the change of the entanglement time constant and the kinetic time constant. Such a possibility seems, however, fortuitous and is, therefore, unlikely. In any case, the data clearly show that a new mechanism becomes effective in region II for the release of strain which is even more effective than the one which operates in region I.

For the mixtures of $C_{14}DMAO$ with $C_{10}OH$ and the CPySal system at the point of zero charge we find the extremely low power law exponent of 1.3 (Figure 11) for the viscosities and an exponent of -1 for the structural relaxation times (Figure 12). The exponent of about 1.3 has also been observed by J. Candau et al. (28) on another system. A detailed explanation has, however, not yet been given. In the following an explanation will be presented for the first time for this low exponent. For the systems for which we find this low exponent, we assume that the cylindrical micelles are indeed very flexible and the expression "wormlike" is, therefore, a very good description of the situation. In such a situation the persistence length would be much shorter than the contourlength between two neighbouring entanglement points.

Furthermore, the persistence length should be independent of the concentration. The diffusion of the rods can, therefore, be described by a constant diffusion coefficient. In order that two arms can collide they have to diffuse a distance x. In order for two neighbouring rods to undergo a bond exchange process they have to diffuse at least the average distance x between two arms. The time constant τ_D for the diffusion should thus be $\propto x^2/D$. Since the meshsize x decreases with the square root of the concentration we finally obtain

$$\tau_s \propto \frac{1}{D \cdot c} \tag{12}$$

This single mechanism explains the experimental data very well. We conclude, therefore, that for situations where $\eta \propto (c/c^*)^{1.3}$ the viscosity is controlled by a diffusion controlled bond interchange mechanism. The absolute values of the viscosities and the time constants for such situations might still vary from system to system because the persistence length of the systems should depend on the particular conditions of the systems. With decreasing chainlength the persistence length should decrease and D increase. For such situations we would expect to find the lowest activation energies for the viscosity. A somewhat different but still similar explanation for the $\tau \propto \Phi^{-1}$ dependence could be based on the assumption that the threadlike micelles are connected or fused together at crosslinks. These crosslinks points could be visualized as disklike micelles from which the threadlike micelles extend. In other words, the transient intermediate species, which are assumed in the various bond intercharge mechanisms, are now assumed to be stable. In the extreme situation all endcaps could be connected, and no free end caps are present in the system. The resulting network could be in the saturated or unsaturated condition. We can then assume that the crosslink point can slide along the threadlike micelles. The sliding along the contour length can be described as a one-dimensional diffusion process with a concentration independent diffusion coefficient. A knot can be dissolved if two network points meet on their random walk. We can further assume that the structural relaxation time would be determined by this random walk process. For a saturated network the mean distance between the knots would depend with $l \propto \Phi^{-1/2}$ on the volume fraction, and the time constant for the diffusion process across this distance should thus be like in equation 12.

We note that this model leads to the same concentration dependence as the previously described diffusion controlled bond intercharge model. Both models, thus, seem to be consistant with the experimental data for which $\eta \propto (\Phi/\Phi^*)^{1.3}$. For both models reptation processes are no longer necessary for the release of strain. The two models can probably be differentiated on the basis of results of self diffusion coefficients for the surfactant molecules. For a connected network a surfactant molecule would be in a similar situation as in a L_3-phase. For such phases it has been shown that the self diffusion coefficient of a surfactant molecule becomes independent of the surfactant concentration (43). For the connected threadlike micelles we would thus expect to find similar behavior. The measurements of Kato et. al. show, however, that the self diffusion of surfactants becomes faster with concentration as does the structural relaxation time. These results indicate that the networks under the θ-condition are not connected and the structural relaxation time is indeed given by a diffusion limited bond interchange mechanism.

Conclusions

Rheological data are reviewed for aqueous viscoelastic surfactant solutions with entangled threadlike micelles. The viscosity of these solutions can be represented by power laws of the form $\eta \propto (c/c^*)^x$ in which the power exponent varies between $x = 8.5$ to 1.3 for different systems. The exponent does not seem to vary continuously but seems to have distinct values. Values that have been observed are 8.5, 4.5, 3.5, 2.3 and 1.3. For many systems the viscoelastic fluids follow simple Maxwell behavior, and the zero-shear viscosities are a product of a single shear modulus and a single structural relaxation time ($\eta^\circ = G^\circ \cdot \tau_s$). For the different systems the shear modulus always follows the same scaling law $G^\circ \propto (c/c^*)^{2.3}$. The distinct values of the power law for the viscosity are, thus, the result of distinct and different mechanisms for the decay of the structural relaxation times.

The exponent 8.5 is observed for charged micelles when the charge is not shielded by excess salt. The high exponent is probably due to growth of the rodlike micelles in the entanglement region above c^* and a pure reptation mechanism for the structural relaxation time. For the lower exponents the structural relaxation time seems to be controlled by a combination of reptation and disentanglement processes as has been proposed by M. Cates. An explanation is given for the very low exponent of 1.3. It is postulated that for systems with this mechanism the structural relaxation time is determined purely by kinetic processes, which are diffusion controlled, and the mechanism is called diffusion controlled bond interchange mechanism. It is assumed that the cylindrical micelles loose their identity at each collision. The structural relaxation time for this case becomes shorter with concentration according to $\tau_s \propto (c/c^*)^{-1}$. The systems which follow this mechanism are close to the L_1/L_α-phase boundary. Data for the alkyldimethylamineoxides show that even uncharged systems can undergo a change in the relaxation mechanism with increasing concentration. A switch in the relaxation mechanism for these systems is also observed when increasing amounts of cosurfactants like n-alcohols (C_xOH) with $x > 6$ are added to the alkyldimethylamineoxides. Cosurfactants increase the size of the micelles but the viscosity passes over a maximum with the cosurfactant concentration.

Literature Cited

1. Hoffmann, H.; Ebert G. *Angew. Chem. Int. Ed. Engl. 27* **1988**, 902-912.
2. Rehage, H; Hoffmann, H. *Molecular Physics* **1991**, *Vol. 74, No. 5*, 933-973.
3. Hoffmann, H.; Löbl, M.;. Rehage, H.; Wunderlich, I. *Tenside Detergents* **1985**, *22, 6*, 290-298.
   Thurn, H.; Löbl, M.; Hoffmann, H. *J. Phys. Chem.* **1985**, *89*, 517-522.
4. Clausen, T. M.; Vinson, P. K.; Minter, J. R.; Davis, H. T.; Talmon, Y.; Miller, W. G. *J. Phys. Chem.* **1992**, *Vol. 96, No. 1*, 474-484.
5. Toshiyuki, Shikata; Hirotaka, Hirata; Tadao Kotaka *Langmuir* **1987**, *3, No. 6* 1081-1086.
   Rehage, H.; Hoffmann, H. *J. Phys. Chem.* **1988**, *Vol. 92, No. 16*, 4712-4719.
   Mishra, B. K.; Samant,S. D.; Pradhan, P.; Mishra, S. B.; Manohar, C *Langmuir* **1993**, *9*, 894-898.
6. Candau, S. J.; Hirsch, E.; Zana, R.; Delsanti, M. *Langmuir* **1989**, *5*, 1225.
7. Hoffmann, H.; Rauscher, A.; Gradzielski, M.; Schulz, S. F. *Langmuir* **1992**, *Vol. 8, No. 9*, 2140-2146.
8. Khatory, A.; Kern, F.; Lequeux, F.; Appell, J.; Porte, G.; Morie, N.; Ott, A.; Urbach, W. *Langmuir* **1993**, *9*, 933.

9. Barnes, H. A.; Hutton J. F.; Walters, K. *An Introduction to Rheology*; Rheology Ser. 3; Elsevier: New York, **1989**.
10. Cates, M. E. Contribution to an ACS Symposium on Polymeric Microemulsions and Polymer-Microemulsion Interactions, New Orleans, August **1987**.
11. Rauscher,A.; Rehage, H.; Hoffmann, H. *Progr. Colloid Polymer Science* **1991**, *Vol. 84*, 99-102.
12. Khatory, A.; Lequeux, F.; Kern F.; Candau, S. J. *Langmuir* **1993**, *Vol. 9, No. 6*, 1456-1464.
13. Jacob N.; Israelachvili, D.; Mitchell J.; Barry W. Ninham *J. Chem. Soc. Faraday Trans. II* **1976**, *72*, 1525.
14. Sumio, Ozeki; Shoichi Ikeda, *J. Colloid Interface Sci.* **1982**, Vol. 87 No. 2.
15. Hoffmann, H.; Rehage, H.; Reizlein K.; Thurn, H. *ACS Symp. Ser. 27,2 Macro-Microemulsions* **1985**, 44-66.
16. Hoffmann, H.; Kalus, J.; Schwandner, B. *Ber. Bunsenges. Phys. Chem.* **1987**, *91*, 99-106.
17. Hoffmann, H.; Krämer, U.; Thurn, H. *J. Phys. Chem.* **1990**, *Vol. 94, No. 5*, 2027-2033.
18. Angel, M.; Hoffmann, H.; Krämer U.; Thurn, H. *Ber. Bunsenges. Phys. Chem*, **1989**, 93, 184-191.
19. Schönfelder, E.; Hoffmann, H.; *Ber. Bunsenges.* **1993** in press.
20. MacKintosh, F. C.; Safran, S. A.; Pincus, P. A. *J. Phys. Condens. Matter* **1990**, *2* 359-364.
    Safran, S. A.; Pincus, P. A.; Cates, M. E.; MacKintosh, F. C. *J. Phys. France* **1990**, 51, 503-510.
21. Leguex, F.; Candau,J. Chapter 3 of this book..
22. Hoffmann, H.; Löbl, M.; Rehage, H. *Progress and Trends in Rheology II* **1988** *Proceedings of the Second Conference of European Rheologists.* *Supplement to Rheologica Acta* **1988**, Vol. 26.
23. Löbl, M.; Thurn, H.; Hoffmann, H. *Ber. Bunsenges. Phys. Chem.* **1984**, *88*, 1102-1106.
24. Drye, T. J.; Cates, M. E. *J. Chem. Phys.* 15. January **1992**, *96, (2)*.
25. Candau, S. J.; Merikhi, F.; Waton, G.; Lemaréchal, P. *J. Phys. France* **1990**, *51*, 977-989.
26. Valiente, M.; Thunig, C.; Munkert, U.; Lenz U.; Hoffmann, H. *J. of Colloid and Interface Sci.* 1. October **1993**, *Vol. 160, No. 1*, 39-50.
27. Imae, Toyoko; Sasaki, Motoi; Ikeda, Shoichi *J. of Colloid and Interface Sci.* February **1989**, *Vol. 127, No. 2*, 511-521.
28. Candau, S. J.; Khatory, A.; Lequeux,F.; Kern, F. *J. Physique IV Proceedings of the workshop on Complex liquid systems*, 6 - 10 July **1992**, Polistena.
29. Hoffmann, H.; Rehage H.; Rauscher, A. *Structure and Dynamics of Strongly Interacting Colloids and Supramolecular Aggregates in Solution*, 493-510.
30. Candau, J.; Hirsch, E.; Zana, R.; Adam, M. *J. Colloid Interface Sci.* **1988**, *112*, 430.
31. Hoffmann, H.; Hofmann, S.; Illner, J. C. in press.
32. Pilsl, H.; Hoffmann, H.; Hofmann, S.; Kalus, J.; Kencono, A. W.; Lindner P.; Ulbricht, W. *J. Phys. Chem.* **1993**, *Vol. 97, No. 11*, 2745-2754.
33. Candau, S. J.; Hirsch, E.; Zana, R. *Phys. of Complex and Supramolecular Fluids*, Ed. by Safran, S. A.; Clar, N. A. **1987**, 569.
34. Ott, A.; Urbach, W.; Langevin, D.; Schurtenberger, P.; Scartazzini, R.; Luisi, P. L. *J. Phys.: Condens. Matter* **1990**, 2, 5907-5912.
    Ott, A.; Bouchaud, J. P.; Langevin, D.; Urbach, W. *Phys. Review Letters*, 22. October **1990**, *Vol. 65, No. 17*.
    Messager, R.; Ott, A.; Chatenay, D.; Urbach, W.; Langevin, D. *Phys. Review Letters*, 4. April **1988**, *Vol. 60, No. 14*.
35. Kato T.; Terao, T.; Tsukada, M.; Seimiya,T. *J. Phys. Chem.* **1993**, *97*, 3910-3917.

36. Nemoto, Norio; Kuwahara, Mitsue *Langmuir* **1993**, 9, 419.
37. Hoffmann, H. *Progr. Colloid Polym. Sci.* **1990**, *Vol. 83*, 16-28.
38. Hoffmann H.; Rauscher, A. *Colloid & Polymer Sci.* 1993, *Vol. 271 No. 4*, 390-395.
39. Hoffmann, H.; Munkert, U.; Thunig, C.; Valiente, M. *J. of Colloid and Interface Sci.* **1994**, *163*, 217-228.
40. Appell, J.; Porte, G.; Khatory, A.; Kern, F.; Candau, S. *J. Phys. II* **1992**, *2*, 1045.
41. Drye, T. J.; Cates, M. E. *J. Chem. Phys.* 1992, *96*, 1367.
42. Turner, M. S.; Marques, C.; Cates, M. E. *Langmuir* **1993**, 9, 695.
43. Ott, A.; Urbach, W.; Langevin D.; Hoffmann, H. *Langmuir* **1992**, *8*, 1045.
44. Hoffmann, H. *Adv. Mater.* 1994, 6, No. 2, 116-129.

RECEIVED August 1, 1994

Chapter 2

Theoretical Modeling of Viscoelastic Phases

M. E. Cates

Cavendish Laboratory, Madingley Road, Cambridge CB3 OHE, England

Viscoelastic surfactant phases usually contain long, semiflexible micellar aggregates; these behave as reversibly breakable polymers. A "reptation-reaction" model, which couples the diffusive disentanglement of the micelles to the kinetic equations describing their reversible breakdown, explains many recent experimental observations. Work on the linear viscoelastic spectrum is reviewed, with an emphasis on extracting quantitative structural and kinetic data from rheological experiments. This can be done by analysing small departures from Maxwell behavior in the spectrum. A rheological theory of nonlinear response in steady shear is also described. The microscopic constitutive equation displays a flow instability, which leads to a plateau in the shear stress under increasing shear rate, and large normal stresses.

Wormlike micelles

In many aqueous surfactant solutions, very long, semiflexible micellar aggregates are formed *(1–4)*. Typical systems involve ionic surfactants in the presence of added salt, though some nonionic surfactants show similar phases. The overlap threshold of the micelles can be very low (a few percent by volume); at higher concentrations the resulting phase is viscoelastic, and strongly resembles a semidilute polymer solution. For example, the osmotic pressure Π and high-frequency shear modulus (plateau modulus) G_o, each vary with volume fraction ϕ approximately as $\phi^{2.25}$. These scaling laws are as predicted by polymer theory *(5)*. However, the micellar "polymers" are intrinsically polydisperse

0097–6156/94/0578–0032$08.00/0

(the chain size distribution is in equilibrium) and break and recombine on experimental time scales. These factors distinguish entangled micellar solutions from conventional polymeric materials.

The chain-length distribution for semidilute self-assembling polymers, such as micelles, can be calculated in a mean-field (Flory-Huggins) approach *(6)*. Let the number per unit volume of chains containing L subunits be $c(L)$. We choose units so that the volume fraction is

$$\phi = \sum_{L=1}^{\infty} L\, c(L) \simeq \int_0^{\infty} L\, c(L)\, dL \tag{1}$$

The packing energy of a chain can be written as $AL + E$, where the term linear in L comes from the body of the chain and the constant term E is that required to create two chain ends (or hemispherical end-caps in the case of micelles). This is typically $5 - 25kT$. Taking units where $kT = 1$ (this is done from now on unless otherwise stated) we have, in the Flory-Huggins approach, the Helmholtz free energy

$$F = \sum_{L=1}^{\infty} c(L)\left[\ln c(L) + E + AL\right] + F_I(\phi) \tag{2}$$

Here the $c \ln c$ terms are contributions to the translational entropy and F_I represents a mean-field interaction term which is assumed to depend only on the total volume fraction.

Minimizing F with respect to $c(L)$ at fixed ϕ yields the equilibrium distribution

$$c(L) \simeq e^{-E}\, e^{-L/\bar{L}} \tag{3}$$

where the mean chain length (or aggregation number) is

$$\bar{L} = \phi^{1/2} e^{E/2} \tag{4}$$

Thus we have a broad distribution of chain lengths with a mean that increases slowly with ϕ, and rapidly with E. These results are independent of the form of the mean-field interaction term $F_I(\phi)$. They can be somewhat modified by polyelectrolyte effects, and also by excluded volume correlations (beyond mean field theory) which we do not discuss here *(7,8)*. It is very difficult to measure $\bar{L}$ directly in experiment; for ordinary polymers one would dilute the system and use light scattering, but this is not appropriate for micelles where $\bar{L}$ varies under dilution.

The main differences between wormlike micelles and conventional polymers lie in the reversibility of the aggregation process. In many systems, reversible breakdown and fusion occur on time-scales comparable to, or more rapid than, those associated with chain diffusion. Hence the dynamics (such as viscoelasticity) are strongly sensitive to the micellar kinetics. Three different model schemes for the kinetics of micellar fusion and breakdown have been analysed *(9–11)*, as follows:

(i) Reversible scission: a chain breaks randomly anywhere along its length. The reverse reaction is end-to-end fusion.

(ii) End interchange: the end of one chain attacks the central part of another. The reverse reaction is the same process.

(iii) Bond interchange: two chains swap a central bond via a four-armed intermediate. The reverse reaction is the same process.

The kinetics can be probed experimentally by the Temperature Jump method *(12)*. In equilibrium we have for the size distribution (rewriting Eq.(3))

$$c(L) = \frac{\phi}{\bar{L}^2} e^{-L/\bar{L}}$$

where $\bar{L} \sim e^{E/2kT}$ is the mean micelle length. (Since we must consider changes in temperature, we suspend temporarily our convention that $kT = 1$.) If T is suddenly changed (e.g. by a capacitor discharge), $\bar{L}$ is altered and $c(L)$ must relax to the new distribution. The relaxation is probed by light scattering which measures some functional $f[c(L)]$ of the size distibution.

Expanding perturbatively for a small jump gives the initial perturbation

$$\Delta c(L, t = 0) \simeq (L - \bar{L}) e^{-L/\bar{L}} \tag{5}$$

To calculate the time evolution, we can substitute this into suitable (mean-field) kinetic equations describing the mechanisms (i–iii) above. For example in reversible scission one finds

$$\dot{c}(L) = -k_1 L c(L) + 2k_1 \int_L^\infty c(L')\, dL' + \frac{1}{2} k_2 \int_0^\infty c(L') c(L - L')\, dL'$$

$$-k_2 \int_0^\infty c(L')\, dL'$$

The first term represents scission of chains of length L; the second, scission of chains of length $L' > L$ (to give a chain of length L); the third (nonlinear) term

is the combination of two smaller chains to make one of length L, and the fourth term is loss of chains of length L by combination with another chain. The rate constants must obey $k_1/k_2 = e^{-E/kT}$ which is fixed by demanding that the steady state solution corresponds to thermodynamic equilibrium. Linearizing the kinetic equation and solving with the appropriate initial condition gives
(10)
$$\Delta c(L,t) = \Delta c(L,0) \exp\left[-2t/\tau_{break}\right]$$

where $\tau_{break} = 1/k_1 \bar{L}$ is the waiting time for a break to occur on a chain of the mean length. The T-jump perturbation relaxes monoexponentially because it is an eigenfunction of the kinetic equation *(10)*. This behavior has been observed unambiguously in experiment *(13)* for chains near the overlap threshold (the signal becomes very small in the true semidilute regime).

End-interchange and bond-interchange have similar kinetic equations. Remarkably, in both cases, T-jump corresponds to a zero mode, and the perturbation $\Delta c(L,0)$ does not decay in time *(11)*. The presence of a zero mode is linked in each case to the fact that the reaction scheme preserves the total number of chains in the system. However, it is unclear why the T-jump experiment should coincide exactly with the relevant mode. All other perturbations to the size distribution relax with a longest decay time of order τ_{break}; this we generically define as the lifetime of a chain of length $\bar{L}$ before undergoing a reaction. The virtual absence of T-jump decay on a time scale of seconds has been observed in the system CTAC/NaSal/NaCl *(14)*, even though in that system τ_{break} is known to be less than 1 second. (This information comes from analysis of the viscoelastic spectrum – see below.) This is consistent with an end-interchange reaction mechanism which was proposed for that system *(14)*.

Linear Viscoelastic Spectra

The most spectacular property of entangled micellar solutions is their viscoelasticity. The relaxation time for stress response to an applied shear is highly variable, but typically about 1 second. Of interest is the function $\mu(t)$, the fraction of shear stress remaining at time t after a small step strain is made. This relaxation function, together with the plateau modulus G_o, determines uniquely all the remaining linear viscoelastic functions *(15)*.

For unbreakable polymers, the relaxation function $\mu(t)$ can be predicted using the reptation model, which applies for chains large compared to the entanglement molecular weight, L_e. In this model, each chain is viewed as trapped in a tube, representing entanglements with its neighbours *(5,15)*. The chain can diffuse only along its own length (reptate); it does so with a curvilinear diffusion constant $D_c \sim 1/L$. (This dependence reflects the fact that the drag on a moving chain increases linearly with its length.) The final escape from the tube is characterised by a time τ_{rep} which varies roughly as $L^2/D_c \sim \phi^2 L^3$. (The exponent for ϕ is slightly model-dependent.) If a step strain is applied at time zero, the chain creeps out of its original tube gradually into a new environment, creating new tube as it goes. The new tube is in equilibrium in the strained environment, and therefore carries no stress. In the reptation model, therefore, $\mu(t)$ is given by the fraction of the original $(t = 0)$ tube through which neither end of the chain has passed before time t.

This can be calculated by casting the problem as a one-dimensional stochastic process (see Doi and Edwards *(15)*). In this process, we imagine a polymer represented as a straight line segment. At time $t = 0$, a particle (representing a tube segment) is released at a random position on the line, and starts to diffuse (Fig.1a). It is killed when it reaches either end of the line (the end points are absorbing boundaries). The resulting stress relaxation function $\mu_L(t)$ for unbreakable polymers of length L is (fairly) close to single exponential. This, along with many other predictions of the reptation model, has been confirmed in experiments on conventional polymers *(15)*. Within the reptation picture it is straightforward to include polydispersity; for an exponential distribution of chain lengths, the result is the weight average:

$$\mu(t) \simeq \phi^{-1} \sum c(L) L \mu_L(t) \simeq \exp[-(t/\tau_{rep})^{1/4}] \tag{6}$$

where from now on τ_{rep} denotes $\tau_{rep}(\bar{L})$. (More sophisiticated treatments of polydispersity yield qualitatively similar results.) This broad relaxation spectrum seems to be approached in some micellar systems, such as CPyCl/NaSal with low amounts of added salt *(16,17)*.

Eq.(6) assumes that the micellar kinetics are so slow as to not influence the reptation process itself, in which case polydispersity is the sole effect of self-assembly. In practice, there is a (much larger) class of micellar systems which show experimentally, instead of (6), behavior approaching a simple Maxwell

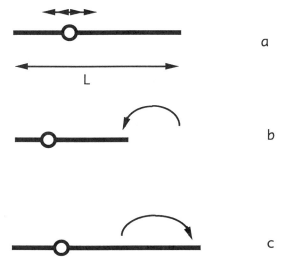

1(a): Reptation model represented as a stochastic process. (b): Motion of a chain end representing chain scission. (c): Motion representing recombination of a chain.

fluid *(16,17,18)*,

$$\mu(t) \simeq e^{-t/\tau} \tag{7}$$

This can be explained by a model in which the reptation process is coupled to the kinetics of micellar breakdown. Chains move by reptation but are also subject to intermittent changes in length, brought about by micellar reactions. The coupled reptation-reaction model can again be cast as a one-dimensional stochastic process. The diffusing particle, representing a tube segment, is launched at time zero on a chain whose length is chosen from the equilibrium distribution; it performs a random walk with the appropriate (curvilinear) diffusion constant. The absorbing boundaries at the chain ends now have their own stochastic dynamics; portions of chain are removed or added (Figs.1(b), 1(c)) with probabilities that are determined from the kinetic equations for the reaction scheme selected. This gives an algorithm from which $\mu(t)$ can be efficiently computed *(9,11)*. The dynamic modulus of the material $G^*(\omega)$ obeys

$$G^*(\omega)/G_o = i\omega \int_0^\infty \mu(t) e^{i\omega t} dt \tag{8}$$

where $G_o \simeq kT\phi/L_e$ is the plateau modulus.

For negligibly slow micellar kinetics ($\tau_{break} \gg \tau_{rep}$), the result (6) is recovered, whereas in the opposite limit of rapid reactions, near-Maxwell behavior, (7), is indeed found. In this limit the Maxwell time τ is given by

$$\tau \simeq \left(\tau_{rep}\tau_{break}\right)^{1/2}$$

for reversible scission and end interchange, and

$$\tau \simeq \tau_{rep}^{2/3}\tau_{break}^{1/3}$$

for bond interchange *(9,11)*. In each case, τ obeys the inequalities $\tau_{break} \ll \tau \ll \tau_{rep}$. The limiting form of a single exponential stress decay can be explained in terms of the fact that the time τ for stress relaxation remains much greater than the breaking time τ_{break}. On a time scale of a few τ_{break}, any given tube segment loses all memory of its initial position relative to the ends of the chain which occupies it. For later times, all tube segments are equivalent, and relax with an equal rate. The resulting relaxation is thus nearly monoexponential, but with departures at short times ($t \leq \tau_{break}$).

Within the model, a quantitative analysis of these short time, high frequency departures may be made. They are most visible when the complex elastic modulus $G^*(\omega)$ is plotted in the Cole-Cole representation: that is, the imaginary part of G^* (the loss modulus, G'') against the real part of G^* (the storage modulus, G') *(15)*. By comparing the departures seen experimentally with those calculated in the model, an estimate of τ_{break} can be deduced from the viscoelastic spectrum. This has been done for a CTAB/KBr system somewhat above the entanglement threshold *(11,19)*; the result of $\tau_{break} \simeq 0.1s$ is very close to that determined independently by T-jump measurements *(13)*. This is strong evidence that the model is correct for the system studied. It shows that the kinetics in this system are of the reversible scission type; as discussed earlier, end-interchange and bond-interchange do not contribute to relaxation of the signal in T-jump. Note that, even for reversible scission, this signal becomes weak in the strongly entangled regime. Hence our viscoelastic method for determining the breaking time has several advantages.

Another interesting estimate can be extracted from the viscoelastic spectrum: that of the mean micellar length. The procedure here is more complicated but again involves analysing the shape of the Cole-Cole plot. In many experiments one finds a turn-up in the plot at the right hand edge (corresponding to very high frequencies). At these frequencies, the chain motion is dominated by fluctuations of short sections of chain within the tube; the resulting contributions to the storage and loss moduli are increasing functions of frequency. The turn-up can be handled theoretically if we take into account the internal relaxation modes of the chain (Rouse or Zimm modes); at high enough frequencies ($\omega \gg \tau_{break}^{-1}$) the physics is the same as for unbreakable chains.

By incorporating these modes within a Poisson renewal model for reversibly breakable chains, Granek *(20)* has shown that the ratio of the minimum of $G''(\omega)$ (at the turn-up point) to its maximum (at the top of the main semicircle) is of order $L_e/\bar{L}$, where L_e is the length of a section of micelle between entanglements. The latter can separately be estimated from the elastic modulus (see above). The resulting estimates for the mean micellar length $\bar{L}$ are typically in the micron range for CTAB/KBr, CTAC/NaSal/NaCl, and CPyCl/NaSal systems *(14)*, but for the most extreme cases (e.g., 15 mM CPyCl + 12.5 mM NaSal *(16)*) estimates for the mean contour length of more than 0.1 mm are obtained. The approach is subject to error from unspecified order unity factors, but no other method is currently available for estimating micellar contour lengths in the entangled regime. As with the breaking time, careful analysis of the linear viscoelastic properties of entangled micelles allow subtle structural and dynamic parameters of these systems to be quantified. For more details on the extraction of length estimates from the Cole-Cole plot, the reader is referred to the paper by Candau et al. in this volume.

In fact, analysis using this method shows that an interesting anomaly in the volume fraction dependence of the mean micellar length $\bar{L}$ arises in certain systems. Theory predicts an increasing trend (Eq.(4)), and this is often indeed seen, for example in the system CTAC /NaSal with a fixed molar ratio in 0.1 M NaCl solution *(14)*. The mean micellar length is in the micron range and increases with concentration of surfactant, as predicted. However, for the same system in 0.25 M NaCl solution, a decreasing trend for $\bar{L}$ is extracted from the fits *(20)*. A possible explanation is that the added salt, as well as inhibiting end-caps (which is expected, since it effectively reduces the head-group size of the surfactant) also promotes formation of labile cross-links between the micelles (for the same reason).

A statistical analysis of the role of crosslinks in micellar systems is given by Drye and Cates *(21)*. These authors distinguished between "saturated" and "unsaturated" networks: in the saturated case, virtually all contacts between chains are converted into links and no entanglements, as such, remain. For an unsaturated network, in contrast, there are many entanglements still present. This should be the first state of the system reached on adding a crosslink promoter (such as salt); the saturated state is likely only close to a phase boundary involving coexistence with excess solvent.

The impact of reversible crosslinks on the reptation-reaction model has been examined by Lequeux *(22)*. His analysis is restricted to the unsaturated case, in which a section of chain between crosslinks can still be considered as confined to a tube. Relaxation again requires the section of chain to break, at a point near enough to a given tube segment that the new end can pass through it before reacting with something else. Lequeux showed that the main effect of cross-linking was through the curvilinear diffusion constant D_c for a chain end. Remarkably, this is enhanced, rather than reduced, by the presence of reversible links:

$$D_c^{eff} \sim 1/\bar{L}_s \qquad (9)$$

where $\bar{L}_s \leq L$ is the mean strand length between crosslinks. The enhancement can be understood in terms of "kinks" or stored length defects. In a linear chain, a kink must travel from one end of the chain to another before a small diffusive displacement of the chain is established. In a network, it has to travel only a distance of order that to the nearest junction. The remainder of the network acts as a stored-length "reservoir" and absorbs the kink.

To incorporate Lequeux's result, we need only make the replacement $\bar{L} \rightarrow \bar{L}_s$ in our previous discussion. (In fact the same applies in the nonlinear viscoelastic theory discussed below.) When crosslinks dominate in the determination of D_c, the viscoelastic spectrum provides a sensitive measure of the cross-link density (just as it provides a sensitive measure of the end cap density, or equivalently $\bar{L}$, when end-caps dominate). The strand length is likely to be a decreasing function of ϕ at fixed salt, which would help explain the anomalous data on CTAC at high salinity mentioned above. Lequeux *(22)* also considers the case when both chain ends and crosslinks are present, and provides an interpolation formula for this intermediate regime.

Nonlinear Viscoelasticity

The strongest challenge to a molecular theory of viscoelasticity lies in the prediction of *nonlinear* behavior; for entangled micelles this can be spectacular in both transient and steady flows *(16,17,23)*. Various studies of transient effects have been made, but only recently have detailed results for the steady-shear response of an entangled micellar solution become available *(17)*. We focus mainly on steady shear in what follows.

We seek a constitutive equation relating the stress tensor $\mathbf{S}$ at time t to the strain history of the sample at earlier times:

$$\mathbf{S}(t) = f[\mathbf{K}(t' < t)]$$

where $\mathbf{K} = \nabla \mathbf{u}$ is the velocity gradient (or rate of deformation) tensor. The deformation of a vector $\mathbf{r}$ due to deformation between times t' and t is written

$$\mathbf{r}(t) = \mathbf{E}_{t't} \cdot \mathbf{r}(t')$$

where the deformation tensor between times t' and t obeys

$$\mathbf{E}_{t't} = \exp\left[\int_{t'}^{t} \mathbf{K}(t'')dt''\right]$$

A constitutive scheme for reversibly breakable chains was proposed in Ref.*(24)*, and has recently been studied numerically *(25)*. We assume that the deformation, which may be large on the scale of a chain, is not so large as to cause mechanical rupture of the micelles. This is likely to be a good model, since for order unity deformations (in the nonlinear range) the chain tension remains of order kT/L_e. The system is overdamped, so elastic energy associated with large length-scale deformations cannot be converted into scission energy. Only when the elastic energy stored locally (within a diameter or two of an incipient breakage) exceeds the thermal energy is any change in scission rates expected. We therefore assume reaction rates are unperturbed by flow; this may have to be modified, however, at very high flow rates.

The (deviatoric) stress tensor for a polymeric system may be written using standard ideas from rubber elasticity theory as *(15)*

$$\mathbf{S} = \frac{15}{4} G_o \left[\mathbf{W} - \frac{1}{3}\mathbf{I} \right]$$

where

$$\mathbf{W} = \langle \mathbf{uu} \rangle$$

is the (tensorial) second moment of the orientational distribution function $P(\mathbf{u})$ for tube segments. This second moment determines not only the stress but the effect of flow on the time evolution of $P(\mathbf{u})$ itself, as shown below.

Consider now the birth and death of tube segments. In the linear region, we have the survival function $\mu(t) = e^{-t/\tau}$ corresponding to a death rate for tube segments $\mathcal{D} = 1/\tau$. By conservation, this is balanced by a birth rate $\mathcal{B} = 1/\tau$. These rates are modified, however, in nonlinear flows. The main effect is called *retraction (15)*. If a random-walk sequence of tube segments is subject to a finite deformation, it increases in length by a factor

$$L'/L = \langle |\mathbf{E} \cdot \mathbf{u}| \rangle_0$$

where the subscript 0 denotes average over an isotropic distribution, $P_0(\mathbf{u}) = 1/4\pi$. This stretching effect is compensated by a fast motion (with time scale $\tau_{rep} L_e/L$) in which the chain shrinks back down its tube to restore the original tube length. To a first approximation, the process can be taken as instantaneous; for a general $P(\mathbf{u})$ it causes tube segments to die at an additional rate

$$v = \frac{1}{L}\frac{\partial L}{\partial t} = \langle |\dot{\mathbf{u}}| \rangle = \langle \mathbf{u} \cdot \partial \mathbf{u}/\partial t \rangle = \mathbf{W} : \mathbf{K}$$

where $\dot{\mathbf{u}} = \mathbf{K} \cdot \mathbf{u}$ implies the last equality. The death rate by retraction depends on the current orientational second moment, $\mathbf{W}$, which in turn depends on the flow history. (The rate v is positive in almost all flows *(15)*.) Treating the two sources of death as independent, we may approximate $\mathcal{B}$ and $\mathcal{D}$ by

$$\mathcal{B} = 1/\tau$$

$$\mathcal{D} = 1/\tau + v(t)$$

These expressions represent a somewhat simplified approximation to a more complex interpolation formula which can be obtained from a more detailed treatment of the stochastic retraction process *(24)*.

To get a constitutive equation, we must find the stress contribution at the present time (t) due to a segment created earlier, at time t'. Such a segment was created isotropically (with probability distribution $P_0(\mathbf{u})$) but has been deformed by the flow. Its new length is $|\mathbf{E}_{t't} \cdot \mathbf{u}|$ and its unit tangent is

$$\mathbf{u}' = \frac{\mathbf{E}_{t't} \cdot \mathbf{u}}{|\mathbf{E}_{t't} \cdot \mathbf{u}|}$$

The mean contribution to $\mathbf{W}$ from this portion of tube, if it survives, is therefore

$$\hat{\mathbf{Q}}_{t't} = \langle \frac{\mathbf{E}_{t't} \cdot \mathbf{u}\, \mathbf{E}_{t't} \cdot \mathbf{u}}{|\mathbf{E}_{t't} \cdot \mathbf{u}|} \rangle_0$$

To calculate the stress at time t, we need only compute $\mathbf{W}(t)$ as the sum over contributions from segments born earlier:

$$\mathbf{W}(t) = \int_{-\infty}^{t} \mathcal{B}(t') \exp\left[-\int_{t'}^{t} \mathcal{D}(t'')dt'' \right] \hat{\mathbf{Q}}_{t't}\, dt'$$

where the exponential factor is the probability of survival between times t' and t. In the fast breaking limit considered here, this survival factor is the same for all tube segments, which means that our constitutive equation for reversibly breakable chains is in fact much simpler than the one for unbreakable polymers *(15)*. (The latter is often simplified using an uncontrolled "independent alignment" approximation *(15)*; there is no need for such a step here.) However, use of the single-exponential memory function means that the model is restricted to flow rates small compared to τ_{break}^{-1}.

For steady flows, the constitutive equation can be solved using standard numerical procedures. The model predicts a shear stress $\sigma = S_{xy}$ that increases smoothly from zero as the shear rate $\dot{\gamma}$ is increased, but then shows a maximum in shear stress $\sigma^* = 0.67 G_o$ at a certain shear rate $\dot{\gamma}_1 = 2.2/\tau$. The decreasing shear stress at $\dot{\gamma} > \dot{\gamma}_1$ suggests that steady shear flow in this region is unstable. However, the predicted decreasing behavior cannot persist to infinite shear rates, and eventually there must be an upturn at flow rates beyond the range for which the model applies (Fig.2). We denote the upper shear rate at which S_{xy} again reaches σ^*, as $\dot{\gamma}_2$. Steady Couette flow in the regime $\dot{\gamma}_1 \ll \dot{\gamma} \ll \dot{\gamma}_2$ can only be supported if the system forms two or more "shear-bands". These are layers of high- and low-shear-rate material (of equal shear stress) which coexist at volume fractions arranged to match the imposed macroscopic shear rate. At the interface(s) between bands, the shear stress, and the component of the normal stress perpendicular to the interface, are both continuous (26).

This flow instability is quite similar to one predicted (though not yet directly observed) in ordinary polymer melts and solutions, where it has been proposed as the origin of the so-called "spurt effect" (26). In pipe flow, the value of σ in the coexisting bands is selected by the interfacial boundary conditions on the normal stress. In cone and plate experiments, however, we expect that, as the shear rate is slowly increased, the flow remains stable up to the maximum shear stress σ^*, this limiting stress fixing the value in the shear-banded region. Thus a horizontal plateau in shear stress should appear at $\sigma = \sigma^* = 0.67 G_o$. Figure 3 shows the predicted shear stress compared with recent data of Rehage and Hoffmann (17) on an aqueous solution of CPyCl / NaSal (100mM/60mM). This well-characterized system has a reasonably good Maxwell spectrum in linear response, indicating $\tau_{break} \ll \tau_{rep}$ as required. The relaxation time $\tau = 8.5$ s and the plateau modulus $G_o = 31.2$ Pa are independently found from linear response measurements, so there are no free parameters in the fit (25). The agreement between theory and experiment is clearly excellent.

The remarkable qualitative behavior predicted by theory, namely saturation of the shear stress with increasing flow rate, seems to be widely observed in entangled micellar solutions (see other papers in this volume). This extreme shear thinning may play a role in several applications (shampoos, etc.) where a gelatinous behavior at low shear must be combined with flow (e.g. in a pump) at higher applied stresses. Note that on theoretical grounds, it is much more informative to plot the shear *stress* (rather than the effective viscosity) against flow rate. For a saturating stress, the conventional viscosity plot shows

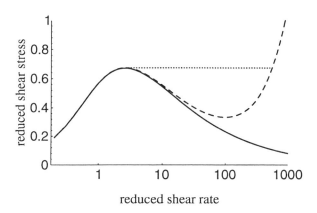

2. Qualitative form of the stress vs. strain rate curve in steady flow. The high shear rate turn-up arises from processes not included in the constitutive equation. Dotted line is the shear-banding plateau (see text).

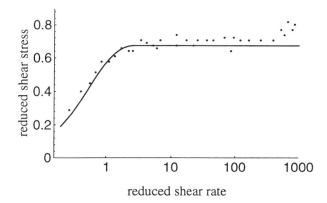

3. Log-linear plot of reduced shear stress σ/G_o against reduced shear rate $\dot{\gamma}\tau$. The solid curve is the prediction of the model with no adjustable parameters. The data is from Ref. *(17)*. Note the apparent turnup at high shear. Theory from Ref. *(25)*.

a slope of -1 on logarithmic axes; but with such a representation it is hard to see whether the shear stress is truly constant or only approximately so. Our model predicts a true plateau for a steady state measurement in which equilibrium is reached. Since the equilibrium involves an inhomogenous flow, this may of course take time to develop.

We have also studied the behavior of the first normal stress difference, $N_1 = S_{xx} - S_{yy}$ (25). Experimentally, this shows a roughly linear increase with shear rate in the plateau region of shear stress. At the upper end of the range, very large normal stresses are obtained compared to those measured for shear rates near $\dot{\gamma}_1$ (where $N_1 \simeq G_o$). The near-linear dependence is quite consistent with the shear banding mechanism, so long as we adopt a model of the high shear phase (of shear rate $\dot{\gamma}_2$) for which the value of N_1 is indeed very large compared to G_o. Well within the banded region, the normal stress will then increase linearly, in proportion to the volume fraction of the high shear phase needed to maintain the imposed overall flow rate.

To model the high shear phase more quantitatively is difficult, and involves extra model assumptions beyond those used above. However, for unbreakable chains, a model has recently been proposed which does predict a form for the shear and normal stresses in the very high shear regime (27). The model involves a dumb-bell type chain which at high shear is constrained to lie in a tube aligned along the flow. (Conventional tube models neglect any fluctuations in the position of the chain about the axis of the tube, and so predict a monotonically decreasing shear stress (28).) We have assumed that the asymptotic results for the modified model, in which transverse fluctuations in chain position are accounted for, can be applied even for reversibly breakable chains (25). With this assumption, we can calculate the high shear branches of the shear and normal stress curves explicitly, and use the results in a fit to the normal stress data (17) (Fig.4). The fit allows us to extract an estimate of the interesting parameter $\bar{L}/L_e$. The experimental data at high shear is somewhat ambiguous, but if we treat the apparent turnup in the shear stress at high shear (Fig.3) as a real effect (signifying the onset of $\dot{\gamma}_2$) we find that $\bar{L}/L_e \simeq 25$. This estimate is in very good agreement with that found from a Cole-Cole analysis of the high frequency linear viscoelastic spectrum, using the method outlined earlier.

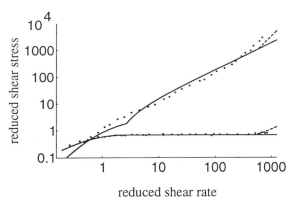

4. Log-log plots of shear and normal stresses, with theoretical fits. Data from Ref. *(17)*, theory from Ref. *(25)*.

Conclusion

The "reptation-reaction" model gives a good account of linear viscoelastic relaxation spectra in several entangled micellar systems. Good estimates of kinetic and structural data can be extracted from the spectra. These include estimates of micellar breakdown rates, and micellar lengths (or strand-lengths in systems where labile crosslinks are more numerous than chain ends). The model also describes the nonlinear flow regime, and results have been presented for steady shear. Using parameters from the linear spectrum, and without further fitting, the results for the shear stress as a function of flow rate are in quantitative agreement with experiment. Those for the normal stress give a tentative estimate of the ratio $\bar{L}/L_e$ which is consistent with that found from the linear spectrum.

In future work we hope to address transient flows in micellar systems. This is potentially very complex in the nonlinear regime; for example in startup, shear-banding effects will dominate the steady response so a correct treatment must allow for the dynamics of the formation of the bands. Moreover, certain types of transient response observed (such as stretched exponential relaxation on cessation of shear *(17)*) have no obvious interpretation within the constitutive equation proposed above; this perhaps suggests that the dependence of micellar kinetics on the strain history cannot be completely ignored.

The entangled micellar phase is, of course, not the only viscoelastic phase of surfactant solutions. Several liquid crystalline phases are strongly viscoelastic; however, their rheology may be controlled by nonequilibrium factors which determine the domain texture. The study of these metastable viscoelastic phases, particularly smectic "emulsions" *(29,30)*, in which multilayer bilayer vesicles arise, is a fascinating field where many advances are likely in the next few years.

Acknowledgments

The author is grateful to his collaborators N. A. Spenley, T. C. B. McLeish, R. Granek and M. S. Turner, whose joint work is reviewed here.

Literature Cited

1. Porte, G. and Appell, J. *J. Phys. Chem.* **1981**, *85*, 2511.

2. Candau, S. J.; Hirsch, E.; and Zana, R. *J. Colloid Interface Sci.* **1985**, *105*, 521.

3. Ikeda, S.; Hayashi, S. and Imae, T. *J. Phys. Chem.* **1980**, *84*, 744.

4. Hoffmann, H.; Kalus, J.; Thurn, H. and Ibel, K. *Ber. Bunsenges. Phys. Chem.* **1983**, *86*, 37.

5. de Gennes, P. G. *Scaling Concepts in Polymer Physics*; Cornell University Press: Ithaca, 1979.

6. Scott, R. L. *J. Phys. Chem.* **1965**, *69*, 261.

7. MacKintosh, F. C.; Safran, S. A. and Pincus, P. A. *Europhys. Lett.* **1990**, *12*, 697.

8. Cates, M. E. *J. Physique* **1988**, *49*, 1593.

9. Cates, M. E. *Macromolecules* **1987**, *20*, 2289.

10. Turner, M. S. and Cates, M. E. *J. Physique* **1990**, *51*, 307.

11. Turner, M. S. and Cates, M. E. *J. Physique II* **1992**, *2*, 503, and references therein.

12. Lang, J. and Zana, R. in *Surfactant Solutions*; Ed. R. Zana, Dekker: New York 1987.

13. Candau, S. J.; Merrikhi, F.; Waton, G. and Lemarechal, P. *J. Physique* **1990**, *51*, 977.

14. Kern, F.; Zana, R. and Candau, S. J. *Langmuir* **1991**, *7*, 1344.

15. Doi, M.; Edwards, S. F. *The Theory of Polymer Dynamics*; Clarendon: Oxford (1986).

16. Rehage, H. and Hoffmann, H. *J. Phys. Chem.* **1988**, *92*, 4217.

17. Rehage, H. and Hoffmann, H. *Molec. Phys.* **1991**, *74*, 933.

18. Shikata, T.; Hirata, H.; and Kotaka, T. *Langmuir* **1987**, *3*, 1081; **1988**, *4*, 354.

19. Kern, F.; Lemarechal, P.; Candau, S. J. and Cates, M. E. *Langmuir* **1992**, *8*, 437.

20. Granek, R. and Cates, M. E. *J. Chem. Phys.* **1992**, *96*, 4758.

21. Drye, T. J. and Cates, M. E. *J. Chem. Phys.* **1992**, *96*, 1367.

22. Lequeux, F. *Europhys. Lett.* **1992**, *19*, 675.

23. Shikata, T.; Hirata, H.; Takatori, E. and Osaki, K. *J. Non-Newtonian Fluid Mech.* **1988**, *28*, 171.

24. Cates, M. E. *J. Phys. Chem.* **1990,** *94*, 371.

25. Spenley, N. A.; Cates, M. E. and McLeish, T. C. B. *Phys. Rev. Lett.* **1993,** *71*, 939.

26. McLeish, T. C. B. *J. Polym. Sci. Polym. Phys. Edn.* **1987,** *25*, 2253, and references therein.

27. Cates, M. E.; McLeish, T. C. B. and Marrucci, G. *Europhys. Lett.* **1993,** *21*, 451.

28. Marrucci, G. and Grizzuti, N. *Gazz. Chim. Ital.* **1988,** *188*, 179.

29. Roux, D. and Cates, M. E. to be published.

30. Hoffmann, H.; Thunig, C.; Munkert, U.; Meyer, H. and Richter, W. *Langmuir* **1992,** *8*, 2629.

RECEIVED July 8, 1994

Chapter 3

Dynamical Properties of Wormlike Micelles
Deviations from the "Classical" Picture

François Lequeux and Sauveur J. Candau

Laboratoire d'Ultrasons et de Dynamique des Fluides Complexes, Unité de Recherche Associée au Centre National de la Recherche Scientifique Numéro 851, Université Louis Pasteur, 4 rue Blaise Pascal, 67070 Strasbourg Cedex, France

The dynamical properties of semi-dilute solutions of worm-like micelles are classically described by a coupled reptation-reaction model that predicts, in particular, scaling behavior to dilution of characteristic rheological parameters. The above model is valid for neutral micelles. However, as the non ionic surfactants do not generally undergo large micellar growth, most experiments were carried out up to now on highly screened micelles. Recently it was observed experimentally that the classical model doesn't apply to systems with an excess of salt and this was interpreted by assuming an intermicellar branching. Also, at low ionic strength and/or large surfactant concentrations, intermicellar orientational correlations affect the rheological behavior. In this paper, we give an overview of the dynamical behavior of worm-like micelles in a large domain of the surfactant-salt diagram, with special emphasis on the effects of branching and of orientational correlations.

Recently, evidence has accumulated about the formation in aqueous surfactant solutions of long flexible cylindrical micelles with a large spread in length and properties resembling those of polymer solutions *(1, 2)*. While the osmotic compressibility and cooperative diffusion constant scale as power laws with the volume fraction of surfactant Φ in the semi-dilute regime, the dynamical properties of these systems are more complex. This is due to the self-assembling nature of the micelles that can be considered as equilibrium polymers undergoing reversible breakdown process. A recent model based on the reptation theory which describes the rheological properties of entangled polymeric chains but including the effects of reversible scission kinetics has been derived by Cates (this book, chapter 2). This model is meant for neutral or highly screened micelles, the dynamical properties of charged micelles being more complex as the electrostatic interactions affect considerably the micellar growth rate *(3, 4)*. Indeed, the experiments performed on systems containing no or little salt showed unambiguously the effect of electrostatic interactions *(2, 5, 6)*. But, unexpectedly, the rheology of the systems containing an excess of salt couldn't be described by the model of linear micelles and it was conjectured that the observed behavior was due to an intermicellar branching *(6, 8, 10)*. The classical model doesn't take into account either the experimental results obtained at high surfactant concentrations, and these deviations are obviously linked to the onset of the orientational correlations.

In this paper, we review recent experimental results concerning the effects of the salt content and of the orientational correlations on the rheological behavior of worm-like micelles. We also propose a quite general schematic diagram [surfactant concentration - salt concentration] describing the different structural features of these systems.

Effects of Surfactant and Salt Concentrations on the Structural Properties of Micellar Solutions

Figure 1 shows a schematic diagram that gathers the experimental observations and some conjectural predictions of these last years. The coordinate axis refer respectively to the surfactant volume fraction Φ and to the molar ratio C_s/C of the salt and the surfactant concentrations. The location of the lines delimiting the different domains in the phase diagram depends strongly on the nature of both the surfactant and the salt and the possibility of observing all the domains for a single system rests in a delicate choice of the surfactant-salt couple. Two domains are of particular interest for the rheological studies.
 - The semi-dilute domain, located between the lines $\Phi^*(C/C_s)$ and $\Phi^{**}(C/C_s)$, characterized by a transient network-like structure.
 - The Onsager domain, located above $\Phi^{**}(C_s/C)$ that corresponds to a regime where the persistence length of the cylindrical micelle is larger than the mesh size of the network, leading to strong orientational correlations.

Within these two domains, the micellar morphology varies considerably upon increasing the salt content.

At very low salt concentration, the micelles are stiffer and smaller but as discussed later, they grow much faster than the neutral micelles.

When the amount of salt is sufficient to screen out the electrostatic interactions, the micelles are linear, their growth is described by the mean field model and the coupled reptation/reaction theory of Cates applies *(1)*.

In presence of an excess of salt, several sets of experimental data suggest that intermicellar branching occurs *(7)*, that can eventually lead at very high salt content to the formation of a saturated network, stable only at sufficiently high surfactant concentrations *(6-10)*.

There are some experimental evidences of phase separation at high salt concentration but it is not clear at the moment whether these phase separations correspond to the formation of a saturated network or to a classical phase separation of polymers in poor solvent conditions. The ordered phases in the upper part of the phase diagram have been observed *(11, 12)*. One also expects to observe in the upper right corner of the diagram a bicontinuous cubic phase.

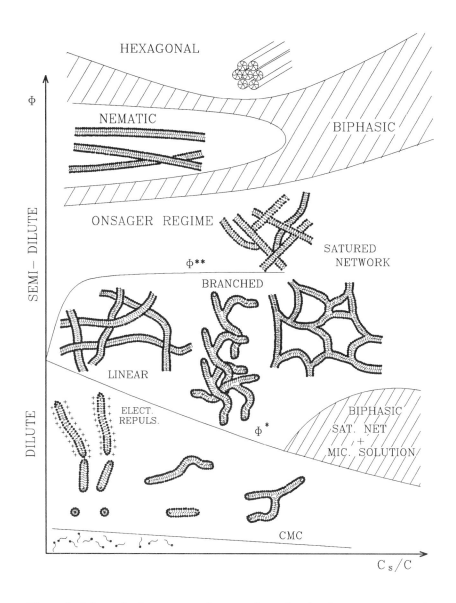

Figure 1. Schematic diagram : surfactant volume fraction Φ versus molar ratio C_S/C of salt over surfactant concentrations

Semi-dilute Regime

Stress Relaxation Function. The main features of the theoretical models describing both equilibrium and dynamic properties of wormlike micelles can be found in the Cates chapter of this book (Chapter 2). Here we simply recall the theoretical results that are needed to discuss our experiments, more specifically regarding the shape of the stress relaxation function and the scaling laws to dilution. Figure 2 shows an example of Cole-Cole plots calculated by Granek and Cates *(13)*.

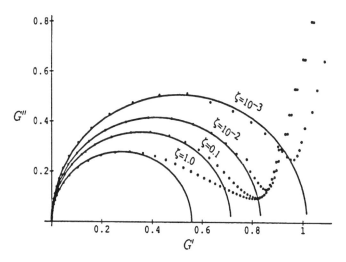

Figure 2. Calculated Cole-Cole plots for the scission-recombination process including the reptation, the breathing and the Rouse regimes for different values of the parameter $\zeta = \tau_{break}/\tau_{rep}$ (as denoted on the plot). $\ell_e/\bar{L} = 0.1$. The lines are semi-circles fitted to the top of dotted curves. (Adapted from ref.13).

At low frequencies the behavior is Maxwellian. Above a frequency of the order of the inverse of the breaking time τ_{break}, one observes a departure from the Maxwellian behavior, the variation of $G''(G')$ being linear with a slope -1. This is followed by the breathing regime with a sharper decay of G''. Eventually there is an upturn associated with the occurrence of the Rouse modes. The minimum of G'' delimits the low frequency range (on the left-hand side of the Cole-Cole plot) that corresponds to modes involving the micellar length and the high frequency range (on the right-hand side of the Cole-Cole plot) controlled by local modes.

This model was extended later on by one of us to the case of branched micelles *(10, 14)*. It was shown that the general features of the stress relaxation remain, provided one replaces the average length $\bar{L}$ of the micelle by a new length $\bar{L}_c$ defined

as
$$\overline{L}_c = \frac{n_2}{n_1 + 2n_3}$$

where n_1 is the concentration of end-caps, n_2 the number density of persistence lengths, proportional to the volume fraction Φ of surfactant and n_3 the concentration of 3-fold connections.

For linear micelles $n_3 = 0$ and $\overline{L}_c = \overline{L} \propto \Phi^{1/2}$.

For saturated networks $n_1 = 0$ and $\overline{L}_c = \overline{L}_s \propto \Phi^{-1/2}$

where $\overline{L}_s$ is the average length of the strand of the network. A determination of $\overline{L}_c$ is provided by the measurements of the minimum G''_{min} of G'' according to the relation *(13)*:

$$\frac{G''_{min}}{G'_\infty} = A \frac{\ell_e}{\overline{L}_c}$$

where G'_∞ is the plateau modulus, ℓ_e the entanglement length and the prefactor A is of the order of 1.

One must remark that the above relation is valid only in the limit where some entanglements remain. For saturated networks ℓ_e has no meaning and one doesn't expect to observe a plateau modulus. Also, for very small τ_{break} (e.g. a small multiple of the entanglement time) a large departure from Maxwell behavior is observed, due to the increasing influence of the Rouse regime at high frequencies (cf. Figure 2). In this regime, the value of G''_{min} has no particular significance and cannot be used to evaluate $\overline{L}_c$.

The analysis of the experimental data, based on a direct comparison with the theoretical Cole-Cole plots calculated by Granek and Cates might be misleading because it involves an adjustable parameter, namely the amplitude of the tube length fluctuations.

The following procedure for obtaining the different parameters characteristic of the micellar systems has been used with some success *(10, 15)*.

First, one determines the ratio $\zeta = \tau_{break}/\tau_{rep}$ where τ_{rep} is the reptation time of a micelle with the mean length $\overline{L}_c$ by a direct superposition of the theoretical Cole-Cole plots calculated by Turner and Cates *(16, 17)* and that disregard non-reptative high frequency effects. To do that, we normalize the data by dividing both G' and G" by G'_{osc}, the radius of the osculating circle of the G' vs G" Cole-Cole plot at the origin. Then we superimpose to the normalized experimental Cole-Cole plots the series of calculated ones with different value of ζ. This procedure is illustrated in Figure 3 relative to a solution of CTAB at a concentration of 0.35 $\underline{M}$ in presence of 1.5 $\underline{M}$ KBr at T = 35°C. From this comparison one obtains the ratio $\zeta = \tau_{break}/\tau_{rep}$. Note the departure in the high frequency range, due to the non-reptative effects.

From the low frequency or low shear measurements one determines the zero shear viscosity η_0. The plateau modulus G'_∞ is obtained by extrapolation of the Cole-Cole data to the horizontal axis. It must be noted that the shallower the minimum of G" in the Cole-Cole plot, the higher the accuracy in the determination of G'_∞. Then the terminal time T_R is determined from the relation $\eta_0 = G'_\infty T_R$. In the asymptotic limit $\tau_{break}/\tau_{rep} \ll 1$, T_R scales like $(\tau_{break}\tau_{rep})^{1/2}$. This scaling was argued on physical grounds *(1)* but it is also found from simulation experiments that provide also the prefactor of the scaling law *(16, 17)*, thus allowing an estimate of τ_{rep}. The significance of this prefactor has been discussed in detail by Granek and Cates *(13)*.

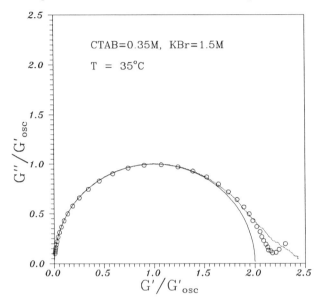

Figure 3. Experimental normalized Cole-Cole plot (open circles) for a CTAB/Kbr solution. The dotted line represent the best fit of the calculated Cole-Cole plots to the experimental data, using the Turner-Cates simulations disregarding the non-reptative effects. This fit is obtained for $\zeta = \tau_{break}/\tau_{rep}$ = 0.06. Also is drawn the osculating semi-circle at the origin (Adapted from ref. 10).

By combining the measures of T_R and ζ, and using the simulation results, one obtains both τ_{break} and τ_{rep}. Finally, the ratio G''_{min}/G'_∞ provides an estimate of the ratio $\ell_e/\overline{L}_c$. The entanglement length can be estimated from the relation :

$$G'_\infty \simeq k_B T/\xi^3 = k_B T/\ell_e^{9/5}$$

where ξ is the correlation length *(13)*.

The values obtained for these different parameters by applying the above procedure to the data of Figure 3 are listed below :

τ_{break}/T_R	τ_{break}/τ_{rep}	η_0	G'_∞	T_R	τ_{break}	G''_{min}/G'_∞	$\overline{L}_c$
0.54	0.06	220 Pa.s	320 Pa	0.69 s	371 ms	0.047	0.67 μm

Scaling Behavior to Dilution. The only well characterized scaling behavior to dilution of the different measured parameters refer to the regime of entangled linear micelles without electrostatic interactions.

In that regime, the growth and the kinetics of the micelles are described by

$$\overline{L}_c \sim \Phi^{1/2} \qquad \tau_{break} = (k_1 \overline{L}_c)^{-1} \sim \Phi^{-1/2}$$

where k_1 is the rate constant per unit time per unit arc length.

Combining the above relationship with the classical scaling laws for polymers, leads to :

$$\eta_0 \sim \Phi^{7/2} \qquad G''_{min}/G'_\infty \sim \Phi^{-7/4}$$

Such a behavior has been observed on several systems *(19, 20)*. But some detailed studies have revealed significant deviations from the above predictions. These are illustrated in Figures 4 and 5 showing the surfactant concentration dependences of respectively η_0, and G''_{min}/G'_∞ for different systems.

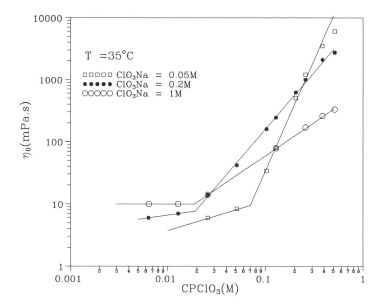

Figure 4. Variations of the zero shear viscosity of CPClO$_3$/ClO$_3$Na solutions as a function of CPClO$_3$ concentration (Reproduced with permission from ref.6. Copyright 1993 Editions de Physique).

Focusing for instance on the behavior of the zero shear viscosity, one notices that the exponent of the power law describing η_0 vs Φ is smaller than 7/2 in presence of an excess of salt. In some systems it becomes larger than 7/2 at very low ionic strength *(6)*. The results obtained at low ionic strength can be understood at least qualitatively by the effect of electrostatic interactions. The main contribution of such an effect to the free energy is the reduction of the effective end cap energy due to the repulsion between surface charges that favors the breaking of the micelles, but to a lesser degree as the concentration is increased *(4)*. Therefore the high values of the exponent obtained for the zero shear viscosity can be explained by an enhanced growth rate. Quantitative comparisons are difficult as the growth rate depends on the energy associated with surfactant packing near the cylinder end caps. The steepest variation of the viscosity with surfactant concentration is obtained for surfactants with an end-cap energy high enough that the micellar growth occurs without addition of salt.

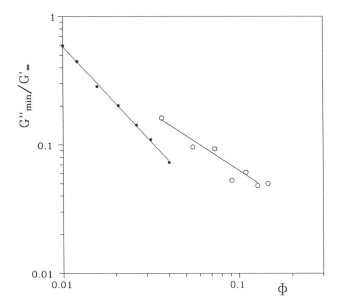

Figure 5. Variations of the ratio G''_{min}/G'_∞ versus surfactant volume fraction
— Open circles : CTAB/1.5 $\underline{M}$ KBr solutions ; T = 35°C. The data are fitted with a straight line with slope −0.88 (Adapted from ref. 10)
— Filled circles : CPyCl/NaSal/0.5 M NaCl solution with [CPyCl]/[NaSal] = 2. T = 25°C. The data are fitted with a straight line with slope −1.44 (Adapted from ref. 19).

It must however be noted that in the absence of salt the semi-dilute domain is very limited, if not suppressed, as the long range electrostatic interactions promote the onset of the Onsager reduced regime at $\Phi \sim \Phi^*$.

In the other limit of high salt content the reduced dependence of the viscosity as a function of the surfactant concentration was attributed to the branching of the micelles. A model based on the coupled reptation/reaction for branched micelles was developed by one of us *(14)*. Again, quantitative fits are difficult in the absence of any information on the respective proportion of end caps and branch points. In the limit where the network is saturated, no reptation process can be invoked. It was suggested that the flow mechanism is based on a process similar to that proposed to describe the rheological properties of the bicontinuous L_3 phase *(18)*, and that involves a sliding of the branch points along "the cylindrical micelles" *(7-9)*.

Concerning the elastic modulus G'_∞ which usually is given by the density of entanglements, it is difficult to relate this parameter to both the density of entanglements and density of connections. At the present stage, we can just infer from the experiments that, for a given surfactant concentration, G'_∞ decreases upon increasing the connection density to eventually vanish as the network saturates.

Finally, the behavior of G''_{min}/G'_∞ shown in Figure 5 upon dilution is also in accord with the theoretical expectations. For the system CPyCl/NaSal with a molar ratio [CPyCl]/[NaSal] = 2 in presence of 0.5 M NaCl which corresponds to the condition where the effect of electrostatic interactions is suppressed, the variation of G''_{min}/G'_∞ with surfactant volume fraction is well described by the theory of linear micelles *(19, 20)*.

On the other hand, for the system CTAB/KBr with an excess of salt the variation of G''_{min}/G'_∞ with Φ is less than expected from such a model *(10)*. Again, this is well taken into account by a volume fraction dependence of $\overline{L}_c$ intermediate between $\Phi^{1/2}$ and $\Phi^{-1/2}$, due to the coexistence of end caps and branch points.

The self-consistency of the above results can be checked by looking to the behavior of the product $\overline{L}_c \, \tau_{break}$. Both parameters $\overline{L}_c$ and τ_{break} can be determined by using the procedure described in the preceding paragraph. This product should be constant upon varying the surfactant concentration and equal to k_1^{-1}. This was indeed observed for the systems CTAB/(1.5 $\underline{M}$ KBr) as shown in Figure 6.

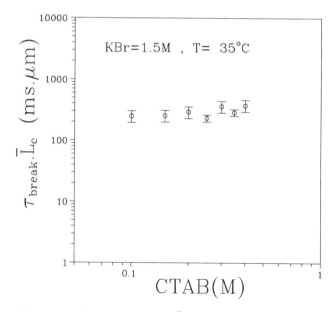

Figure 6. Variation of the product $\tau_{break} \overline{L}_c$ versus surfactant concentration for CTAB/1.5 $\underline{M}$ KBr solutions ; T = 35°C.

Onsager Regime

The signature of this regime is the presence of a peak in the static structure factor *(11, 12)*. Figure 7 shows an example of a scattering curve obtained by small angle neutron scattering on an aqueous micellar solution of $CPClO_3$ (0.05 $\underline{M}$ $NaClO_3$) with a surfactant volume fraction equal to 0.2.

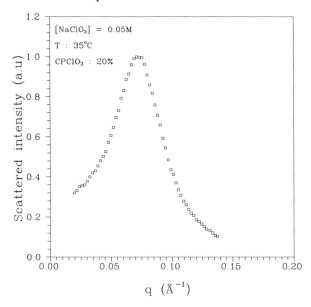

Figure 7. Scattering function obtained by S.A.N.S. on a 20 % $CPClO_3$ solution (Adapted from ref. 21)

In the surfactant concentration range where a peak is observed, the variation of the position of the peak with concentration is consistent with a $\Phi^{1/2}$ power law, thus indicating that the cylinders are highly packed. The peak corresponds to a locally hexagonal packing of cylinders, as inferred from an analysis involving the position of the peak, the diameter of the micelles and the volume fraction *(21)*. The system remains isotropic, even though strong orientational correlations are observed in a rather large concentration domain :

$0.15 \leq \Phi \leq 0.345$ for the $CPClO_3/0.05$ $\underline{M}$ $NaClO_3$ system.

The effect of orientational correlations appears also in the high frequency local motions *(12)*. The typical frequency behavior of the viscoelastic modulus is similar to that observed in the classical semi-dilute regime but with two significant differences :

- the elastic modulus G'_{osc} (corresponding to the osculating semi-circle) decreases upon increasing the surfactant concentration. This effect can be associated with the orientational correlations that decrease the stress-strain susceptibility as in the case of solutions of rod like polymers *(23)*

- the high frequency modes become slower and slower as Φ increases, thus invading the low frequency relaxation, due to an increase of the range of the orientational correlations with the surfactant concentration (cf. Figure 8).

Hence, the evolution of the Cole-Cole diagram with the surfactant concentration clearly shows the role of the orientational correlations, the signature of which is given by the SANS experiments.

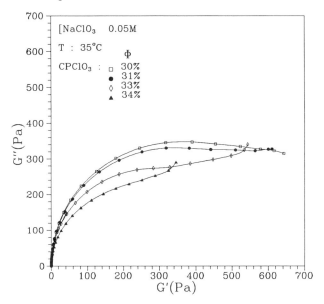

Figure 8. Viscoelastic behavior near the isotropic-nematic transition in the isotropic domain for $CPClO_3/ClO_3Na$ solutions : as the surfactant concentration increases, the radius of the osculating circle decreases and the high frequency modes invade the low frequency half circle region. The $CPClO_3$ volume fractions are denoted in the plot. A phase separation occurs at Φ = 34.5 %.

Conclusions

The versions of the reptation model extended to include the effect of reversible breakage appear to offer a promising framework for understanding the linear viscoelastic measurements on entangled linear or branched worm-like micelles. In particular, they provide an explanation of the general features of the stress relaxation. Other predictions, such as the scaling behavior to dilution of the different rheological parameters determined experimentally are less confirmed experimentally. The data suggest that salt concentration and specific counterion effects are important in determining the dependence of average micellar size and branching degree on surfactant concentration which complicates the interpretation of the measurements. Furthermore, the strong orientational correlations that appear at low salt content and/or high surfactant concentration give rise to local collective modes, thus perturbing the high frequency behavior of the stress relaxation.

From the dynamical experiments, the different micellar structures and organizations encountered can be drawn in a schematic diagram of surfactant concentration-salt concentration, but more structural studies are required to confirm this still rather conjectural diagram.

Literature Cited

1. See for instance : Cates, M.; Candau, S.J. *J. Phys. Condens. Matter*, **1990**, *2*, 6869-6892 and references therein. See also this book chapter 2.
2. Hoffmann, H. this book, chapter 1. See also : Rehage, H.; Hoffmann, H. *Molecular Physics*, **1991**, *74*, 933-973 and references therein.
3. Safran, S.; Pincus, P.; Cates, M.; Mackintosh, F. *J. Phys. (Paris)*, **1990**, *51*, 503-509.
4. Mackintosh, F.; Safran, S.; Pincus, P. *Europhys. Lett.*, **1990**, *12*, 697-702.
5. Delsanti, M.; Moussaid, A.; Munch, J.P. *J. Coll. Interf. Sci.*, **1993**, *157*, 285-290.
6. Candau, S.J.; Khatory, A.; Lequeux, F.; Kern, F. *J. Phys. IV*, **1993**, *3*, 197-209.
7. Porte, G.; Gomati, R.; El Haitamy, O.; Appell, J.; Marignan, J. *J. Phys. Chem.*, **1986**, *90*, 5746-5751.
8. Appell, J.; Porte, G.; Khatory, A.; Kern, F.; Candau, S.J. *J. Phys. (France) II*, **1992** 2, 1045-1052.
9. Khatory, A.; Kern, F.; Lequeux, F.; Appell, J.; Porte, G.; Morie, N.; Ott, A.; Urbach, W. *Langmuir*, **1993**, *9*, 933-939.
10. Khatory, A.; Lequeux, F.; Kern, F.; Candau, S.J. *Langmuir*, **1993**, *9*, 1456-1464.
11. Porte, G. Private communication.
12. Schmitt, V.; Roux, D.; Lequeux, F. Private communication.
13. Granek, R.; Cates, M. *J. Chem. Phys.* **1992**, *96*, 4758-4767.
14. Lequeux, F. *Europhys. Lett.*, **1992**, *19(8)*, 675-681.
15. Kern, F.; Lemarechal, P.; Candau, S.J.; Cates, M. *Langmuir*, **1992**, *8*, 437-440.
16. Cates, M., Turner, M. *Europhys. Lett.*, **1990**, *11*, 681-686.
17. Turner, M.; Cates, M. *J. Phys. (Paris)*, **1990**, *51*, 307-318.
18. Snabre, P.; Porte, G. *Europhys. Lett.*, **1990**, *13*, 641-645.
19. Berret, J.F. ; Appell, J.; Porte, G. *Langmuir*, to be published.
20. Koehler, R.; Kaler, E.; this book chapter 7.
21. Schmitt, V.; Lequeux, F.; Pousse, A.; Roux, D. submitted.
22. Richetti, P.; Roux, D. Private communication.
23. Doi, M., Edwards, S., In *The Theory of Polymer Dynamics*, Clarendon Press, Oxford, **1986**.

RECEIVED July 8, 1994

Chapter 4

Interesting Correlations Between the Rheological Properties of Rod-Shaped Micelles and Dye Assemblies

H. Rehage

Institute of Environmental Chemistry, University of Essen, Universitätsstrasse 5–7, W–45141 Essen, Germany

Gel-like supermolecular structures are often observed in biological or macromolecular systems, but they can even occur in dilute solutions of surfactants or dyes. These molecules, under suitable conditions, assemble reversibly into large aggregates of rod-like geometry. In the quiescent state, the dynamic properties are mainly controlled by breaking and recombination reactions. At these experimental conditions very simple scaling laws are observed. At finite values of the shear rate, the anisometric aggregates are aligned in the direction of flow, and this process affects all rheological properties. The present paper gives some insight into the correlations between orientation and stress, and it describes some interesting flow phenomena observed in both types of association colloids.

Many surfactant and dye solutions, even at rather high concentrations, behave as simple Newtonian liquids with viscosities only slightly above that of water. This is often observed in concentration regimes where the molecules are completely dissolved or at conditions where spherical aggregates are present. Rod-shaped particles, however, exhibit more complicated flow phenomena. This has been known for more than 50 years, but the molecular processes underlying these phenomena have only been explored in recent time. It turns out that some solutions exhibit interesting relaxation properties. This occurs at experimental conditions where the average lifetime of the rod-shaped particles is small in comparison to the translation properties of the whole aggregate. Micelles are in thermal equilibrium with single monomers. In a given time interval a certain number of monomers are leaving the micelle by diffusion processes, but at the same time other molecules are just entering the aggregate. In true equilibrium, the net exchange is equal to zero, and each micelle can be characterized by a certain aggregation number. This holds, at least, if we average over long times and many aggregates. One should keep in mind, however, that such a micelle is a

0097–6156/94/0578–0063$08.36/0

fluctuating, dynamic particle that can change its size and shape continuously. Micelles are fragile, dynamic objects that are constantly formed and destroyed by the addition and loss of surfactant monomers. They have, hence, a characteristic lifetime or breaking time, which depends in a very complicated way on the chain lengths, the ionic strength, the aggregation number and other parameters. Under experimental conditions where the average life time of the anisometric micelles is much smaller than the diffusion constants of the whole aggregate (reptation), there are numerous breaking and reformation processes within the time scale of observation. As a consequence, an applied shear stress relaxes through chemical pathways. This leads to monoexponential stress decay when investigating a large ensemble of micelles. In recent years, these processes were extensively investigated by M. Cates (1-3). Starting from a pure reptation model, where relaxation occurs by curvilinear diffusion processes, Cates introduced the concept of limited lifetimes, in order to include breakage and healing reactions. It is nowadays well established that the theoretical model proposed by M. Cates describes the basic features of viscoelastic surfactant solutions (4-10). This holds especially in the linear viscoelastic regime. At these conditions the response at any time is proportional to the value of the initiating signal. Doubling the stress will double the strain (superposition principle). In practice it is found that most materials show linear time dependent behavior even at finite deformations as long as the strain remains below a certain limit. This threshold varies from sample to sample and is thus a property of the material. In rubber elastic systems, the linear stress-strain relation is observed up to high deformations of a few hundred percent, whereas in energy elastic systems the transition occurs already at very small shear strains of only a few thousandths. On a molecular scale, this phenomenon is due to orientation effects and reversible structure breakdown. In the theory of linear viscoelasticity all material parameters are functions of the time or, in oscillatory experiments, functions of the angular frequency ω (the frequency dependence results from the Fourier transformation of the time domain). At conditions where the linear relationship between stress and strain are no longer valid, new phenomena occur. It is, for instance, well known that the relaxation process changes at these experimental conditions. This phenomenon is often observed in solutions of entangled macromolecules where the superposition of shear flow on the dynamic characteristics leads to modifications of the long-time part of the relaxation spectrum. It is generally assumed that with increasing flow the slow relaxation processes are suppressed. In a first approximation one could say that a certain shear rate $\dot{\gamma}$ influences the relaxation times $\tau = 1/\dot{\gamma}$ (here τ denotes the relaxation time). In a more precise model it should be taken into account that in fact a larger region of the relaxation spectrum is shifted. Such typical processes are even observed in surfactant solutions. It is, therefore, interesting to study the dynamic properties of these entangled rod-shaped aggregates at large deformations as they occur during flow. Besides these changes of relaxation processes there is still another phenomenon that becomes important at large values of the velocity gradient. During flow, anisometric particles will be oriented along the streamlines. This is a typical dynamic process because the suspended aggregates rotate in the fluid. In laminar flow the anisotropic particles are forced to rotate with a non-uniform angular velocity $\omega(\phi)$. The rotational diffusion rate attains a minimum at $\phi=0°$ and a maximum at $\phi=90°$. The Brownian motion, however, tends to counteract the hydrodynamic alignment by causing random fluctuation. The competition of both influences leads to the flow induced decrease of the orientation angle. At low values of the velocity gradient, the average orientation of the longest axis is at 45°, which corresponds to a random distribution of aggregates. With increasing shear rate this value is gradually shifted towards zero degrees. This orientation process of the particles leads to changes of rheological properties. It is often observed that the shear viscosity, the relaxation modulus, the storage modulus or the loss modulus are strongly decreasing whereas the first normal stress difference or the flow birefringence are increasing because of the alignment processes. At high values of the velocity gradient, the orientation is certainly one of the most important parameters affecting all rheological properties. Whereas the quiescent state of viscoelastic surfactant solutions is mainly controlled by fluctuations of the micellar aggregates, the behavior during flow is additionally influenced by orientation processes, which also cause modifications of the relaxation spectra. The non-Newtonian behavior of these solutions is of great practical interest, and it is, therefore, important to understand the principal phenomena that occur during flow. The present paper is an attempt in such a direction and gives a detailed description of the relationship between shear stress and orientation. The basic ideas and concepts have a general, universal character, which may lead to a deeper insight into dynamic properties of viscoelastic liquids.

Materials

Viscoelastic Surfactant Solutions. In surfactant solutions, the phenomenon of viscoelasticity can be induced by addition of specific compounds to some amphiphilic molecules. Up to now, three different types of added molecules are known to give the desired effects: a second, oppositely charged surfactant, organic counterions like the salicylate counterion and uncharged compounds like esters or aromatic hydrocarbons. Typical examples exhibiting strong viscoelastic properties at very low concentrations are the cationic surfactants cetylpyridiniumchloride (CPyCl) or cetyltrimethylammoniumbromide (CTAB) combined with different amounts of sodium salicylate (NaSal). Another interesting surfactant system is dimer acid betain, which gives the desired effects already without adding a second component. The complicated structure of dimer acid betain is shown in Figure 1. In contrast to anionic or cationic surfactants, where the mechanical properties depend on the counterion concentration, the zwitterionic dimer acid betain nearly behaves as a noncharged system and the results are not influenced by ionic strength. From a theoretical point of view, this corresponds to an ideal case which is rarely observed in other types of viscoelastic surfactants. The dimeric acid betain solutions can be used as simple model systems for studies of fundamental principles of flow. A detailed description of the rheological properties is given in reference 10.

Viscoelastic Solutions of Chromophores. For about a century it has been known that many dyes, especially noncharged molecules, tend to form different types of aggregates in aqueous solutions (*11, 12*). Specifically for the case of cyanines and related chromophores, it was observed that this aggregation process leads to dramatic changes of the optical properties of the dilute solutions. The best known molecular assemblies that are present in these solutions are called Jelley or Scheibe aggregates in honor of the two authors who first discovered this interesting phenomenon (*13-16*). The presence of J-aggregates usually leads to the observation of an intense narrow absorption band that is shifted to longer wavelengths relative to the monomer adsorption. Aggregates with absorption maxima at shorter wavelength are designated as H-aggregates. On the basis of an extended dipole model, Kuhn and coworkers have proposed a brickwork like arrangement of cyanine chromophores to explain the J-band (*17*). In solution, aggregates have a threadlike, linear structure whereas in monolayers they are arranged in a planar fashion (*17*). J-aggregates were observed in aqueous solutions, at the surfaces of crystals, at oppositely charged polymers or at the surface of monolayers. A typical molecule that forms rod-shaped aggregates in solution is pseudoisocyanine chloride (1,1'-diethyl-2,2'cyanine chloride). The molecular structure of this compound is given in Figure 2. Quantum-mechanical calculations of Kuhn and coworkers and comparison with absorption spectra showed that the rod-shaped aggregates in solutions have a sandwich-like structure (*17*). Such a molecular arrangement is schematically shown in Figure 3. In monolayers, surface active chromophores can form similar closest packing structures, where the molecules are in a brick stone work arrangement. A typical dye showing this behavior at the surface between water and air is shown in Figure 4.

According to investigations of Kuhn et al. these molecules aggregate at the water surface in a brick stone work arrangement (*17, 18*). This special two-dimensional structure is schematically shown in Figure 5. As in micellar solutions, the aggregates formed by chromophores are in thermal equilibrium with single monomers. We should, hence, expect a similar type of stress relaxation. This interesting problem will be discussed further in the text.

Dye molecules forming J-aggregates can be synthesized following a method first developed by Sondermann (*19*). Surfactants that are able to build-up rod-shaped aggregates, are commercially available (Fluka, Merck, Aldrich). The details of the synthesis of dimer acid betain are summarized in Ref. 10.

Rheological Methods and Fundamental Equations of Flow

Linear Viscoelastic Properties. The properties and dynamic features of supermolecular structures observed in solutions of surfactants or chromophores can easily be studied by means of rheological measurements. There are in principal four different types of tests that can be performed to measure the principal material properties. In transient tests, a stepwise transition is used from one equilibrium state to another. In these experiments, a certain shear stress σ, a certain deformation γ or a certain shear rate $\dot{\gamma}$ is suddenly applied at t=0 and then held constant thereafter. Although the initial extension is infinitely fast, there is a certain time response of the viscoelastic material which can be measured to get the desired rheological functions. In the relaxation test, a step function shear strain is applied at t=0. The resulting stress is time dependent, and this quantity is measured after the amount of deformation has occurred. From these data, the relaxation modulus $G(t,\gamma)$ can be calculated by the relation:

$$G(t,\gamma) = \sigma(t,\gamma)/\gamma \qquad (1)$$

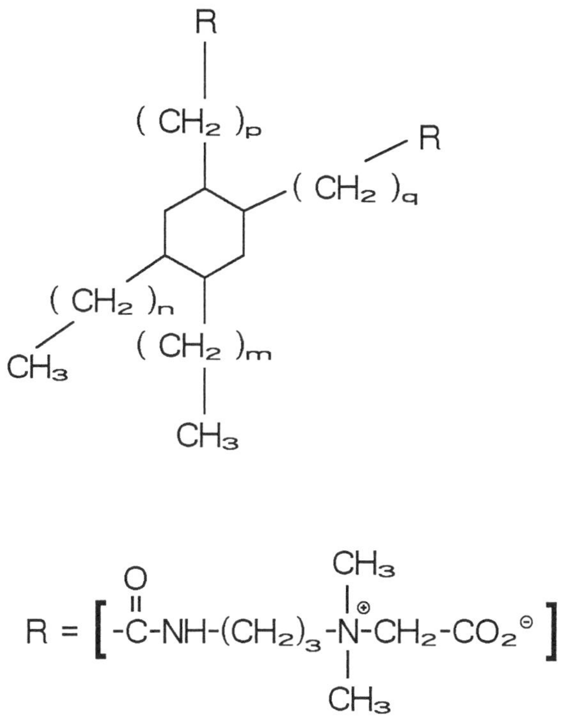

Figure 1. Dimer acid betaine. m, n, p and q are natural numbers which can be varied during the chemical synthesis.

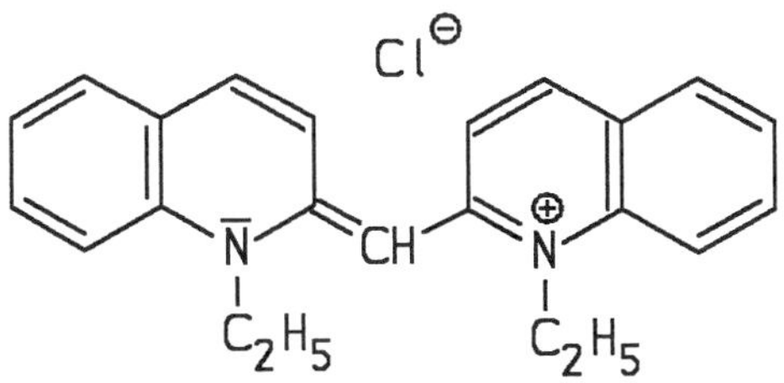

Figure 2. Dimer acid betaine molecular structure of pseudoisocyanine chloride.

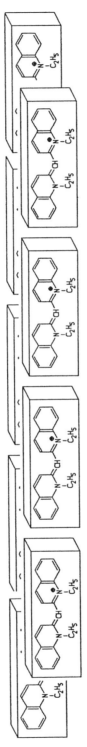

Figure 3. Threadlike arrangement of pseudoisocyanine chloride molecules.

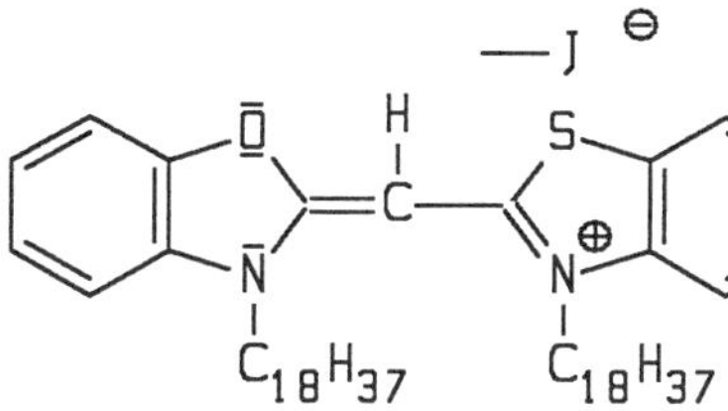

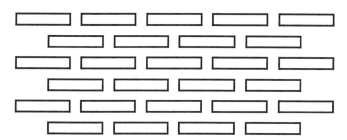

Figure 4. Molecular structure of a surface active dye molecule (3-octadecanoyl-benzthiazol-(2)[3-octadecanoyl-benzoxazol-(2)]-monomethincyanin-iodide.

Figure 5. Brick stone work arrangement of surface active chromophore molecules at the surface of water. The dye molecule is approximated by a rectangle with dimensions 0.6*0.4 nm (bird's-eye view).

In the limit of the linear stress-strain relationship, the relaxation modulus does not depend on γ. The second variable describes the fact that at higher values of the shear strain the relaxation process depends upon the initial amount of deformation (this process is due to changes of the relaxation spectrum). As long as the data are obtained in the linear region, the second variable may be neglected.

From the above expression it is clear that the relaxation modulus describes the stress relaxation after the onset of shear strain. In dilute solutions of surfactants or dye molecules, an applied shear stress always relaxes to zero after infinitely long periods of time. This does not hold for more concentrated regimes where anisotropic phases occur. Very often, these systems have yield values, and at these conditions the shear stress does not relax to zero.

Beside these transient methods there is another type of test that can be performed to provide information on the viscoelastic properties of the sample: periodic or dynamic experiments. In this case the shear strain is varied periodically with a sinusoidal alternation at an angular frequency ω. A periodic experiment at frequency ω is qualitatively equivalent to a transient test at time $t=1/\omega$. In a general case, a sinusoidal shear strain is applied to the solution. The response of the liquid to the periodic change consists of a sinusoidal shear stress which is out of phase with the strain. The shear stress can be split into two different components. The first component is in phase with the deformation and the second one is out of phase with the strain. From the phase angle δ, the amplitudes of the shear stress $\hat{\sigma}$, and the amplitude of the shear strain $\hat{\gamma}$ it is possible to calculate the storage modulus G' and the loss modulus G''.

$$G'(\omega,\gamma) = [\hat{\sigma}(\omega,\gamma/\hat{\gamma}]\cos(\delta) \qquad (2)$$

$$G''(\omega,\gamma) = [\hat{\sigma}(\omega,\gamma/\hat{\gamma}]\sin(\delta) \qquad (3)$$

The storage modulus describes the elastic properties of the sample, and the loss modulus is proportional to the energy dissipated as heat (viscous resistance). It is convenient to express the periodically varying functions as a complex quantity which is termed the magnitude of the complex viscosity $|\eta^*|(\omega,\gamma)$ This quantity can be calculated from the following equation:

$$|\eta^*|(\omega,\gamma) = \frac{\sqrt{G'^2(\omega,\gamma) + G''^2(\omega,\gamma)}}{\omega} \qquad (4)$$

It can be shown that for most viscoelastic solutions there exists a simple correlation between dynamic and steady-state flow characteristics (20). In a first approximation the complex viscosity $|\eta^*|(\omega,\gamma)$ at certain angular frequency ω can be compared with the steady-state value of the shear viscosity $\eta(\infty,\dot{\gamma})$ at the corresponding shear rate $\dot{\gamma}$. Details of these correlations between linear viscoelastic functions and non linear effects will be discussed further in this text.

The rheological properties of viscoelastic samples can be represented by simple mechanical models. In the regime of linear stress-strain relations, it is assumed that the elastic properties are given by a Hookean spring, and that the viscous phenomena can be represented by a Newtonian dash pot. The simplest model that can describe a visco-elastic surfactant solution is called the Maxwell material. It consists of a spring and a dash pot connected in series (Figure 6).

The elastic spring corresponds to a shear modulus G_o and the dash pot represents the constant viscosity η_o. The dynamic properties of the Maxwell element can be represented by linear differential equations, the solution of which give the desired material functions. The behavior under harmonic oscillations can be obtained from the following formulas:

$$G'(\omega) = \frac{G_o \omega^2 \tau^2}{1 + \omega^2 \tau^2} \tag{5}$$

$$G''(\omega) = \frac{G_o \omega \tau}{1 + \omega^2 \tau^2} \tag{6}$$

$$|\eta*|(\omega) = \frac{\eta_o}{\sqrt{1 + \omega^2 \tau^2}} \tag{7}$$

In the regime of linear viscoelasticity the material functions do not depend upon γ. As a consequence, they are only functions of the angular frequency ω. In transient experiments the Maxwell material can be described by single exponential relaxation processes:

$$G(t) = G_o e^{-t/\tau} \tag{8}$$

The Maxwell material can be used for the theoretical description of monoexponential stress decay because there is only one mechanism to store elastic energy (just one spring). Multiexpontential stress relaxation processes can be described by a combination of different Maxwell materials where each model corresponds to one separate relaxation time (generalized Maxwell model). The behavior under harmonic oscillations can be obtained from the following formulas:

$$G'(\omega,\gamma) = \sum_{i=1}^{n} G_i(0,0) \frac{\omega^2 \tau_i^2}{1 + \omega^2 \tau_i^2} \tag{9}$$

$$G''(\omega,\gamma) = \sum_{i=1}^{n} G_i(0,0) \frac{\omega \tau_i}{1 + \omega^2 \tau_i^2} \tag{10}$$

The relaxation modulus is similarly given by:

$$G(t,\gamma) = \sum_{i=1}^{n} G_i(0,0) e^{-t/\tau_i} \tag{11}$$

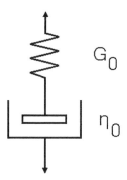

Figure 6. Schematic representation of the Maxwell fluid.

The dynamic features of the generalized Maxwell material can be characterized by a set of discrete relaxation times and relaxation moduli. Such representation of experimental values is called a discrete relaxation spectrum. It is easy to show that in the simplest viscoelastic liquid, represented by the Maxwell material, the spectrum is reduced to the existence of only one single point. If a discrete distribution of relaxation processes occurs, the relaxation spectrum gives detailed information on the dynamic features of the viscoelastic sample. Relaxation times with high values of the shear modulus correspond to mechanisms of strong stress release. It can easily be shown that each experimentally observed stress relaxation process can in principle be fitted with any desired degree of accuracy to a series of terms as in equation 11, if n is sufficiently large. In this way it is possible to determine the discrete spectrum of lines, each with a location τ_i and an amplitude G_i. Real treatment of experimental data, however, shows that it is usually very difficult to resolve more than a few lines. This is mainly due to the experimental error. It turns out that only contributions with the longest relaxation times can be obtained from transient or dynamic experiments. Beyond this, the empirical choice of parameters G_i and τ_i is largely arbitrary. This difficulty can be avoided by introducing continuous spectra. For this case, equation 11 becomes:

$$G(t,\gamma) = \int_0^\infty F(\tau,\gamma)\,e^{-t/\tau}\,d\tau \tag{12}$$

If the number of elements in the generalized Maxwell model is increased to infinity, the expression $F(\tau,\gamma)\,d\tau$ can be interpreted as the shear modulus corresponding to a relaxation time having a value between τ and $\tau+d\tau$. Thus $F(\tau)$ is actually the spectrum of the distribution of shear moduli. The spectrum itself is usually not accessible by direct experiment. It is, however, possible to calculate it from measurements of G', G'', or $G(t)$ by numerical or graphical differentiation or by the use of finite difference methods. We shall not go into more details at this point. The interested reader is referred to the detailed text of Tschoegl et al., which gives an excellent introduction to the different types of spectrum calculations (21). An alternative distribution function, $H(\tau,\gamma)$ is sometimes used in place of $F(\tau,\gamma)$.

$$H(\tau,\gamma) = \tau\,F(\tau,\gamma) \tag{13}$$

It is evident, that the function $H(\tau,\gamma)$ is the corresponding distribution function of viscous moduli. A certain distribution of relaxation times is, hence, responsible for a corresponding distribution of shear moduli and viscosities. Of course, only two of these three distributions are independent. Thus, an alternative expression for the relaxation modulus is given by:

$$G(t,\gamma) = \int_0^\infty H(\tau,\gamma)\,e^{-t/\tau}\,d\ln\tau \tag{14}$$

It is often observed, that $H(\tau,\gamma)$ has a shape rather similar to that of $G''(\omega,\gamma)$ reflected in the modulus axis. The maxima represent concentrations of relaxation processes. In viscoelastic solutions of dye molecules or surfactants, $H(\tau,\gamma)$ vanishes at long times, whereas in real cross-linked samples a finite value results. Equations 12 and 14 may be interpreted as integral constitutive equations for viscoelastic materials. They are often used to calculate appropriate linear viscoelastic properties, such as the relaxation modulus, from other types of excitation, e.g., from sinusoidal oscillations. Thus, once the spectrum of the viscoelastic constants is known, it is possible to generate the response to any desired type of excitation. At conditions where the relaxation process is discrete, the use of these integral equations is possible but inconvenient because one has to operate with delta functions in their practical utilization. The relaxation spectrum is then not continuous but it consists of separate points (lines). Particularly in viscoelastic solutions of dyes and surfactants, one often observed discrete relaxation processes that can easily be described by a series of Maxwell materials.

Non-Linear Viscoelastic Properties. It is well known that the anisometric aggregates of micellar solutions can be oriented under the influence of a velocity gradient. This happens in such a way that the probability of finding the longest axis in the direction of flow is a maximum. The short axes are aligned in the direction of the velocity gradient. One of the most important optical characteristics of a viscoelastic sample is the refractive index. It is determined by the polarizability of atomic groups. Dye and surfactant molecules are optically anisotropic, and the rod-shaped aggregates they are forming have, consequently, unequal refractive indexes in different directions. In the quiescent state the aggregates are randomly oriented in space and they form optically isotropic materials. Under the action of a velocity gradient, however, the sample becomes anisotropic and the optical properties must be represented by the refractive index tensor. This property and the average angle of orientation can be measured with a flow birefringence apparatus. Details of the measuring technique are extensively described in a recent book of Janeschitz-Kriegl (22).

It is often observed that the anisotropic distribution of shear stresses coincides with the alignment of the anisometric aggregates. This holds for a large group of macromolecules but also for concentrated solutions of rod-shaped particles. From a physical point of view that means that there exists a linear relationship between the shear stress and birefringence tensor. Such characteristic properties are quantitatively described by the stress optical rule, which is sometimes called the stress optical law. Such a relation was first proposed by Lodge and postulated in recent years by Doi and Edwards for solutions containing rigid, overlapping rods (23,24). From the stress optical law, thus defined, the following equations can be derived:

$$\Delta n(\infty,\dot{\gamma})\sin[2\chi(\infty,\dot{\gamma})] = 2C\sigma_{21}(\infty,\dot{\gamma}) \tag{15}$$

$$\Delta n(\infty,\dot{\gamma})\cos[2\chi(\infty,\dot{\gamma})] = C\{\sigma_{11}(\infty,\dot{\gamma}) - \sigma_{22}(\infty,\dot{\gamma})\} \tag{16}$$

In these equations $\Delta n(\infty, \dot{\gamma})$ denotes the steady-state value of the flow birefringence. This quantity describes the optical anisotropy of the streaming solution. $\Delta n(\infty, \dot{\gamma})$ can be calculated together with $\chi(\infty, \dot{\gamma})$ from experimental values of flow birefringence (*22*). In many surfactant solutions, the orientation of rod-shaped micelles can be described by a simple equation, which was first proposed by S. Hess (*25,26*). The term $\sigma_{11}(\infty, \dot{\gamma}) - \sigma_{22}(\infty, \dot{\gamma})$ describes the first normal stress difference. The anisotropic character of the flowing solutions gives rise to additional stress components, which are different in all three principal directions. This phenomenon is called the Weissenberg effect or the normal stress effect. According to the stress optical law, the normal stresses are zero at $\chi = 45°$, and they become more and more important with increasing shear rate. It is convenient to express the mechanical anisotropy of the flowing solutions by the first normal stress difference $N_1(\infty, \dot{\gamma}) = \sigma_{11}(\infty, \dot{\gamma}) - \sigma_{22}(\infty, \dot{\gamma})$ and the second normal stress difference $N_2(\infty, \dot{\gamma}) = \sigma_{22}(\infty, \dot{\gamma}) - \sigma_{33}(\infty, \dot{\gamma})$. The normal stresses are due to the fact that finite elastic strains are developing in the streaming liquid. On a molecular scale, these forces describe the anisotropic character of the oriented systems. In viscoelastic solutions at rest, entropic forces determine the random distribution of the rod-shaped aggregates. During flow, a dynamic orientation process occurs, which is induced by hydrodynamic forces. This phenomenon is associated with the formation of anisotropic restoring forces. It can easily be shown that the force in the direction of flow is greater than the one in the direction of the velocity gradient. For this reason, the first normal stress difference is usually positive in sign ($N_1 > 0$). As normal stresses are associated with non-linear effects, they do not appear explicitly in the description of the Maxwell material. In the regime of small deformations, they are equal to the isotropic ambient pressure. At elevated shear rates, however, the normal stress differences become more and more important because they are usually increasing with $\dot{\gamma}^2$. The term $(\sigma_{11}(\infty, \dot{\gamma}) - \sigma_{22}(\infty, \dot{\gamma}))/(2\dot{\gamma}^2)$ denotes the steady state values of the first normal stress coefficient. The constant C in equation (24) is called the stress optical coefficient. In solutions of macromolecules, this constant has been derived on the basis of the Kuhn-Grün theory (*27*).

$$C = \frac{(2\pi/45)(n^2 + 2)^2(\alpha_1 - \alpha_2)}{nkT} \qquad (17)$$

n is the mean value of the refractive index, α_1 is the polarizability of the molecule in the direction of the main axis, and α_2 is the polarizability of the molecule in the normal direction. $(\alpha_1 - \alpha_2)$ denotes the optical anisotropy of the flexible chains. The above expression is only valid when the average refractive index of the particles is equal to the one of the surrounding medium ($dn/dc \approx 0$, matching solvents). Equation 17 is derived from the theory of ideal networks in rubber-elastic polymers and does not include intermolecular interactions between the chains. In viscoelastic surfactant solutions, the stress optical law usually holds, and the close relationship to the phenomena observed in polymeric systems suggests a critical test of equation 17. From measurements of the electric birefringence it is possible to obtain the difference $(\alpha_1 - \alpha_2)$ (*9*). On the other hand, this term can also be calculated from the polarizabilities of the paraffin chains, the polar head groups, and the condensed counter-ions assuming that about 8% of the monomers are oriented perpendicular to the surface of the aggregate. It is generally accepted that the monomeric molecules do not have a strictly parallel arrangement in the micellar core. The internal phase of micelles is, without doubt, more liquid-like than solid-like. This fact was recognized by Papenhuizen et al., who used flow birefringence measurements to get information on the size and the shape of micellar aggregates (*28,29*). In this context it is interesting to note that in viscoelastic surfactants the constant C usually attains the same value which does not depend upon the surfactant or salt concentration. The interpretation of equation 17 leads to the conclusion that this might be due to a constant term $(\alpha_1 - \alpha_2)$. In surfactant solutions, the difference of the polarizability depends mainly on the optical parameters of the alkyl chains, and this remains nearly constant if the other experimental conditions are changed. Typical values are of the order of $-C = 2 \cdot 10^{-7} \text{ Pa}^{-1}$ (*9*). The negative sign is due to the fact that the surfactant molecules are perpendicularly oriented to the surface of the anisometric aggregates. It turns out that the stress optical coefficient is one of the most important parameters describing the optical anisotropy of the cylindrical micelles.

Linear Rheological Properties of Viscoelastic Dye and Surfactant Solutions.

As discussed before, some viscoelastic surfactant solutions can be characterized by monoexponental stress decay. This holds at experimental conditions where the average breaking time of the rod-shaped aggregates is short in comparison to the diffusion process of the whole aggregate. A typical example of such behavior is shown in Figure 7. The points are experimental data and the drawn lines are theoretical fits according to equations 5 and 6. It is evident that the rheological properties of the solutions can be represented by a monoexponential stress decay. Many cationic surfactant solutions combined with different types of counterions show such typical properties. It is, however, not always possible to observe such simple features. An excellent example representing more complicated relaxation phenomena is shown in Figure 8.

In this experiment a step function shear strain is applied at t=0. The relaxation modulus $G(t,\gamma)$ was calculated from the time dependent shear stress $\sigma(t,\gamma)$. The experimental data show a striking analogy with entangled polymer solutions. There exists a broad spectrum of different relaxation times, which can be represented by the spectrum $H(\tau)$ (Figure 9).

According to the theory of Cates, such relaxation phenomena correspond to processes where the diffusion time of the elongated aggregates is short in comparison to the breaking time. It is interesting to note that the relaxation spectrum obtains two different maxima which might correspond to the two main processes causing stress relaxation. One could, hence, argue that one extreme value describes the average breaking time and the other one the reptation properties of the rod-shaped micelles. At the present state, we cannot prove this assumption, but it is certainly an interesting problem which we will study in more detail.

In solutions of cationic surfactants, multiexponential relaxation functions are generally observed at low values of the surfactant concentration and counter-ion excess. In the non-charged solutions of dimer acid betaine, however, the average lifetime of the micelles is always much longer than the diffusion constant, and even by varying the concentration, changing the temperature, or adding salt it is not possible to observe real

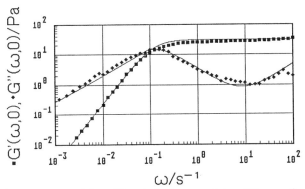

Figure 7. The storage modulus G' and the loss modulus G" as a function of the angular frequency ω for a solution of 100 mmol CPyCl and 60 mmol NaSal at T=20°C.

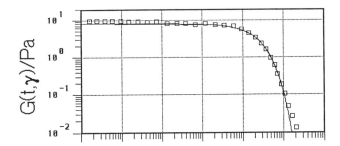

Figure 8. The relaxation modulus $G(t,\gamma)$ as a function of time for a dimer acid betaine solution of 80 mmol at T=20°C. The drawn line corresponds to monoexponential stress decay. Deviations occur at short and at long times.

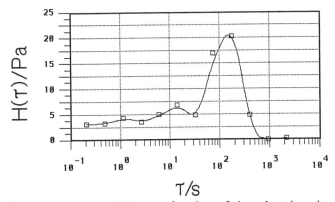

Figure 9. The line spectrum $H(\tau)$ as a function of the relaxation time τ for a solution of 100 mmol dimer acid betaine at T=20°C.

monoexponential relaxation functions (*10*). Very similar properties are also observed in solutions of dye molecules. This is shown if Figure 10, which gives some insight into the dynamic features of these solutions.

It is evident that the dynamic features of these dye solutions cannot be described by monoexponential relaxation functions. This is clearly shown in Figure 11, which gives some information on the relaxation modulus of viscoelastic chromophore gels.

From Figure 11 it becomes evident that the stress relaxation process after a step function shear strain cannot be described by the theoretical equations of a Maxwell material. The calculation of the relaxation spectrum yields Figure 12.

In viscoelastic solutions of dye molecules, the stress decay is characterized by a broad distribution of relaxation times. It is interesting to note, that the spectrum does not change as a function of the concentration. This phenomenon is similar to the dynamic features observed in dimer acid betaine solutions. As the molecular processes of dye aggregation are not well understood at the present state, it is difficult to relate the kinetic phenomena to breaking times or diffusion processes. In analogy to surfactant solutions, these processes might be very similar

Non-Linear Viscoelastic Properties

The phenomenon of non-linear response was discovered already at the earliest stages of rheological research. As soon as viscometers became available, many departures from Newtonian behavior were found which are due to orientation effects or to changes in the relaxation spectra. A typical example showing non-linear relaxation processes is given in Figure 13. The transient rheological properties of surfactant solutions can be investigated by stepwise transitions from the quiescent state to the deformed one. In this experiment, a step function shear strain is applied at t=0 and the relaxation modulus can be calculated from the time dependent shear stress according to equation 1. At low values of the shear strain, the relaxation modulus can often be described by monoexponential relaxation functions (Figure 13). At higher values of the shear strain, the stress decay is accelerated and this process can be characterized by stretched exponential relaxation functions. This is very similar to corresponding phenomena observed in solutions of entangled macromolecules. In solutions of polymer molecules it is often observed that the viscosity decreases with increasing shear rate. This typical behavior is called shear thinning or pseudoplastic. In former investigations we have shown that the decrease of the viscous resistance with increasing shear is due to the orientation of the anisometric aggregates (*9*). It is generally observed that the shear thinning behavior of $\eta(\infty,\dot\gamma)$ and $|\eta^*|(\omega,0)$ are similar in character. Among several variants of correlations between dynamic measurements and steady-state characteristics, the empirical method of Cox and Merz has attracted the greatest interest (*20*). A simple relationship between linear viscoelastic functions and non-linear processes can be obtained by comparing $\eta(\infty,\dot\gamma)$ and $|\eta^*|(\omega,0)$ at $\omega=\dot\gamma$. According to the observations of Cox and Merz, these functions coincide for entanglement networks of polymers (*20*). Deviations occur, however, if besides the pure mechanical contacts there are other types of forces such as hydrogen bonds or ionic interactions, which also contribute to the cross-linking process. In general, the steady state shear viscosity of these solutions is much smaller than the corresponding dynamic value because the number of these additional contacts

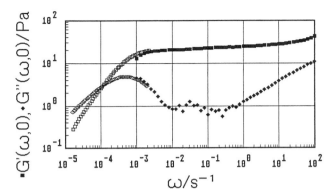

Figure 10. The storage modulus and the loss modulus for a viscoelastic solution of 40 mmol pseudoisocyanine chloride at T=20°C. Filled symbols are obtained from dynamic measurements and open symbols are calculated from measurements of the relaxation modulus.

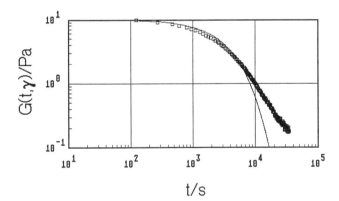

Figure 11. The relaxation modulus as a function of time for a solution of 40 mmol pseudoisocyanine chloride at T=20°C. Open symbols represent experimental data and the drawn line corresponds to monoexponential stress decay.

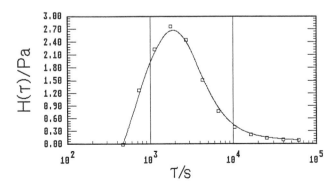

Figure 12. The relaxation spectrum as a function of the relaxation time for a solution of 40 mmol pseudoisocyanine chloride at T=20°C.

is reduced during flow, whereas it remains unchanged in oscillatory experiments. In viscoelastic surfactant solutions, the strong decrease of the steady-state shear viscosity is clearly correlated with the orientation of the anisometric micelles (9). This simple explanation, however, cannot be used for sinusoidal oscillations. In these experiment, the extinction angle is of the order of 45°. Due to the completely different orientation processes, the molecular interpretation of the Cox-Merz rule may be rather complicated. For surfactant solutions, one usually observes satisfactory overlap between these two functions. This is clearly demonstrated in Figure 14, which gives a direct comparison of dynamic and steady state flow characteristics.

The existence of a simple correlation between the components of the complex viscosity and the steady state values of the stress tensor underlines once more the close relationship to corresponding phenomena observed in polymer solutions. The slope of 1 points to the existence of a limiting shear stress. This phenomenon is often observed in surfactant solutions and is clearly uescribed in Figure 15.

Similar phenomena are often observed in solutions of entangled macromolecules. When the shear stress exceeds a critical value, the flow behavior is changed to plug flow, with slippage along the wall. This process is often termed the "spurt" effect (22). In viscoelastic surfactant solutions, this phenomenon was extensively investigated by M. Cates (31). The theoretical model predicts an instability of flow, which occurs at a critical value of the shear rate $\dot{\gamma}$. As a consequence, the observed plateau value of the shear stress is related to orientation effects, which occur at high values of the velocity gradient. It is interesting to note that the theoretical model predicts a simple correlation between the maximum shear stress and the shear modulus:

$$\sigma_{max} = 0.67 \, G'(\infty, 0) \tag{18}$$

Experimental studies of different surfactant systems show that this relationship holds at conditions where the relaxation process is close to single exponential (7,9,10). This corresponds exactly to the theoretical predictions. In concentration regimes with multiexponential stress decay, large deviations occur (7,10). Recent experiments with optical rheometers (rheoscopes) seem to indicate that this phenomenon is due to vortex formation (32). It is interesting to note that there exist some simple relations between linear viscoelastic properties and non-linear processes. One of the most popular observations was first postulated by Gleissle (33,34). The first mirror relationship defined by Gleissle is of an empirical nature. It is based on the experimental observation that the shear viscosity coincides with the transient viscosity measured in start-up flow. In this test, a step function shear rate $\dot{\gamma}$ is suddenly applied at t=0 (start-up flow). The shear stress growth coefficient $\eta^+(t, \dot{\gamma})$ can be calculated by:

$$\eta^+(t, \dot{\gamma}) = \sigma(t, \dot{\gamma}) / \dot{\gamma} \tag{19}$$

In viscoelastic solutions this function is exponentially increasing. A typical example showing this type of behavior is given in Figure 16. The symbols correspond to experimental data and the drawn line describes a monoexponential relaxation process according to the theoretical equation of a Maxwell material:

$$\eta^+(t, \dot{\gamma}) = \eta^+(\infty, \dot{\gamma})(1 - e^{-t/\tau}) \tag{20}$$

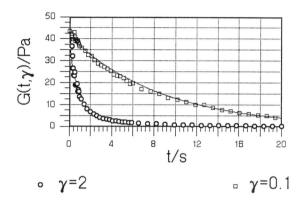

Figure 13. The relaxation modulus as a function of the time for two different deformations γ for a solution of 100 mmol CPyCl and 60 mmol NaSal at T=20°C. The drawn line corresponds to monoexponential stress decay.

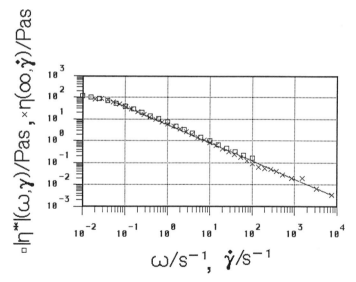

Figure 14. The steady state values of the shear viscosity $\eta(\infty,\dot{\gamma})$ as a function of the shear rate and the magnitude of the complex viscosity $|\eta^*|(\omega,0)$ as a function of the angular frequency for a solution of 3 mmol dimeric acid betaine at T=20°C.

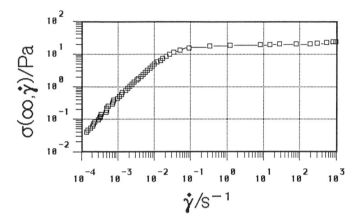

Figure 15. The steady state shear stress as a function of the shear rate for a 3 mmol solution of dimeric acid betaine at T=20°C.

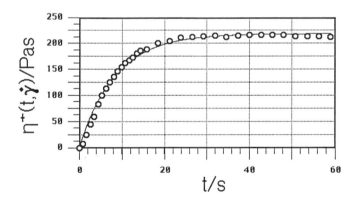

Figure 16. The shear stress growth coefficient $\eta^+(t,\dot{\gamma})$ as a function of the shear time for a solution of 100 mmol CPyCl and 60 mmol NaSal at T=20°C ($\dot{\gamma} = 0.05\,\mathrm{s}^{-1}$).

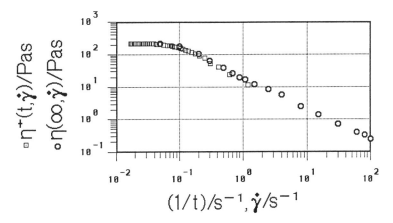

Figure 17. Comparison of the shear stress growth coefficient as a function of $1/t$ and the stationary values of the shear viscosity as a function of the shear rate for a solution of 100 mmol CPyCl and 60 mmol NaSal at T=20°C.

According to the observations of Gleissle, there exists a simple relationship between the shear stress growth coefficient and the stationary values of the shear viscosity (*33,34*).

$$\eta^+(t,\dot{\gamma}) = \eta(\infty,\dot{\gamma}) \qquad \text{for } t = 1/\dot{\gamma} \tag{21}$$

One should, hence, obtain the shear viscosity from transient data by plotting the shear stress growth coefficient as a function of $1/t$ in a double logarithmic plot. Such treatment of experimental data is shown in Figure 17.

It is evident that the mirror relation holds very well. This implies that there exists a simple correspondence between the material functions in the linear regime and the non-linear region. It is easy to show that, within the limits of experimental error, the first relation of Gleissle can only apply to fluids whose viscosity varies at high values of the shear rate as $1/\dot{\gamma}$. In a double logarithmic plot, such a curve has a slop of -1. This holds, as we have discussed before, for a large number of surfactant solutions. In this case, the mirror relation can successfully be used to calculate non-linear phenomena from linear viscoelastic constants or inverse.

Conclusions

In the preceding sections we have seen that the dynamic properties of rod-shaped micelles and dye aggregates lead to specific phenomena that are typical for the presence of "entanglement" networks. The quiescent state is largely influenced by the kinetic processes, whereas the streaming solutions are mainly controlled by orientation effects. The simple relationships observed between all rheological functions look particularly promising in view of the growing interest in theoretical work on complex liquids. Viscoelastic surfactants or dye solutions may, therefore, be used as simple model systems in order to get a broader and deeper insight into the fundamental principles of flow.

Acknowledgments
The author has benefited from intensive conversations and collaborations with numerous colleagues. Financial support by grants of the "Deutsche Forschungsgemeinschaft" DFG (SFB 213), and the "Fonds der Chemischen Industrie" are gratefully acknowledged. Special thanks are due to Prof. G. Platz for his valuable suggestions and comments concerning the physical chemical properties of viscoelastic dye solutions.

Literature Cited

1. Cates, M.E. *Macromolecules* **1987**, *20*, 2289.
2. Cates, M.E. *J. Phys. France* **1988**, *49*, 1593.
3. Turner, M.S.; Cates, M.E. *Langmuir* **1991**, *7*, 1590.
4. Candau, S.J.; Hirsch, E.; Zana, R. *J. Colloid Interface Sci.* **1985**, *105*, 521.
5. Candau, S.J.; Hirsch, E.; Zana, R.; Adam, M. *J. Colloid Interface Sci.* **1988**, *122*, 430.
6. Candau, S.J.; Hirsch, E.; Zana, R.; Delsanti, M. *Langmuir* **1989**, *5(5)*, 1225.
7. Khatory, A.; Lequeux, F.; Kern, F.; Candau, S.J. *Langmuir* **1993**, *9*, 1456.
8. Rehage, H.; Hoffmann, H. *J. Phys. Chem.* **1988**, *92*, 4712.
9. Rehage, H.; Hoffmann, H. *Molecular Physics* **1991**, *74*, 933.
10. Fischer, P.; Rehage, H.; Grüning, B. *Tenside Surf. Det.* **1993**, *31*, 99.
11. Vickerstaff, T. *The Physical Chemistry of Dyeing*, 2nd ed.; Oliver and Boyd: London, 1954.
12. Mason, S.F. *J. Soc. Dyers Colourists* **1968**, *84*, 604.
13. Jelley, E.E. *Nature* **1936**, *138*, 1009.
14. Jelley, E.E. *Nature* **1937**, *139*, 631.
15. Scheibe, G. *Angew. Chemie* **1936**, *49*, 563.
16. Scheibe, G. *Angew, Chemie* **1937**, *50*, 212.
17. Czikkely, V.; Forsterling, H.D.; Kuhn, H. *Chem. Phys. Lett.* **1970**, *6*, 11.
18. Bucher, H.; Kuhn, H. *Chem. Phys. Lett.* **1970**, *6*, 183.
19. Sondermann, J. *Liebigs Ann. Chem.* **1971**, *749*, 183.
20. Cox, W.P.; Merz, E.H. *J. Polym Sci.* **1958**, *28*, 619.
21. Emri, I.; Tschoegl, N.W. *Rheologica Acta* **1993**, *32*, 311.
22. Janeschitz-Kriegl, H. *Polymer Melt Rheology and Flow Birefringence*; Springer Verlag: Berlin, 1977.
23. Lodge, A.S. *Trans. Faraday Soc.* **1956**, *52*, 120.
24. Doi, M.; Edwards, S.F. *J. Chem. Soc. Faraday Trans.* **1978**, *74*, 418.
25. Hess, S. *Phys. Rev. A.* **1971**, *25*, 614.
26. Thurn, H.; Lobl, M.; Hoffmann, H. *J. Phys. Chem.* **1985** *28*, 517.
27. Kuhn, W.; Grun, F. *Kolloid-Z.* **1942**, *101*, 248.
28. den Otter, J.L.; Papenhuijzen, J.M.P. *Rheologica Acta* **1971**, *10*, 457.
29. Janeschitz-Kriegl, H.; Papenhuijzen, J.M.P. *Rheologica Acta* **1971**, *10*, 461.
30. Vinogradov, G.V.; Malkin, A. Ya. *Rheology of Polymers*; Springer Verlag, Berlin, 1980.
31. Spendley, N.A.; Cates, M.E.; McLeish, T.C.B. *Phys. Rev, Lett.* **1993**, *71*, 939.
32. Leonhard, H.; Rehage, H. to be published.
33. Gleissle, W. *Proc VIII International Congress on Rheology* Naples, **1980**, 457.
34. Gleissle, W. *Rheol. Acta* **1982**, *21*, 484.

RECEIVED September 23, 1994

RHEOLOGY AND STRUCTURE

Chapter 5

Microstructure of Complex Fluids by Electron Microscopy

S. Chiruvolu, E. Naranjo, and J. A. Zasadzinski

Department of Chemical Engineering, University of California, Santa Barbara, CA 93106

Understanding the relationship between molecular organization and macroscopic properties of emulsions, lyotropic and thermotropic liquid crystals, colloidal dispersions, and other such "complex fluids" is generally a first step towards understanding rheology. As the trend in both science and technology is toward dilute, mixed surfactant solutions, light scattering, X-ray diffraction, and various spectroscopy techniques are not sufficient to determine complex fluid microstructure. Modern rapid-freezing methods followed by freeze-fracture or cryoelectron microscopy are much better suited to visualize the three-dimensional structure of the aggregates that make up the dispersion, while simultaneously revealing particle orientation and distribution. The direct structural information obtained by microscopy requires no model-dependent interpretation as is the case in scattering or spectroscopy data. However, the loss of ambiguity in interpretation can be more than made up for by artifacts of sample preparation.

Most attempts at understanding the rheology of surfactant solutions and complex fluids begin with a more or less detailed model of molecular organization and interactions in solution (*1-5*). Light, X-ray, and neutron scattering, NMR spectroscopy, fluorescence quenching, flow birefringence, and rheological techniques have provided a wealth of information on the *implications* of surfactant microstructure (*6-20*). However, especially in dilute or mixed surfactant systems, the microstructure is often so unexpected as to make it difficult to construct a model; or worse, a model built on simple structural concepts leads to erroneous interpretation of experimental results (*21,22*). Electron microscopy is becoming a necessary complement to these techniques as the structural information provided is model independent. Information at all relevant length scales, from 1 nm to 100 μm is available in the same experiment, allowing not only the structural characterization of the individual surfactant aggregate, but the organization of the aggregates in solution.

Two methods have evolved, each with specific benefits, to image surfactant solutions with electron microscopy; they are direct imaging of the frozen, hydrated material in the electron microscope, also known as cryo-TEM, and freeze-fracture replication (*23, 24*). Although these techniques have been applied to microstructured

0097-6156/94/0578-0086$08.00/0

fluids for more than three decades (*25-31*), they are only now coming into wider acceptance in the complex fluids community. This reluctance to embrace microscopy is undoubtedly due to the many possible artifacts associated with the non-equilibrium aspects of rapid freezing (*24,32-37*) or staining and drying (*38-48*) procedures.

In many early investigations, the cooling rates employed were too low to prevent structural rearrangements caused by crystal formation and phase separation. If the solvent phase crystallizes as distinct grains, the advancing crystal front expels any dispersed or dissolved species to the grain boundaries, thereby destroying the structure and spatial organization of the material. More than a decade ago, in reviewing the state of electron microscopy as applied to colloidal dispersions, Menold, Lüttge, and Kaiser (*36*) concluded that solvent crystallization caused by slow cooling usually led to unacceptable distortions, segregation, and reorganization of dispersed particles. The general consensus at that time was that, although intuitively appealing, electron microscopy was not a valid means of determining fluid microstructure due to freezing artifacts. However, in more recent times microscopy has been essential to settling long-standing questions such as the bicontinuous structure of oil-water-surfactant microemulsions (*49,50*), the structure of thermotropic blue phase (*51-53*) and twisted smectic liquid crystals (*54*), the organization of bilayers in dilute lamellar and isotropic L_3 phases (*55-57*), the organization of worm-like (*58-63*) and disc micelles (*58,64*), vesicle-micelle transitions (*65-67*), vesicle-lamellar phase transitions (*68-70*), the microstructure of lyotropic smectic and nematic phases (*64,71-75*), and aqueous polymer gels (*76,77*).

The obvious alternatives to microscopy for detailed structural characterization in more ordered lyotropic and thermotropic liquid crystals have been light, X-ray, and neutron scattering (*20,86,87*). Although each of these techniques have specific limitations, most samples can be examined in the fluid state without any special precautions. Unfortunately, the price of simplified sample preparation is increased difficulty in data interpretation. There are three principal problems in interpreting scattering data; the first is the averaging of structure over a macroscopic scattering volume, the second is the difficulty in uniquely inverting the Fourier scattering pattern to determine real space structure, and the third is the limited spatial dimensions available from a particular experiment.

Typical scattering experiments are constructed so that information is collected and averaged over a macroscopic volume (from mm^3 to cm^3) which contains a large number of individual particles, domains, etc. This averaging can be useful to assigning overall structural features and symmetries, but the detailed characteristics of the microstructure are averaged out. These characteristics, which may include structural features such as the number and distribution of polymer crosslinks or entanglements (*76,77*), membrane geometry and topology (*49,50,72-75*), liquid crystalline defects (*26, 51-54, 88*), particle shape and size polydispersity (*66-71*), grain and domain boundaries (*89, 90,91*), can be important, even dominant factors in macroscopic physical phenomena. For instance, the importance of dislocations and grain boundaries in the deformation of crystalline solids is well documented (*92*).

The more limiting feature of scattering data is that the information provided is indirect. A scattering experiment can be described mathematically as the Fourier transform of some density, atomic number, etc. distribution in the sample (*87,93*). To understand the experiments, the researcher has two options. The first is to uniquely invert and interpret the experimentally limited amount of Fourier data, a difficult if not impossible task. The second option is to invoke a mathematical model of the anticipated intra- and inter- particle structures and distributions and compare the theoretical scattering from the model with the experimental data. Although this second option is used almost exclusively, the physical limitations and assumptions inherent to the model can bias the data. Often, several conflicting models, and even unphysical models, can account for the same experimental data (*93*). An important example of this has been in investigations of middle phase microemulsions. Scattering patterns are not detailed enough to distinguish between several proposed structural models due to the polydispersity and random structures of the microemulsions (20); however freeze-fracture images by Jahn and Strey (49,50) clearly show that the microemulsion consists of interconnected, bicontinuous oil and water domains as originally proposed by Scriven (94).

Structure on both an average and discrete basis can be seen simultaneously and directly with electron microscopy. For freeze-fracture, image contrast is generated by simple mass-thickness contrast (24). For direct imaging, the thin samples typically prepared are nearly pure phase objects and the dominant contrast is phase contrast (23). In both cases, it is possible to relate the structures observed in the images to the structures in the sample with a minimum of modeling or interpretation. It is usually much easier to interpret the real space structure from an image than from scattering data in Fourier space; however, an added benefit of the image is that the Fourier space information can be readily obtained from the image by either optical diffraction from the TEM negative, or by modern digital image analysis (24).

The third benefit of microscopy is that structural studies of colloidal systems often need to cover a wide range of length scales simultaneously to determine the details of the particles making up the dispersion and their three-dimensional orientation and distribution in the suspending fluid. This is perhaps the greatest benefit of either freeze-fracture or cryo-TEM. In a given image, information is available at length scales from the resolution limit of about 2 nm to roughly 100 microns, or about 6 decades in length scale. For a typical scattering experiment, information can only be obtained over $2 - 3$ decades in length scale.

The main drawback to TEM investigations of microstructured fluids has been the fear that freezing-induced artifacts cause unacceptable changes in the chemical or physical state of the sample, leading to uncertainty and error in the interpretation of the TEM images. Recent advances in the science and art of sample freezing, including the jet-freezing technique pioneered by Müller, Meister, and Moor (95), and controlled plunge freezing of thin sandwich samples (96) have allowed for rapid advances in the applications of freeze-fracture to microstructured fluid systems. With cryo-TEM, it has been shown that water and other simple liquids can be frozen quickly enough to prevent crystallization (80,97). Studies of blue phase liquid crystals have shown that phases stable over as little as 0.5°C can be captured by rapid freezing (52,53).

With the expanding possibilities for investigations of microstructured fluids via high resolution microscopy, this review attempts to present the state of the art of freeze-fracture and cryo-TEM as it is applied to surfactants. We will present what we believe to be the necessary and sufficient conditions for reliable sample freezing, a discussion of fracture in frozen liquids, the limitation and resolution of replication, and a discussion of image resolution in freeze-fracture and cryoelectron microscopy.

Experimental Methods

Rapid Freezing. The most important step in any TEM investigation is the initial rapid freezing or quench step. The goal of rapid freezing is to remove heat at such a rate that 1) the details of the individual particle structure are retained and 2) the distribution and orientation of the particles are not disturbed. For most microstructured fluid systems, the second criterion is much more difficult to achieve and usually requires that the solvent or continuous phase be vitrified. In freeze-fracture, vitrification generally is taken to mean that the solvent phase contains no recognizable crystals larger than the typical resolution of the images. In cryo-TEM, it is easy to see if vitrification has occurred by examining the diffraction pattern of the solvent phase. Simple plunge freezing into liquid propane or ethane has been sufficient to vitrify water films up to about $50 - 200$ microns in thickness.

However, if the solvent phase does crystallize, dispersed particles and solutes are swept to the crystalline grain boundaries and information on the original orientation and distribution of the particles in the solvent is lost (see Figure 1). The distribution of the colloidal particles reflects the crystallization behavior of the solvent rather than any of the original properties of the system. Theoretical estimates of the cooling rates necessary to vitrify water range from 10^4 to 10^{10} K/sec (98), although experimentally, the minimum cooling rate appears to be in the range of 10^4 to 10^5 K/sec (97).

For freeze-fracturing a broad variety of liquid samples of any viscosity such as suspensions, solutions, and emulsions, the most practical method of rapid freezing is sandwiching the sample in a rigid container that is then contacted with a liquid cryogen. A wide variety of sample holders have appeared in the literature (99); however, the most commonly used are variations on the "copper sandwich" holders developed by Gulik-Krzywicki (100) and commercialized by Balzers (BUO-12-056T and variations; Hudson, New Hampshire). Typically, from $0.1 - 0.5$ ᵐl of sample liquid is pipetted onto one of the planchettes, then a second planchet is used to spread the liquid to form a thin, 10 - 50 micron thick film. A variant of this sample holder with an annular opening in the top planchette is used by Jahn and Strey (50). By far the most common method of rapid freezing is immersing the sample sandwich into a liquid cryogen, typically with a spring-loaded device to increase the relative specimen-cryogen velocity ($50,101$).

For cryo-TEM, the sample liquid is spread as a thin film (<0.15 μm) on polymer coated electron microscope grids prior to plunge freezing (23). The most difficult part of sample preparation for cryo-TEM is to ensure that the sample is thin enough (<150 nm) for electrons to pass through the sample. A small droplet of sample is placed on a grid, then blotted to form a thin film. This entire procedure is done within a temperature and humidity controlled chamber to eliminate evaporation of the solvent (23). Practice is required to form the thin films, and the high shear associated with the thinning process can create or destroy certain microstructures. This technique produces higher cooling rates due to the thinner samples than the freeze-fracture samples, although it is limited to fairly dilute, low viscosity (<10 cp) materials.

Freezing a multicomponent, structured fluid sample sandwiched between metal sheets in a liquid cryogen is a complicated heat transfer process that involves convection at the cryogen-sandwich boundary, conduction through the sandwich, and conduction and possible phase changes, crystallization, etc. within the fluid specimen (101). A simplified model of cooling shows that the cooling rate of samples is limited by convection from the cryogen to the specimen surface ($101,102$). The important criterion is that the Biot modulus of the specimen, hd/2k is $<< 1$; (see (102) for discussion) h is the heat transfer coefficient from the cryogen to the sample, d is the sample thickness, and k is the average sample thermal conductivity; for a typical freeze-fracture sample of thickness $100 - 200$ microns, the Biot number, Bi, is 0.05 to 0.5. For the thinner cryoelectron samples, Bi is even smaller. An important physical consequence of convection limited cooling (Bi < 1) is that the temperature is spatially uniform within the sample during freezing (102). In this approximation, the cooling rate of a sample of area A, volume V, heat capacity C_p, and density ρ, is:

$$\frac{dT}{dt} = \frac{-A}{V} h(T - T_C) \frac{1}{\rho C_p} \tag{1}$$

T is the sample temperature and T_C is the liquid cryogen temperature. Experimental measurements of the average cooling rate of a wide variety of specimens using various cryogens can be correlated using this simplified model (101). The cooling rate is proportional to the ratio of specimen surface area to volume, and inversely proportional to the thermal density, ρC_p, of the sample, which varies little between the typical materials encountered in complex fluids. Surprisingly, the cooling rate is independent of the sample thermal conductivity, and hence, virtually all samples of the same characteristic dimensions freeze at the same rate.

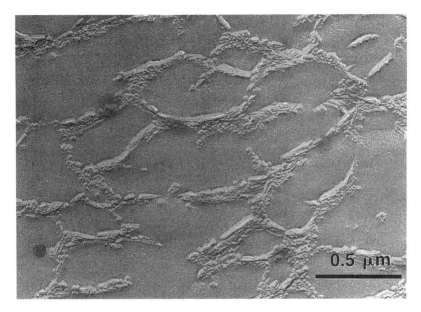

Figure 1a. Freeze-Fracture TEM image of poorly frozen dispersion of tobacco mosaic virus in water. The slow cooling rate allows the formation of characteristic ice patches, surrounded by particles forced to boundaries between crystals.

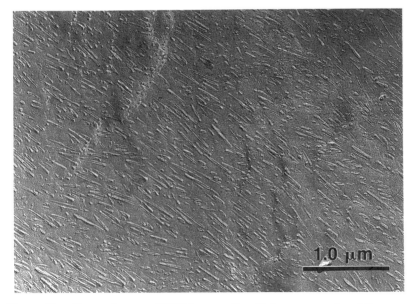

Figure 1b. Freeze-fracture TEM image of well frozen sample of the same dispersion as Figure 1a. The orientation and distribution of the nematically aligned specimen is identical to room temperature material.

Optimization of the freezing process can be achieved by maximizing h, the heat transfer coefficient, while minimizing T_C, the cryogen melting point, by proper choice of cryogen and the velocity at which the cryogen contacts the sample. A limited amount of cryogen boiling enhances the heat transfer coefficient; too much boiling results in the formation of a vapor film around the sample that drastically reduces the rate of heat transfer (*103*). Liquid nitrogen, and other cryogens at their boiling points should be avoided for this reason. The best practical cryogen appears to be ethane or propane cooled to near its freezing point by liquid nitrogen (*98-103*). The distribution and orientation of dispersed macromolecules in water has been preserved using both controlled plunge freezing in liquid propane and propane jet freezing, although it is impossible to say if the water surrounding the colloidal particles is amorphous or microcrystalline (see Figure 1).

Many of the phases of interest in the study of microstructured fluids are either above or below room temperature. This requires that a temperature and environment controlled chamber be coupled to the freezing apparatus. Sealed, two stage ovens have been used to observe temperature dependent phase transitions in lyotropic nematic micellar phases (*64*) and phospholipids (*105*). Bellare (*23*) has constructed a temperature and humidity controlled cell in which cryo-TEM samples can be equilibrated prior to plunge freezing. This chamber has been modified to do temperature jump experiments by focusing a high intensity mercury lamp onto the thin film sample (*79,80*). Jahn and Strey (*49,50*) have used a similar configuration to examine "middle-phase" microemulsions. Thermotropic liquid crystalline phases stable over less than 1°C have been resolved using quick freezing techniques followed by freeze-fracture (*51-53*), provided that the equilibrium phase transition occurs by nucleation and growth. Thermotropic smectic phases have been successfully quenched from above 100°C for freeze-fracture investigation (*54*).

Because the chemical and physical properties of a microstructured fluid cannot be optimized for rapid freezing by chemical or physical cryoprotectants without changing the structure, a judicious choice of systems to investigate usually is the difference between success and failure. As avoiding solvent crystallization is of primary importance to successful images, it is useful to understand the solvent properties that affect crystallization. Crystallization consists of two steps, nucleation of small crystals of a critical size, then the growth of these crystals. Except for extremely pure liquids, nucleation occurs heterogeneously at insoluble impurities; such impurities are often the colloidal particles we wish to study. The crystal growth velocity, u, is the more important quantity to minimize by appropriate choice of sample properties. The crystal growth velocity is proportional to the degree of supercooling, ΔT, the entropy of fusion, β, and the fraction, f, of acceptable sites on the interface (which reflects the steric constraints involved in packing solvent molecules into a different configuration), and inversely proportional to the solvent viscosity, μ:

$$u \, \alpha \frac{f\beta}{\mu} \frac{\Delta T}{T_m} \tag{2}$$

T_m is the equilibrium melting point. Steric effects, are generally more pronounced than are viscous effects in Equation 2. Branched hydrocarbons such as isohexane ($(CH_3)_2CH(CH_2)_2CH_3$) freeze at a much lower temperature (−153.7°C) than does straight chain n-hexane ($CH_3 (CH_2)_4CH_3$) (−94°C), indicating that the steric restrictions imposed by the branching make crystallization much more difficult. Cyclohexane, on the other hand, adopts a fairly rigid conformation in the liquid phase that is easy to pack into a crystalline lattice and freezes at about 6°C (*106*). Clearly, as many of the physical properties of these solvents are similar, to optimize the system for rapid freezing, isohexane, which is sterically hindered, is a much better choice than cyclohexane, which crystallizes readily. In aqueous solutions, salts, solutes and macromolecules that tend to disrupt water structure by hydrogen bonding can hinder crystallization. For instance, ice formation in polyacrylamide gels is suppressed to below -17°C (*107*).

The Freeze-fracture Technique. Once frozen successfully, cryo-TEM specimens are inserted directly into the TEM via a "cryo-transfer holder" which keeps the sample temperature below $-140°C$, the recrystallization temperature of water (*23*). The original sample is examined in the TEM, which presents opportunities and problems that are discussed in the next section. In the freeze-fracture technique, the sample is replaced by a metal "replica" that is compatible with the TEM environment. The fracture, etching, and replication steps of the freeze-fracture technique are carried out at low temperature and high to ultrahigh vacuum. Typically, the "copper sandwich" samples are loaded under liquid nitrogen into a hinged brass block fracture stage. The fracture stage has sufficient thermal mass that the specimens do not heat up significantly during the brief time they are exposed to air during transfer into a vacuum chamber. The fracture stage is clamped to a temperature controlled coldfinger within the vacuum chamber. The specimen stage fractures the sandwiches on opening; a sharp, quick break is preferable to a long, steady pull for reasons discussed below. The stress on the specimen is primarily tensile.

Griffith (*110*) proposed that, even prior to fracture, a brittle material contains a population of small cracks at impurities, boundaries, and other imperfections. When placed under sufficient stress, one or more of these cracks spreads into a brittle fracture, thereby decreasing the elastic energy at the expense of increasing the surface area, and hence the surface energy of the material. A crack will spread when the decrease in elastic energy is at least equal to the energy required to create the new crack surface. The surface energy can be thought of as the product of the new surface area created and the specific energy per unit area of the fracture surface. The Griffith theory states that the fracture will follow the path of least resistance (smallest specific surface energy or smallest molecular cohesion) provided that the fracture area created is not too large. In most two or more phase colloidal dispersions, the fracture surface propagates along the interface between the two phases, usually at particle-solvent boundaries (see Figures 1,2). Apparently, solvent-solvent cohesion and particle-particle cohesion are larger than solvent-particle cohesion. Alternately, small differences in the volume contraction on cooling between the solvent phase and the particles could lead to debonding prior to fracture, or to the formation of cracks at the particle-solvent interface. In either case, the weak zone appears to be at the interface and interpretation of freeze-fracture images is greatly simplified. Branton (*111*) has shown that the weak zone in bilayer membranes is along the hydrocarbon interior of the membrane; this also appears to be true for lyotropic lamellar phases in general.

Etching. The controlled sublimation of the solvent, known as etching, can be used to enhance the topographic variations in a fractured, microstructured fluid. However, removing too much of the solvent can alter the apparent location and distribution of dispersed particles, hide evidence of crystallization induced reorganization, and make the replicas difficult to pick up and clean. If the dispersed phase is entangled and self-supporting, as are polymer solutions and gels, or bicontinuous L_3 or microemulsion phases, a limited amount of etching can bring out the network structure (*76,77*) (see Figure 2). The important parameter in etching a sample is temperature. The sublimation pressure, hence the sublimation rate, is set once the sublimation temperature is fixed. The sublimation rate, S, in nanometers per second from a surface under vacuum can be obtained from gas kinetic theory:

$$S = \frac{P_s - P_v}{\rho_c} \left(\frac{M_c}{2\pi RT} \right)^{1/2} \times 10^7 \tag{3}$$

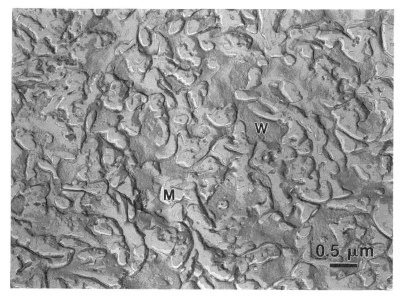

Figure 2. Freeze-fracture image of L_3 phase of cetyltrimethylammonium bromide-sodium 1-octane sulfonate-water mixture. The sample has been etched to bring out the relief between the water (W) and membrane (M) bicontinuous structures.

in which P_s is the saturation vapor pressure and P_v the background pressure of the sublimating phase in dynes/cm^2, M_c is the molecular weight (gr/mole) and ρ_c (gr/cm^3) is the density of the sublimating phase. R is the gas constant, and T is the sublimation temperature (K). The saturation vapor pressure for most solvents can be found in general engineering handbooks. If the background pressure of the solvent is greater than the vapor pressure at the temperature chosen, material will condense from the vacuum onto the sample, obscuring surface details. Hence, care is necessary to understand the relative composition of the residual gases in the vacuum chamber.

Replication. The goal in the replication process is to reproduce the fracture surface as accurately as possible with an electron opaque shadowing layer backed by a continuous, electron transparent, backing layer. The resolution in freeze-fracture electron microscopy is limited by imperfections in replication. Ideally, the evaporated metal atoms, which are usually a mixture of platinum and carbon, stick exactly where they land and form a structureless layer. However, the surface energy of the metal layer is much higher than that of the original fracture surface of water or hydrocarbons; hence, the metal film does not spread or "wet" the surface, but aggregates into small droplets (*112*). The aggregates grow in size, eventually merging with neighboring aggregates to form a continuous film. For most electron microscopy applications, the evaporation is stopped prior to the formation of a continuous metal layer. In our lab, about 1.5 nm of platinum carbon followed by about 15 nm of carbon backing gives optimum results. For proper interpretation of the replica, the sample material must be completely removed from the replica before viewing. In our lab, the cleaning method of Fetter and Costello (*113*) has always given the best results.

Contrast Mechanisms in Electron Micrographs

The contrast mechanism, that is, how the final image is related to the original microstructure, is quite different for freeze-fracture and cryo-TEM, as should be expected from the quite different chemical and physical characteristics of the specimens. An electron micrograph is the intensity distribution of an incident plane electron wave transmitted through the specimen and detected at the viewing plane. The plane of the specimen is denoted by $\mathbf{r} = (x,y)$ and z is the direction of the electron beam. Electrons can be scattered, absorbed, or phase shifted by the sample, all of which lead to contrast in the final image that must be related to structural features in the specimen.

The phase shift and absorbtion are generally described separately. In traversing a specimen of local thickness $t(\mathbf{r})$ and local potential distribution $V(\mathbf{r},z)$, the incident electron wavefunction, Ψ_t, is altered:

$$\Psi_t = \Psi_i \exp\{-i\eta(\mathbf{r})\} \tag{4}$$

corresponding to a phase shift of $\eta(\mathbf{r})$ in the transmitted wave, Ψ_t, relative to the incident wave. $\eta(\mathbf{r})$ is related to the local column average potential in the specimen:

$$\eta(\mathbf{r}) = \sigma \int_0^t V(\mathbf{r},z)\, dz \tag{5}$$

σ is a constant $= \pi/\lambda E_0$ for electrons of wavelength λ, and energy E_0 (*114*).

The incident wave is also scattered within the specimen; that part of the wave that scatters outside the objective aperture of the microscope will not be available for image formation and can be considered absorbed. Electrons are scattered elastically by the Coulomb potential and inelastically by plasmon and inner shell ionizations. The decrease in amplitude of the incident wave can be described as (*114, 115*):

$$\Psi_t = \Psi_i \exp \{ -S_p (E_0, \alpha)\rho(r) \, t(r)\} \qquad (6)$$

in which $S_p(E_0, \alpha)$ is dependent only on the electron energy and the aperture used (hence on the microscope and its operating conditions (*114*)) and $\rho(r)$ is the local, column average mass density within the specimen:

$$\rho(r) = \int_0^t \rho \, (r,z) \, dz \qquad (7)$$

$S_p\rho t$ can be considered the absorption of the specimen; it is directly proportional to the product of the mass and thickness of the sample. In general, the incident wave will undergo both a phase shift and absorption such that the transmitted amplitude is:

$$\Psi_t = \Psi_i \exp \{ \, i\eta(r) - S_p (E_0, \alpha)\rho(r) \, t(r)\} \qquad (8)$$

Specimens are classified as phase objects if the η term dominates the exponent, or as mass-thickness objects if the ρ term dominates. For a typical cryo-TEM sample, the small density differences make the specimen a phase object, and the primary contrast mechanism is phase contrast. For freeze-fracture replicas, the plantinum film scatters electrons very strongly, and the specimens are mass-thickness dominated. Hence, the information obtained from the two techniques is complementary.

Cryo-TEM and the Weak Phase Object Approximation. A further approximation known as the "weak phase object" is often used to help relate the sample features to the image. If $\eta(r) \ll 1$, and density variations can be neglected, the linearized form of Equation 8 is:

$$\Psi_t = \Psi_i \, [1 - i\eta(r)] \qquad (9)$$

The first term in the brackets corresponds to the unscattered beam, and the remaining terms to the scattered beam. (What is actually more important to the contrast are variations from the mean phase shift $\{\eta_{av} - \eta(r)\}$; hence, even though the absolute density or phase shift might be large, if the variance is small, the object can be considered a weak phase object.) However, detectors do not record the amplitude and phase of an image, but its intensity. Hence, a weak phase object would not have any contrast in an ideal image because

$$|\Psi_t|^2 = |1 - i\eta(r)|^2 \approx 1 \qquad (10)$$

where η^2 terms have been neglected in keeping with the weak phase object approximation. Phase contrast can only be realized in an image by further changing the relationship of the scattered beam with respect to the unscattered beam. In optical microscopy, a phase plate is introduced at the focal plane. This plate produces a phase shift of $\pi/2$ between the scattered and unscattered beams resulting in a transmitted beam amplitude of:

$$\Psi_t = \Psi_i \, [1 - i\eta(r)(\exp (-i\pi/2)], \qquad (11)$$

the image intensity, $|\Psi_t|^2 = 1 - 2\eta(r)$, and the image contrast is directly related to the projected potential of the specimen. However, phase plates are generally unavailable for electron microscopes and phase contrast is generated by the combined effects of apertures, lens defects and defocusing the image, which can be calculated using the "transfer theory" of imaging (*114-118*).

TEM images are complicated by the interaction of the transmitted beam amplitude with the "microscope transfer function" that describes the combined effects of imperfections in the electron optics and defocusing (*116,117*). The effect of apertures, aberrations and defocus are lumped into the "point spread function," $P(r)$, which can be thought of as the distortion induced by the microscope on the image amplitude and phase of an ideal point object. The real image amplitude, Ψ_{real}, is then related to the ideal image amplitude, Ψ_{ideal}, by a convolution of the ideal image amplitude with the point spread function (the ideal amplitude is given by Equation 9):

$$\Psi_{real} = \Psi_{ideal} * P(r) \tag{12}$$

where the convolution of two functions $f(r)$ and $g(r)$ is defined in the usual way:

$$f(r) * g(r) = \int_{-\infty}^{\infty} f(r') \, g(r - r') \, dr' \tag{13}$$

It is simpler and more conventional to use the two-dimensional Fourier space representation of both the point spread function and the image amplitude to describe these effects. $S(q)$ is defined as follows:

$$S(q) = \int_{-\infty}^{\infty} \Psi_t(r) \exp(-2\pi i \, (r \cdot q)) \, dr = \Psi_i[\delta(q) - iA(q)] \tag{14}$$

$A(q)$ is the two-dimensional Fourier transform of $\eta(r)$; $\delta(q)$ is the Dirac delta function and represents the unscattered beam. It is simpler to calculate the contributions of apertures, aberrations, and defocus to the Fourier transform of $P(r)$, which is known as the transfer function or $T(q)$. Each imperfection contributes an optical path difference between the actual and ideal waves. These path differences, when integrated over the lens area yields the transfer function, which describes the phase change in the incident wave induced by the imperfections (114). Spherical aberration, defocus and axial astigmatism contribute to the transfer function:

$$T(q) = \exp\{-i\phi(q)\}B(q) \tag{15}$$

where the phase shift $-\phi(q)$ introduced by the lens imperfections is given by:

$$\phi(q) = \frac{\pi}{2}\left[C_s\lambda^3 q^4 + 2\Delta f\lambda q^2 - 2\lambda C_a(q_x^2 - q_y^2)\right] \tag{16}$$

in which Δf is the defocus (positive for overfocus or a strongly excited lens), C_s is the coefficient of spherical aberration and C_a is the astigmatism coefficient of the objective lens. $B(q)$ is the aperture function, $B(q) = 1$ for $|q| < \alpha_o$, and zero everywhere else, where α_o is the acceptance angle of the objective aperture.

The maximum spatial frequency $q_{max} = \alpha_o/\lambda$ determines the best resolution that can be achieved in the image. For example, if α_o is .01 rad (a typical 40 μm aperture for an objective lens of focal length 2 mm), $q_{max} = 2.7$ nm^{-1} for 100 kV electrons and the resolution in the image is $1/q_{max}$ or .37 nm. The real image amplitude, Ψ_{real} is calculated from the inverse Fourier transform of $S_i(q) = S(q)T(q)$:

$$S_i(q) = [\delta(q)\exp[-i\phi(q)]B(q)] - i[A(q)\cos[\phi(q)]B(q)] - [A(q)\sin[\phi(q)]B(q)]; \tag{17}$$

$$\Psi_{real} = 1 - i\left[\eta(r) * F(r)\right] - \left[\eta(r) * G(r)\right] \tag{18}$$

where $*$ is the convolution integral as before (Equation 13) and $F(r)$ and $G(r)$ are the inverse Fourier transforms of $\cos[\phi(q)]B(q)$ and $\sin[\phi(q)]B(q)$ respectively. The convolutions in Equation 18 describe the blurring effect of the microscope lens aberrations on the projected potential of the specimen, $\eta(r)$. The image intensity is:

$$|\Psi_{real}|^2 = 1 - 2\eta(r) * G(r) \tag{19}$$

where the squared terms are neglected, consistent with the original weak phase object approximation. Equation 19 shows that the image intensity is linearly related to $\eta(r)$, and the image represents a distorted version of the structure of the specimen due to the lens aberrations and defocus. If the image contrast is defined as

$$C(r) = \frac{|\Psi_i|^2 - |\Psi_{real}|^2}{|\Psi_i|^2} \qquad (20)$$

where Ψ_i is the incident wave amplitude, then

$$C(r) = 2\eta(r) * G(r). \qquad (21)$$

An ideal microscope (for weak phase specimens) has no aperture ($B(q) = 1$) and introduces a uniform phase shift of $\phi = \pi/2$ and the ideal image contrast is:

$$C_{ideal}(r) = 2\eta(r) = 2\sigma \int_0^t V(r,z) \, dz. \qquad (22)$$

An ideal image of a weak phase object is the projected atomic potential of the specimen.

The behavior of the transfer function over the **q** range resolved in the image determines the image. Of the factors that influence $\phi(q)$, two are under the control of the microscopist - defocus and astigmatism. In any modern microscope, small lenses are available to minimize the effects of astigmatism; hence the third term in Equation 16 can be neglected. What the microscopist does is try to maximize $\sin[\phi(q)]B(q)$ for the **q** range in the specimen by adjusting the degree of defocus. Images are never recorded overfocus, because the spherical aberration and defocus terms are additive and cause rapid oscillations in $\sin[\phi(q)]$ (see Equation 16). The defocus where $\sin[\phi(q)] \approx 1$ over as large a range as possible is known as the optimum defocus. Although the transfer function for a typical 100 kV microscope is optimized for $\Delta f \approx 100$ nm, low frequency (low resolution) information up to 2 nm^{-1} (> 0.5 nm in real space) is transferred to the image with a relatively small amplitude. For cryo-TEM, radiation damage to the specimen limits the useful magnification to about 20,000 times, and the smallest features of interest such as micelles, bilayers, etc. are often 3 - 10 nm in extent and much larger underfocus conditions are needed to bring out this information. The approximate defocus required to optimize phase contrast for a particular resolution can be calculated from Equation 16. Provided **q** is small, the defocus term in Equation 16 is significantly greater than the spherical aberration term. Thus for a spatial frequency **q** the transfer function can be optimized by choosing the Δf value that makes $\sin[\phi(q)] = 1$, or $\phi = \frac{\pi}{2}$, that is

$$\Delta f = r^2/2\lambda \qquad (23)$$

Hence, for a typical rod micelle of diameter 5 nm, with $\lambda = 3.7$ pm for 100 kV electrons, $\Delta f = 3.4$ μm. A drawback to these large defocus values is that $\sin[\phi(q)]$ oscillates widely for higher resolution information in the image, making this information unreliable. That is, an artificial granularity is created by the oscillations in the transfer function at these large defocus values. The net effect is that even a featureless substrate can appear to have a great deal of texture when imaged under these conditions. Care must be taken in distinguishing structural features such as spherical micellar aggregates from the background granularity. This is further complicated by radiation damage to the specimen by the electron beam that can also generate small differences in thickness or potential that can appear to be real microstructure. A general rule for imaging surfactant microstructure is that rod-like micelles and bilayers are relatively easy to visualize due to their large extent in two dimensions, while spherical aggregates and micelles, especially less than 10 nm, are much harder to distinguish from the background, and should be examined with caution (see Figures 3,4).

Mass-Thickness Contrast from Freeze-fracture Replicas. The three-dimensional contours of the fracture surface intuitively appear to give rise to the contrast variations in a freeze-fracture micrograph. However, the surface itself is not

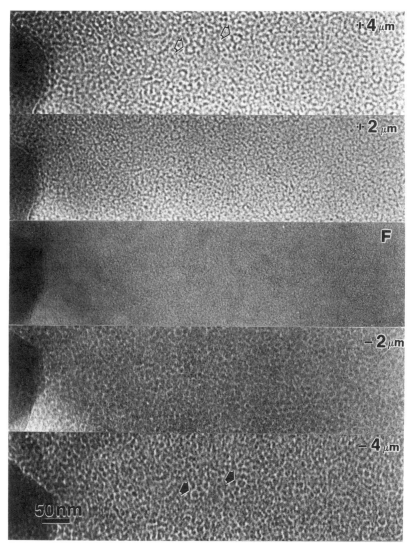

Figure 3.   Sequence of cryo-TEM images of a dilute dispersion of 7 nm diameter silica particles. The images are of the same area, but taken at different amounts of defocus, from +4 μm overfocus to –4 μm underfocus. The image taken at focus (F) is essentially featureless, consistent with the weak phase object approximation. However, distinct features emerge at higher defocus and coarsen with increasing defocus. The arrows at top and bottom show the contrast reversal common to phase contrast imaging. It is difficult in any of the images to unambiguously determine the location of the silica particles relative to the background graininess of the image. Spherical micelles and other small aggregates would be equally difficult to distinguish.

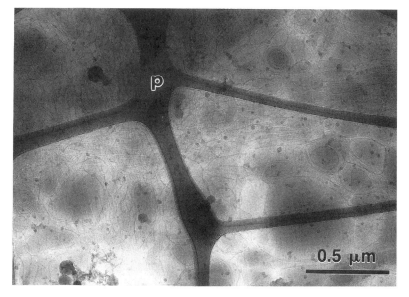

Figure 4. Cryo-TEM image of wormy vesicle phase of dimyristoylphosphatidylcholine and geraniol in water. In this image the bilayers are simple to see as they form the undulating, intertangled network of vesicles. P marks the holey polymer grid upon which the sample is suspended.

examined in the microscope, rather a metal shadowed replica of the surface. If the height (relative to some origin) of the fracture surface can be represented as a single valued function, $f(\mathbf{r})$, the metal thickness, $t_m(\mathbf{r})$, at that point in an unidirectionally shadowed replica, is given by (114):

$$t_m(\mathbf{r}) = d + f(\mathbf{r} - \mathbf{a}) - f(\mathbf{r}), \tag{24}$$

d is the thickness of metal deposited on a horizontal, flat surface, and $\mathbf{a}$ is the horizontal projection of the shadowing metal thickness in the direction of the evaporating metal beam. This expression is only meaningful if $t_m(\mathbf{r}) \geq 0$; the metal film thickness can never be negative. (The regions where Equation 24 is negative correspond to the areas of the specimen where there is no metal film.) If the metal is deposited normal to the fracture surface, $\mathbf{a}$ is zero and no thickness variations are present in the metal film, and hence no contrast variations are developed in the image. This is the case for the carbon reinforcing film used to strengthen freeze-fracture replicas. In a typical experiment, the total metal thickness, Δ, deposited on an oscillating quartz crystal monitor normal to the shadowing direction, and θ, the shadow angle from the horizontal, are measured. In these terms:

$$d = \Delta \sin\theta, \quad |\mathbf{a}| = \Delta \cos\theta \tag{25}$$

Fourier analysis of the thickness distribution is the most accurate method of determining the fracture surface contour, $f(\mathbf{r})$ from a freeze-fracture electron micrograph. The Fourier transform of $t_m(\mathbf{r})$ gives $T_m(\mathbf{q})$:

$$T_m(\mathbf{q}) = d\delta(\mathbf{q}) - [\, 1 - \exp(-2\pi \, \mathbf{q} \cdot \mathbf{a})] \, F(\mathbf{q}) \tag{26}$$

in which $F(\mathbf{q})$ is the Fourier transform of the surface profile $f(\mathbf{r})$. From Equation 26, $T_m(\mathbf{q})$ is zero along the line $\mathbf{q} \cdot \mathbf{a} = 0$; in physical terms, there is no information about fracture surface contours along the direction perpendicular to the shadow direction (114). An optical diffractogram of the freeze-fracture negative shows the line of zero information as a dark zone; this the most accurate determination of the shadowing direction. The fracture surface, $f(\mathbf{r})$, can be determined from the inverse transform of:

$$F(\mathbf{q}) = \frac{d\delta(\mathbf{q}) - T_m(\mathbf{q})}{[\, 1 - \exp(-2\pi\mathbf{q} \cdot \mathbf{a})]} \tag{27}$$

for $\mathbf{q}, \mathbf{q} \cdot \mathbf{a} \neq 0$.

Although Equations 24–27 are exact, they do not provide much insight into the relationship between surface contours and image contrast. A more intuitive description can be obtained by a linear approximation to Equation 24. If the shadowing direction is along the x direction of a Cartesian coordinate system, $\mathbf{r}$ can be written as an ordered pair (x,y) and $f(\mathbf{r})$ is the z coordinate corresponding to (x,y): $z = f(x,y)$. With these assumptions, Equation 24 becomes:

$$t_m(x,y) = \Delta \sin\theta + f((x - \Delta \cos\theta),y) - f(x,y) \tag{28}$$

Equation 25 was used to replace d and $\mathbf{a}$. If Equation 28 is expanded in a Taylor series,

$$f((x - \Delta \cos\theta),y) \approx f(x,y) - \Delta \cos\theta \frac{\partial f}{\partial x} + O[\, (\Delta \cos\theta)^2 \frac{\partial^2 f}{\partial x^2}] \tag{29}$$

If only linear terms are considered:

$$t_m(x,y) = \Delta \sin\theta + (f(x,y) - \Delta \cos\theta \frac{\partial f}{\partial x}) - f(x,y) \tag{30a}$$

$$t_m(x,y) = \Delta (\sin\theta - \cos\theta \frac{\partial f}{\partial x}) \tag{30b}$$

In this approximation, the metal thickness profile is simply related to gradients in the fracture surface profile along the shadow directions. Once the metal thickness distribution is known along the shadow direction, Equation 30b can be integrated to give the fracture surface profile directly (119,120). The error in this approximation can be appreciable at areas where the slope of the fracture surface changes rapidly (where $\partial^2 f / \partial x^2$ is large).

Image contrast for replicas depends on differential scattering of electrons from regions of the specimen with differing $t_m(x,y)$. For a composite replica made up of n layers of different materials, Equation 8 becomes:

$$\Psi_t = \Psi_i \ \exp\left(\sum_{i=1}^{n} -S_p\rho_i t_i(x,y)\right) \tag{31}$$

For electron image film, there exists an approximate relationship between the number of electrons striking the film, N, and the optical density, D:

$$D \approx D_{max}CN \tag{32}$$

D_{max} is the saturation density of the film and C is a constant that depends on the characteristics of the film and developer. Combining Equations 12 and 13 gives the relation between the density of the negative and the film thickness:

$$D \approx D_{max}C \ \Psi_i \ \exp\left(\sum_{i=1}^{n} -S_p\rho_i t_i(\mathbf{r})\right) \tag{33}$$

At first glance, Equation 33 seems to have an inordinate number of parameters that are difficult to measure. However, if ratios of densities, rather than the absolute densities are considered, the result is considerably simplified. In an image of an area of the replica devoid of metal with only the carbon backing layer, the optical density of the negative is:

$$D_c = D_{max}C \ \Psi_i \ \exp\left(-S_p\rho_c t_c\right) \tag{34}$$

At a point $\mathbf{r}$, the metal layer thickness is given by $t_m(\mathbf{r})$ as given in Equation 24, in addition to the carbon layer thickness, t_c, which is uniform over the entire replica. Taking the ratio of the density at the point $\mathbf{r}$ to the reference carbon film density gives:

$$\frac{D_m(\mathbf{r})}{D_c} = \frac{D_{max}C\Psi_i \exp\left(-S_p\rho_c t_c - S_p\rho_m t_m(\mathbf{r})\right)}{D_{max}C\Psi_i \exp\left(-S_p\rho_c t_c\right)} = \exp -S_p\rho_m t_m(\mathbf{r}) \tag{35}$$

and only a single parameter is required to relate the metal thickness to the optical density. Therefore, a single negative contains all the information necessary to determine the fracture surface profile. The carbon density, D_c, is the maximum density recorded on the negative. This simple relationship of metal thickness to optical density, when combined with Equations 27 or 30b, show that there is a direct relationship between the fracture surface profile and the images recorded with the electron microscope. Absolute heights might be determined directly from the electron microscope images; however, recent advances in imaging freeze-fracture replicas with the scanning tunneling microscope allow for a much simpler and direct determination of the three-dimensional fracture profiles with better resolution (123,124).

Conclusions

The early disappointments of TEM investigations of microstructured fluids have undoubtedly limited the number of colloid and materials scientists using this very promising experimental technique. These early investigations were plagued by slow cooling rates that led to unacceptable distortions of fluid structure. With the major advances in ultra-rapid freezing, such problems have now been eliminated. The benefits of direct visualization of fluid structure with near molecular resolution are only slowly being realized. As more investigators come to understand the potential of being able to study the real-space structure of complex dispersions, emulsions, gels, and solutions, the pace of progress will accelerate. Freeze-fracture electron and cryo-TEM microscopy have made it possible to see the three-dimensional structure and organization in a wide range of systems that have previously thwarted analysis; the future promises to be equally bright.

Acknowledgments

We would like to thank M.J. Costello, Y. Talmon, and R. Strey for their comments and continuing discussions on the role and possibilities of freeze-fracture and cryo-TEM. Financial support was provided by the donors of the Petroleum Research Fund and by NSF Grant #CTS - 9102719 and NIH Grant 47334.

Literature Cited

1. Cates, M.E.; Candau, S.J., *J. Phys.: Condens. Matter* **1990**, *2*, 6869.
2. Turner, M.S., Marques, C., and Cates, M. E., *Langmuir*, **1993**, *9*, 695.
3. Hoffmann, H. and Ebert, G., *Angew. Chem. Int. Ed. Engl.*, **1988**, *27*, 902.
4. Langevin, D. *Annu. Rev. Phys. Chem.*, **1992**, *43*, 341.
5. Granek, R. and Cates, M.E., *J. Phys. Chem.*, **1992**, *96*, 4758.
6. Candau, S., Hirsch, E. and Zana, R., *J. Colloid Interface Sci*, **1985**, *105*, 521.
7. Porte, G., Appell, J., *Europhys. Lett.*, **1990**, *12*, 190.
8. Hoffmann, H., et al., *Ber. Bunsen-Ges. Phys. Chem.*, **1981**, *25*, 877.
9. Kato, T., Anzai, S., Seimiya. T., *J. Phys. Chem.*, **1987**, *91*, 4655.
10. Magid, L.J., *Colloids Surf.*, **1986**, *19*, 129.
11. Hirata, H., Sakaiguchi, Y. , *J. Colloid Interface Sci.*, **1988**, *121*, 300.
12. Olsson, U., Söderman, O, Guéring, P., *J. Phys. Chem.*, **1986**, *90*, 5223.
13. Makloufi, R., et al., *J. Phys. Chem.*, **1989**, *93*, 8095.
14. Shikata, T., Hirata, H., Kotaka, T., *Langmuir*, **1987**, *3*, 1081; **1988**, *4*, 354.
15. Khatory, A., et al., *Langmuir*, **1993**, *9*, 933.
16. Anet, F.A.L., *J. Am. Chem. Soc.*, **1986**, *108*, 7102.
17. Hoffmann, H., et al., *Adv. Colloid Interface Sci.*, **1982**, *17, 275*.
18. Bayer, O., et al., *Adv. Colloid Interface Sci.*, **1986**, *26*, 177.
19. Brown, W., Johansson, K., Almgren, M., *J. Phys. Chem.*, **1989**, *93*, 5888.
20. Kahlweit, M., et al., *J. Colloid Interface Sci.*, **1987**, *118*, 436.
21. Gabriel, N.E., Roberts, M.F., *Biochemistry*, **1986**, *25*, 2812 .
22. Eum, K.M., et al., *Biochemistry*, **1989**, *28*, 8206.
23. Bellare, J.R., et al., *J. Elect. Microsc. Tech.*, **1988**, *10* , 87.
24. Zasadzinski, J.A.N.; Bailey, S. M., *J. Elect. Microsc. Tech.*, **1989**, *13*, 309.
25. Buchnall, A., et al., *Mol. Cryst. Liq. Cryst.*, **1969**, *7*, 215.
26. Kléman, M., et al., *Phil. Mag.*, **1977**, *35*, 33.
27. Goodman, J.F. and Clunie, J.S. In *Liquid Crystals and Plastic Crystals*, ed. by G. Gray and P. Winsor, Wiley and Sons, NY, **1974**, Vol. 2.
28. Bachmann, L. and Talmon, Y., *Ultramicrosc.*, **1984**, *14*, 211.
29. Talmon, Y., and Miller, W. G., *J. Colloid Interface Sci.*, **1978**, *67*, 284.
30. Talmon, Y., et al., *Rev. Sci. Instrum.*, **1979**, *50*, 698.
31. Falls, A.H., et al., *Biochim. Biophys. Acta*, **1982**, *693*, 364.
32. Biais, J. et al., *J. Microsc.*, **1981**, *121*, 169;
33. Biais, J., et al., *C. R. Acad. Sci. Paris*, **1977**, *285* , C213.
34. Eley, D.D., et al., *J. Colloid Interface Sci.* , **1976**, *54*, 462.
35. Geymayer, W.F., *J. Polymer Sci. Symp.* **1974**, *44*, 25.

36. Menold, R., et al., *Adv. Colloid. Interface Sci.,* **1976**, *5,* 281.
37. Wade, R.H., et al., *J. Colloid Interface Sci.,* **1986**, *114,* 442.
38. Kunitake, T. and Okahata, Y., *J. Amer. Chem. Soc.,* **1977**, *99,* 3860.
39. Deguchi. K. and Mino, J. *J. Colloid Interface Sci.,* **1978**, *65,* 155.
40. Shikata, T., et al., *J. Colloid Interface Sci.,* **1987**, *119,* 291.
41. Shikata, T., et al., *J. Electron Microsc.,* **1987**, *36,* 168.
42. Shikata, T., Hirata, H., and Kotaka, T., *Langmuir,* **1988**, *4,* 354.
43. Imae, T. Kamiya, R., and Ikeda, S., *J. Coll. Interface Sci.,* **1984**, *99,* 300.
44. Talmon, Y., *J. Colloid Interface Sci.,* **1983**, *366,* 366.
45. Kilpatrick, P., et al., *J. Coll. Int. Sci,* **1985**, *107,* 146.
46. Kilpatrick, P., et al., In *Surfactants in Solution,* K. L. Mittal and P. Bothorel, Eds., Plenum, NY, **1986**, Vol 4, pp. 489-500.
47. Vinson, P.K., and Talmon, Y., *J. Colloid Interface Sci.,* **1989**, *133,* 288.
48. Zasadzinski, J.A. *J. Colloid Interface Sci.,* **1988**, *122,* 288.
49. Jahn, W. and Strey, R., In *Proceedings of the Workshop on Physics of Amphiphilic Layers,* J. Merneir, D. Langevin, and N. Boccara, eds. Springer-Heidelberg, Berlin, **1987**.
50. Jahn, W. and Strey, R., *J. Phys. Chem.,* **1988**, *92,* 2294.
51. Costello, M., et. al., *Phys. Rev. A,* **1984,** *29,* 2957.
52. Zasadzinski, J.A.N., et al., *Phys. Rev. Lett.,* **1986**, *57,* 364.
53. Berreman, D.W., et al., *Phys. Rev. Lett.,* **1986**, *57,* 1737.
54. Ihn, K.J., et al., *Science,* **1992**, *258,* 275.
55. Strey, R., et al., *Langmuir,* **1990**, *6,* 1635.
56. Hoffmann, H., et al., *Langmuir,* **1992**, *8,* 2629.
57. Hoffmann, J., Thunig, C., and Valiente, M., *Coll. Surfaces,* **1992**, *67,* 223.
58. Vinson, P. K., et al., *J. Colloid Interface Sci.,* **1991**, *142,* 74.
59. Cochin, D., et al., *Macromolecules,* **1992**, *25,* 4220.
60. Clausen, T. M., et al., *J. Phys. Chem.,* **1992**, *96,* 474.
61. Magid, L. J., Gee, J., and Talmon, Y., *Langmuir,* **1990**, *6,* 1509.
62. Lin, Z., Scriven, L. E., and Davis, H. T., *Langmuir,* **1992**, *8,* 2200.
63. Kaler, E.W., et al., *J. Phys. Chem.,* **1992** *96,* 6698.
64. Sammon, M., et al., *Phys. Rev. Lett.,* **1987**, *57,* 2834.
65. Kamenka, N., et al., *Colloids and Surfaces,* **1992**, *67,* 213.
66. Herrington, K. L., et al., "Phase Behavior of Aqueous Mixtures of DTAB and SDS," *J. Physical Chemistry,* **1993**, in press.
67. Walter, A., et al., *Biophys. J.,* **1991**, *60,* 1315.
68. Kaler, E.W.; et al., *Science* **1989**, *245,* 1371.
69. Murthy, A.K., et. al., *J. Coll. Interface Sci.* **1991**, *145,* 598.
70. Talmon, Y.; Evans, D.F.; Ninham, B.W. *Science,* **1983**, *221,* 1047.
71. Zasadzinski, J.A. and Meyer, R.B., *Phys. Rev. Lett.,* **1986**, *56,* 636.
72. Zasadzinski, J., et al., *Phil. Mag. A,* **1985**, *51,* 287.
73. Zasadzinski, J. A. N., *Biophys. J.,* **1986**, *49,* 1119.
74. Zasadzinski, J.A.N., et al., *J. Elect. Microsc. Tech.,* **1986**, *3,* 385.
75. Dubois, M., Gulik-Krzywicki, T. and Cabane, B., *Langmuir,* **1993**, *9,* 673.
76. Zasadzinski, J., et al., *Macromolecules,* **1987**, *19,* 2960.
77. Zasadzinski, J.A.N., et al., *Chem. Eng. Comm.,* **1987**, *52,* 283.
78. Heuser, J.E. and Reese, T. S., *J. Cell Biol.,* **1976**, *70,* 357a.
79. Talmon, Y., et al., *J . Electron Microsc. Tech.,* **1990**, *14,* 6.
80. Chestnut, M.H., et al., *J. Microsc. Tech.,* **1992**, *20,* 95.
81. Steere, R.L., *J. Electron Microsc. Tech.,* **1989**, *13,* 159.
82. Gebhardt, C., et al., *Z. Naturforsch.,* **1977**, *32C,* 581.
83. Hatfield, J.C., **1978**, Ph.D. Thesis, University of Minnesota, Minneapolis, Minnesota, unpublished.
84. Sjöblom, E. and Friberg, S., *J. Colloid Interface Sci.,* **1978**, *67,* 16.

85. Robards, A.W. and Sleytr, V.B., *Low temperature methods in biological electron microscopy*. In *Practical Methods in Electron Microscopy*. A.M. Glauert, ed. Elsevier Press, Amsterdam, **1985**, Vol. 10, pp. 5-133.
86. Luzzati, B. and Tardieu, A., *Ann. Rev. Phys. Chem.*, **1974**, *25*, 79.
87. Berne, B. J. and Pecora, R., Dynamic Light Scattering. J. Wiley and Sons, NY, **1976**.
88. Allain, M., *J. Phys., Paris*, **1985**, *46*, 225.
89. Knoll, W., et al., *J. Chem. Phys.*, **1983**, *79*, 3439.
90. Gaub, H., et al., *Biophys. J.*, **1984**, *45*, 725.
91. Gaub. H., et al., *Chem. Phys. Lipids*, **1985**, *37*, 19.
92. Friedel, J., *Dislocations*, Pergamon Press, London, **1964**.
93. Karle, J., *Proc. Nat. Acad. Sci. USA*, **1977**, *74*, 4704.
94. Scriven, L.E., *Nature (London)*, **1977**, *283*, 123.
95. Müller, M., Meister, N. and Moor, H., *Mikroscopie (Wien)*, **1980**, *36*, 129.
96. Costello, M.J., Fetter, R. and Corless, J.M., In *Science of Biological Specimen Preparation*, SEM Inc., AMF O'Hare, Chicago, IL, **1984**, pp. 105-115.
97. Adrian, M., et al., *Nature*, **1984**, *308*, 32.
98. Mayer, E., *J. Microsc.*, **1985**, *140*, 3.
99. Costello, M.J. and Corless, J.M., *J. Microsc.*, **1978**, *112*, 17.
100. Gulik-Krzywicki, T. and Costello, M.J., *J. Microsc.*, **1978**, *112*, 103.
101. Zasadzinski, J.A., *J. Microscopy*, **1988**, *150*, 137.
102. Bailey, S.M. and Zasadzinski, J.A., *J. Microscopy*, **1991**, *163*, 307.
103. Bennett, C.O. and Myers, J.E., *Momentum, Heat, and Mass Transfer* 3rd edn. McGraw-Hill, NY, **1982**, Ch. 19.
104. Gilkey, J.C. and Staehelin, L.A., *J. Elect. Micros. Tech.*, **1986**, *3*, 177.
105. Zasadzinski, J., and Schneider, M., *J. Phys., Paris*, **1987**, *48*, 2001.
106. Roberts, J.D. and Caserio, M.C. *Basic Principles of Organic Chemistry*. W.A. Benjamin, Inc. Menlo Park, CA, **1977**, Ch. 12.
107. Tanaka, T., Ishiwata, S. and Ishimoto, C., *Phys. Rev. Lett.*, **1977**, *38*, 771.
108. Brüggeller, P. and Mayer, E., *Nature (London)*, **1980**, *288*, 569.
109. Israelachvili, J.N., et al., *Quart. Rev. Biophys.*, **1981**, *13*, 121.
110. Dieter, G., *Mechanical Metallurgy*, 2nd Ed. McGraw-Hill, NY, **1976**, Ch. 7.
111. Branton, D. In: *Freeze-etching, Techniques and Applications*, E. L. Benedetti, and P. Favard, Eds. Societé Française de Microscope Électronique, Paris, **1973**, Ch. 10.
112. Adamson, A.W., *Physical Chemistry of Surfaces*, 5th Ed. John Wiley and Sons, Inc., NY, **1990**, Ch. 4.
113. Fetter, R.D. and Costello, M. J., *J. Microsc.*, **1985**, *141*, 277.
114. Misell, D.L. *Image Analysis, Enhancement and Interpretation*. In: *Practical Methods in Electron Microscopy*. A.M. Glauert, ed. North Holland, Amsterdam, **1978**, Vol. 7, Ch. 3, 7.
115. Bellare, J., Ph.D. Thesis, University of Minnesota, Minneapolis, MN, **1964** unpublished.
116. Hanszen, K.J., *Z. Angew. Phys.*, **1966**, *20*, 427.
117. Hanszen, K.J., In: *Advances in Optical and Electron Microscopy*, Vol. 4, ed. by R. Barer and V. E. Coslett, Academic Press, London, (1971) 1.
118. Lenz, F.A., In *Electron Microscopy in Materials Science*, ed. by U. Valdre, Academic Press, New York, **1971**, 541.
119. Krbecek, R., et al., *Biochim. Biophys. Acta*, **1979**, *554*, 1.
120. Rüppel, D. and Sackmann, E., *J. Physique*, **1983**, *44*, 1025.
121. Gruler, H., *Acta. Histochem. Suppl. Band*, **1981**, *23*, 55.
122. Rasigni, M., et. al., *J. Opt. Soc. Am.*, **1981**, *71*, 1124.
123. Zasadzinski, J.A. et al., *Science*, **1988**, *239*, 1014.
124. Woodward, J., et al., *J. Vac. Sci. Tech.*, **1991**, *B9*, 1231.

RECEIVED July 8, 1994

Chapter 6

Cryo-Transmission Electron Microscopy Investigations of Unusual Amphiphilic Systems in Relation to Their Rheological Properties

D. Danino[1], A. Kaplun[1], Y. Talmon[1], and R. Zana[2]

[1]Department of Chemical Engineering, Technion-Israel Institute of Technology, Haifa 32000, Israel
[2]Institut Charles Sadron, Centre National de la Recherche Scientifique, 6 rue Boussingault, 67083 Strasbourg Cedex, France

In this paper we report studies of two types of unusual amphiphiles: the dimeric amphiphiles, alkanediyl-α,ω-bis(alkyldimethylammonium bromide) (I) and the polyamphiphiles poly(disodium maleate-*co*-alkylvinylether) (II), by fluorescence probing, viscometry and cryo-transmission electron microscopy. We show that in amphiphiles (I) the linking of the two surfactant moieties at the level of the head-group by a short alkanediyl spacer results in a strongly enhanced tendency to micelle growth with respect to the corresponding monomeric surfactants, and in a peculiar rheological behavior. The polyamphiphile (II) with a hexadecyl side chain forms long entangled thread-like micelles by end-to-end linking of polymer chains; its solutions are visually viscoelastic. The addition of an anionic or a nonionic surfactant breaks these links and results in the disappearance of the viscoelasticity. The results demonstrate the wealth of information that the coupling of cryo-TEM and other techniques can provide on complex systems such as the ones described here.

Surfactants self-assemble in aqueous solution to form micelles of varied shapes: spherical or spheroidal, elongated (prolate), or disk-like (oblate). The study of the means to control the shape of micelles has always been an important topic in surfactant research, of both academic and applied interest, as the micellar shape determines to a large extent the properties, and also some applications, of surfactants in solution. The introduction of the so-called surfactant packing parameter by Israelachvili et al. (*1*), has provided a criterion to help predict the shape micelles of a given surfactant will adopt in aqueous solution, at concentrations well above the critical micellar concentration (cmc), where micellar growth has taken place. The packing parameter, P, is:

$$P = v/a_M \, l \qquad (1)$$

where v and l are the volume and length of the surfactant hydrophobic moiety, one or two alkyl chains in most instances, and a_M the surface area occupied by a surfactant head-group at the interface between water and the micelle hydrophobic

0097–6156/94/0578–0105$08.00/0

core. The packing parameter has been a tool in the understanding of the effect of various parameters such as surfactant chain length, branching of the alkyl chain, head-group size, ionic strength, temperature, nature of the counterion, etc., on the shape of micelles of conventional surfactants. Recall that l and v can be calculated using Tanford's equations (2). For a conventional surfactant (one alkyl chain and one head-group), v/l is close to 21 $\mathring{A}^2$, the cross-section of the alkyl chain, when the alkyl chain contains 12 or more carbon atoms.

Recently, we have proposed (3) a new way for controlling the shape of micelles, by linking the amphiphiles at the level of the head-group, thus generating oligomeric (dimers, trimers, tetramers, etc.) amphiphiles and polyamphiphiles (when the degree of polymerization is large, typically above 50-100). Figure 1 represents the chemical formulae of the dimeric (I) and polymeric (II) amphiphiles that we have started to investigate (3-10).

In amphiphiles of type I two surfactant moieties are linked at the level of the head group by a polymethylene chain (spacer), containing s carbon atoms. These surfactants, also called "gemini", or "bipolar" surfactants, can be formally considered as the dimers of the two unequal- chains surfactants $C_mH_{2m+1}(C_{s/2}H_{s+1})N^+(CH_3)_2Br^-$. The polyamphiphiles of type II are alternated copolymers of disodium maleate and alkylvinylether, often referred to as "polysoaps".

The idea underlying the use of oligo- and polyamphiphiles is now described. In conventional surfactants the head-groups are randomly distributed on the surface separating the aqueous phase and the micelle hydrophobic core, with a distribution of distances between head-groups peaked at the thermodynamic distance d_T (Figure 2A), determined by the opposite forces at play in micelle formation. d_T is about 7 to 10 $\mathring{A}$, in view of the reported values of the surface area per head group. When using dimeric amphiphiles such as those in Figure 1 for instance, the distribution of distances may become bimodal. Indeed, if the spacer is short enough (small s-value, say, less than 6) it is probably fully stretched at the water-core interface (5). The head-group distance distribution function then exhibits a narrow maximum at the distance d_s, corresponding to the extended length of the spacer(11), and another maximum at the thermodynamic equilibrium distance d_T (Figure 2B). d_T is not expected to differ much from its value for conventional surfactants. A similar situation exists for oligoamphiphiles and polyamphiphiles. This modification of the distribution of head-group distances, as well as the effect of the chemical link between head-groups on the packing of surfactant alkyl chains in the micelle core, is expected to affect strongly the curvature of surfactant layers and, thus, the micelle shape. At the outset, it should be realized that as s increases, d_s tends towards d_T. This occurs at s = 6-7. For such dimeric surfactants the microstructure is not expected to differ much from that of the corresponding conventional surfactants.

With this in mind we have undertaken a study of the microstructure of aqueous solutions of a series of dimeric amphiphiles of type I and of polyamphiphiles of type II (PS10, and PS16), using for this purpose the time-resolved fluorescence quenching (TRFQ) technique (12, 13) and transmission electron microscopy at cryogenic temperature (cryo-TEM) (14). The first method permits the determination of the micelle aggregation number. Cryo-TEM gives direct visualization of the micelles. These studies have been performed in relation with the visual observation of the rheological behavior of the solutions for amphiphiles of type (I). Some viscosity measurements have been performed for the solutions of polyamphiphiles.

In this paper we give first a brief description of TRFQ and cryo-TEM, followed by a review of some reported results (7-10), as well as new ones.

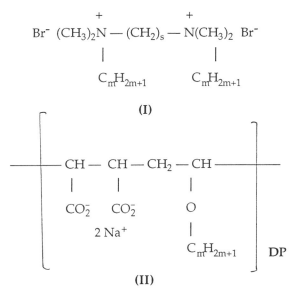

Figure 1. (I) m-s·m amphiphiles: s = 2 - 20; m = 8, 10, 12, 14, and 16; (II) polyamphiphiles: m = 10 is PS10; m = 16 is PS16.

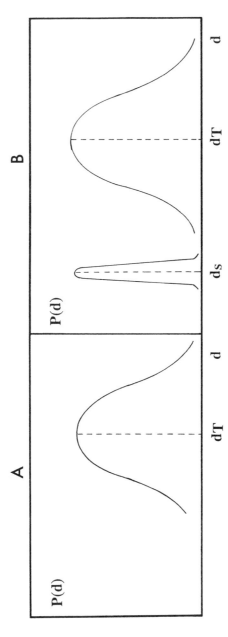

Figure 2. Schematic representation of the distribution of distances between head-groups in micelles of conventional (A) and dimeric (B) surfactants.

Experimental.

Materials. The surfactants (I) were the same as in previous investigations (3, 7), or synthesized and purified similarly. The polyamphiphiles PS10 and PS16 of polymerization degrees DP = 1000 and 4000, respectively, were gifts from Dr. R. Varoqui and E. Pefferkorn (Institut C. Sadron, Strasbourg). They originated from the same batches as in previous studies (7-10).

Time-resolved fluorescence quenching. This method uses a fluorescence probe, here pyrene, and a quencher of this probe fluorescence, here the tetradecylpyridinium ion in its chloride salt. Probe and quencher are selectively solubilized in the micelles. The decay of the probe fluorescence is determined first in the absence of quencher, at a low (< 0.05) [probe]/[micelle] molar concentration ratio, and then in the presence of quencher, at a still low [probe]/[micelle], and with a [quencher]/[micelle] molar concentration ratio close to 1. The time dependence of the fluorescence intensity in both situations is given respectively by (15, 16):

$$I(t) = I(0)\exp(-t/\tau) \tag{2}$$

$$I(t) = I(0)\exp\left\{-A_2t - A_3\left[1 - \exp\left(-A_4t\right)\right]\right\} \tag{3}$$

I(0) is the intensity at time zero, and A_2, A_3, and A_4 are three time-independent constants. In the case of amphiphiles (I) we consistently found $A_2^{-1} = \tau$. This indicates that the probe and quencher distributions between micelles are frozen on the fluorescence time scale. Then, A_3 and A4 are given by (15, 16):

$$A_3 = [quencher]/[micelle] \quad \text{and} \quad A_4 = k_Q \tag{4}$$

where k_Q is the rate constant for intramicellar quenching. From A_3 one can obtain the molar micelle concentration ([micelle]). The micelle aggregation number, N, is calculated from:

$$N = A_3 (C - cmc)/[quencher], \tag{5}$$

where C is the molar surfactant concentration, or the polyamphiphile concentration expressed in moles of repeat unit per liter.

The fluorescence decay curves, plots of I(t) vs. t were recorded using a single-photon counting setup (8).

Cryogenic Temperature Transmission Electron Microscopy (Cryo-TEM). Most analytical techniques are indirect in the sense that microstructural information is obtained only by analysis of the experimental data using a preconceived model. Thus, direct imaging is needed to produce unambiguously that physical model. Transmission electron microscopy is the obvious choice for direct imaging of microstructured systems, because of its high resolution and relative ease of data interpretation, even (in most cases) without additional image analysis. Image analysis may be used to improve data quality (contrast enhancement, three-dimensional reconstruction, etc.) and to quantify it. To examine a complex fluid by TEM, it must be made compatible with the instrument. While the older staining-and-drying techniques cannot preserve the original microstructure of these labile systems, ultra-rapid thermal fixation, that leads to water vitrification, does preserve it. In fact, because of the high tendency of water molecules to rearrange into a hexagonal crystalline solid, they serve as a built-in indicator of the success of fixation. If water is vitrified, then, in all probability, the much heavier molecules of the other components have not rearranged. Vitrified specimens are then examined in the TEM at cryogenic temperature. This is direct imaging cryo-TEM.

Specimens for direct-imaging cryo-TEM are prepared under controlled humidity (usually 100% relative humidity) and temperature (typically between 0 to 80 °C) in the controlled environment vitrification system (CEVS), as described by Bellare et al. (*14*): thin liquid films (ca. 0.2 micrometer thick), spread over holes in a perforated carbon film, supported on 200 mesh electron microscope copper grids, are quenched into liquid ethane at its freezing point, and transferred into a Gatan cooling holder in its 'work-station'. The holder is then inserted into the transmission electron microscope. Ultra-rapid cooling and the precautions taken prior to quenching assure that the structures observed in these cryo-specimens are indeed those found in the original systems studied by this technique (*3, 9, 10, 17-19*).

Results and Discussion

Dimeric Surfactants. Figure 3 shows the plots of micelle aggregation number, N, against surfactant concentration, C, for the homologues of the 12-s-12 series with s = 3, 5, 6, and 8. These results demand two remarks.

(1) The plots appear to extrapolate at low C, close to the cmc, to values of N that are only slightly dependent on the spacer carbon number, s. These values increase from about 20 to 25, as s decreases from 8 to 3. Such values of N correspond to 40 to 50 dodecyl chains per micelle. These numbers are close to those found for the corresponding "monomeric" surfactants $C_{12}H_{25}(C_{s/2}H_{s+1})N^+(CH_3)_2Br$, for which a small decrease of N upon increasing s was also reported (*20*). These values are also close to those calculated from Tanford's equations (*2*) for the aggregation number of a spherical micelle with a hydrophobic core of a radius equal to the length of a dodecyl group in a fully extended conformation (i.e., 16.6Å). Thus, close to the cmc the micelles of dimeric surfactants are nearly spherical.

(2) N increases with C in the case of the four surfactants we have studied. This increase becomes more pronounced as s decreases. For 12-6-12 and 12-8-12, the increase of N with C is rather small, and the micelles are thus expected to remain spherical or spheroidal in the whole concentration range. In this respect these dimeric surfactants behave much like their corresponding "monomers", the propyl- and butyl- dodecyldimethylammonium bromides(*20*). The increase of N with C is more important, but still modest, for 12-5-12. It becomes very important for 12-3-12, indicating a change of micelle shape; the corresponding monomer shows only a small increase of N with C (*20*). The tendency to micellar growth of dimeric surfactants is thus much stronger than that of the corresponding monomers, when the spacer carbon number is small enough. The large increase of N with C for 12-3-12 indicates that the micelles of this surfactant become strongly polydisperse as they grow (*1*).

The aggregation number of 12-2-12 was not determined. However, the rheology of the solutions of this surfactant was very peculiar. 1.3% solution was normally fluid when gently shaken after a long rest (12 hours or more), but became visually viscoelastic when shaken strongly for a sufficient time (shear - thickening), with a recoil of trapped air bubbles when their motion was stopped. A 4% solution was viscoelastic and almost gel-like, whereas a 15% solution was again fluid, although quite viscous (*7*). These observations are similar to those reported by Hoffmann for other surfactants (*This book, Chapter 1*), and suggested the presence of long thread-like micelles in 12-2-12 solutions.

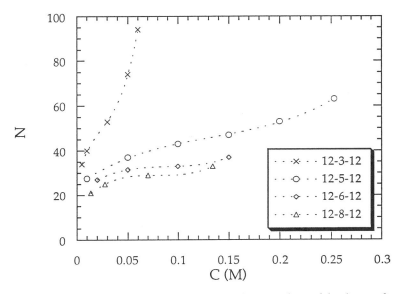

Figure 3. Variation of the micelle aggregation number with the surfactant concentration for the 12-s-12 series.

This expectation was verified by the cryo-TEM investigation of the solutions of this surfactant and of its homologues with longer spacers. Figure 4 shows typical cryo-transmission electron micrographs for solutions of the surfactants 12-2-12, 12-3-12, 12-4-12, and 12-8-12. The 1.3% 12-2-12 solution micrograph clearly shows entangled thread-like micelles, that explain its viscoelasticity. The thread-like micelles are still present in the micrograph of the 7% 12-3-12 solution (note bottom part of the micrograph!), but are not observed in the 2% solution of this surfactant (data not shown). The last solution had a viscosity comparable to that of water, but the 7% solution was extremely viscous. The micrographs of 5% (i.e., below 0.1 M) solutions of 12-4-12 and 12-8-12 show only densely packed spherical or spheroidal micelles. The viscosity of these solutions was low. Thus, direct cryo-TEM visualization of the microstructure of dimeric surfactant solutions supports the conclusions inferred from the N vs. C plots.

In separate experiments (3) we have observed that the addition of DTAB (dodecyltrimethylammonium bromide), which can be considered as the monomer of 12-2-12, to a 2% solution of 12-2-12 resulted in rapid disappearance of the viscoelasticity of the system, already at a DTAB mole fraction of 0.065. Cryo-TEM showed that the initially thread-like 12-2-12 micelles became shorter, and transformed progressively into spherical mixed DTAB/12-2-12 micelles, as the DTAB mole fraction was increased. DTAB micelles are known to remain spherical even at high concentration or ionic strength (21, 22). DTAB incorporation into 12-2-12 micelles is thus expected to inhibit the formation of thread-like micelles by the latter, as observed (data not shown).

Figures 5a and 5b show typical micrographs of solutions of two dimeric surfactants with a very long spacer, the 12-16-12 (2.4%) and 12-20-12 (1.2%). Vesicles, often double-lamellar, are observed. In the 12-20-20 samples we also observe lens-like objects (L), open "membranes" (M), and occasionally tubular liposomes (not shown here). Figures 5c and 5d demonstrate that the fluid microstructures of solutions of the corresponding "monomers", the octyl- (3.0%) and decyl- (0.8%) dodecyldimethylammonium bromides, retain the same state of aggregation as dimeric surfactants, namely, vesicles.

From the above it appears that the dimeric surfactants with a dodecyl chain, form thread-like micelles when the spacer is short enough, spherical micelles for medium length spacers and, vesicles with very long spacers. The corresponding monomers $C_{12}H_{25}(C_{s/2}H_{s+1})N^+(CH_3)_2Br^-$ form only spherical micelles in the same concentration range, when s is small, below 10. In view of this behavior, it appeared interesting to investigate the effect on the microstructure of the solution of forming the dimer of a surfactant that has a tendency to form elongated micelles. One would expect the dimer of such a surfactant to form aggregates having the shape coming next to thread-like micelles in the hierarchy of structures, that is, vesicles, if the spacer is made short enough. We therefore investigated the 16-s-16 series. Recall that hexadecyltrimethylammonium bromide is known to form elongated micelles at high concentration (23). One would thus expect 16-2-16 and probably also homologues with a longer, but not too long spacer, to also form elongated micelles and, eventually, vesicles. This expectation is verified by the micrographs shown in Figure 6. Thread-like micelles (t), vesicles (v) and bilayer membrane fragments (m) can be seen in the micrograph of a 0.9% 16-3-16 solution (Figure 6a). The micrograph of a 3.4% 16-4-16 solution (6b) presents a convincing example of entangled thread-like micelles. Some open membranes (m) are present in this micrograph too, as well as some spheroidal micelles (arrows). And last, micrograph 6d of a 4.0% 16-6-16 solution shows only spherical or very short rod-like micelles.

In conclusion, the results summarized above clearly show that the linking of cationic surfactants two by two at the level of the head-group by a polymethylene space, to form so-called "dimeric" or "bipolar" surfactants, can strongly affect the

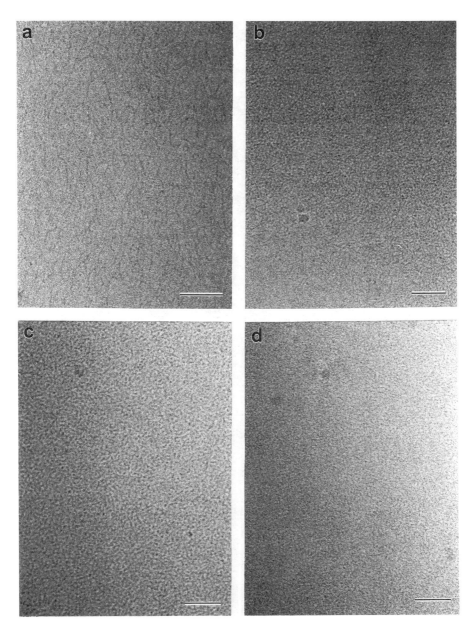

Figure 4. Typical electron micrographs of vitrified aqueous 12-s-12 solutions: (a) s = 2, 1.3%; (b) s = 3, 7.0%; (c), s = 4, 5.0%; (d), s = 8, 5.0% . Bar = 100 nm.

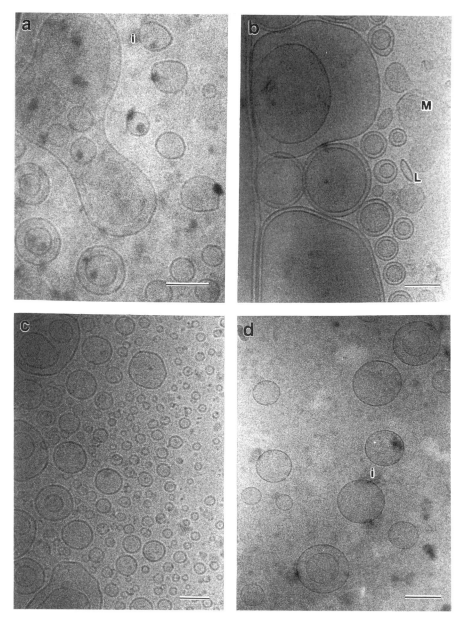

Figure 5. Cryo-transmission electron micrographs of vitrified solutions of: (a) 2.4% 12-16-12 (i denotes a small cubic ice area); (b) 1.2% 12-20-12 (L denotes a lens-shaped vesicle, M an open membrane); (c) 3.0% octyldodecyldimethyl-ammonium bromide (d) 0.8% decyldodecyldimethylammonium bromide. Bar = 100 nm.

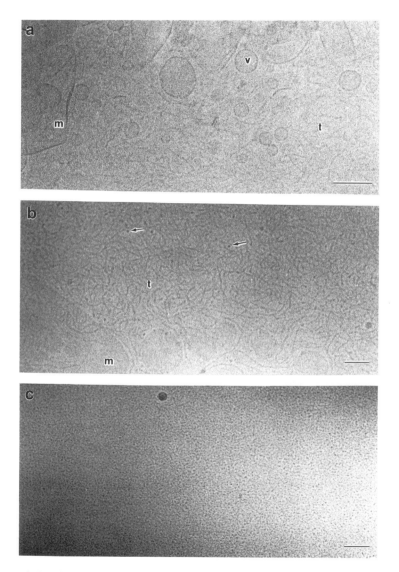

Figure 6. Typical electron micrographs of vitrified 16-s-16 solutions: (a) 0.9%
16-3-16 (m - membrane, v - a closed vesicle, t - thread-like micelles); (b) 3.4%
16-4-16 (a large open membrane is denoted by "m"; arrows indicate spheroidal
micelles); (c) 4.0% 16-6-16. Bar = 100 nm.

nature of the surfactant aggregates, if the spacer is short enough. In general, upon "dimerization", one goes from one type of aggregate to the one next in the hierarchy of structures upon increasing value of the packing parameter.

Polyamphiphiles. Typical electron micrographs obtained with PS10, and PS16 in aqueous solutions were published elsewhere (*9*). Spheroidal aggregates were seen with PS10, whereas the micrographs of PS16 showed long, entangled, thread-like micelles, very much like those in Figure 4a of 12-2-12, or in Figure 6b of 16-4-16. The solutions of PS10 were of low viscosity, but those of PS16 were visually viscoelastic at the concentration of 9 mM used. The diameter of the PS10 aggregates, as estimated from the electron micrographs, was consistent with that expected for spherical micelles made of a surfactant with a decyl chain. Separate fluorescence probing studies of PS10 solutions had revealed the presence of hydrophobic microdomains made of 15 to 20 repeat units, and with about one microdomain per 100 repeat units (*8, 24*). The average distance between microdomains on a given polymer chain was calculated to be between 10 and 40 nm (depending on whether the polymer segment connecting two successive microdomains is in a coiled or fully extended conformation). Microdomains located on different polymer chains would be at about the same range of distances under the experimental conditions used. This situation may very well result in the microstructures seen in the micrographs of PS10, with somewhat randomly distributed microdomains (*9*).

The diameter of the PS16 thread-like micelles was consistent with the length of a surfactant with a hexadecyl chain (*9*). However, the micelles were much longer than expected for a single PS16 chain. In fact, the micrographs showed very few micelle ends, observed typically as black dots. This led us to conclude that PS16 thread-like micelles are formed by end-to-end linking of a large number of PS16 chains. Indeed, the ends of these chains are not organized in microdomains for lack of neighboring repeat units, and tend to connect one to another, to prevent energetically costly contacts between water and hexadecyl side-chains. As pointed out (*10*), these features of PS16 solutions make them a choice material for studying the interaction between PS16 and surfactant because the binding of surfactant molecules to PS16 micelles may induce changes in the microstructure and rheology of the system.

We have shown through viscosity measurements and cryo-TEM investigations that PS16 interacts with the anionic surfactant sodium dodecylsulfate (SDS). This interaction leads to the progressive break-up of the thread-like micelles into the individual polymer chains and the disappearance of the viscoelasticity of the system (*10*).

A similar study has been performed, where SDS was replaced by either of the two nonionic surfactants hexaethyleneglycol monodecyl or monododecyl ether ($C_{10}E_6$ and $C_{12}E_6$). Figure 7 shows that the addition of the nonionic surfactant to a 3 mM PS16 solution leads to a rapid decrease of the solution viscosity, which appears to depend only little on the surfactant chain length. Similar experiments performed with a 9 mM viscoelastic PS16 solution showed that the viscosity decrease is accompanied by the disappearance of the viscoelasticity at surfactant concentrations as low as 0.9 mM, that is, below the cmc of $C_{10}E_6$ (1 mM). Thus, nonionic surfactants are much more effective than SDS in modifying the rheology of PS16 solutions. This difference probably reflects the absence, in the case of the nonionic surfactants, of repulsive electrostatic interaction that SDS molecules must overcome to incorporate into a PS16 microdomain, which bears a net electrical charge of the same sign. The plot in Figure 7 goes through a shallow minimum at a [surfactant]/[PS16] molar concentration ratio of about 0.3, then the viscosity increases slowly with surfactant concentration. These experiments, repeated in the

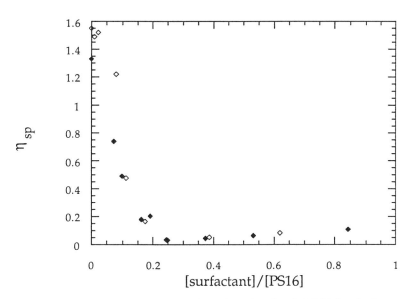

Figure 7. Variation of the specific viscosity of a 3 mM PS16 solution upon addition of $C_{10}E_6$ ($\lozenge$) and $C_{12}E_6$($\blacklozenge$) with the [surfactant]/[PS16] molar concentration ratio.

absence of PS16, showed no change of solution viscosity. This effect is thus characteristic of the system, and may be associated with some conformational change of the PS16 chain induced by the binding of surfactant.

Figure 8 shows electron micrographs of the PS16/$C_{10}E_6$ system at a polymer concentration of 9 mM(monomer)/L and various values of the molar concentration ratio R = [surfactant]/[PS16]. In Figures 8a and 8b, at R = 0.06, the micrographs show long thread-like micelles, somewhat oriented, probably as a result of sample preparation. In Figure 8c, where R = 0.13, many micellar ends are clearly visible as black dots, indicating that the fragmentation of the long micelles has already started, as the result of the incorporation of the nonionic surfactant into the microdomains. Only short elongated micelles can be seen at R = 0.36 (Figure 8d). These micelles have lengths in the 100 nm range, a value consistent with the PS16 molecular weight (9). The sequence of micrographs in Figure 8 is very similar to that for the system SDS/PS16 (10). In both instances cryo-TEM provides direct visual evidence of the disruption, upon addition of anionic or nonionic surfactants, of the long thread-like micelles of PS16 into their constituting elements, i.e., individual polymer chains. This disruption is clearly the result of the incorporation of the surfactant into the PS16 microdomains. The question that then arises is whether the surfactant distributes in a somewhat uniform manner along the thread-like micelles, or tends to concentrate in some specific spots, such as the microdomains corresponding to the junctions between polymer chains. Indeed, the latter microdomains are probably not as well organized as those away from the polymer chain ends. Measurements of surfactant self-diffusion in the SDS/PS16 (25) system showed that the amount of bound dodecylsulfate ion to PS16 is small, 15% or less of the added SDS. This result, and that very low concentrations of nonionic surfactants are sufficient to disrupt the thread-like micelles, suggest that the bound surfactant is located mainly at the junctions between polymer chains. Hence, it drastically affects micelle length even at very low content.

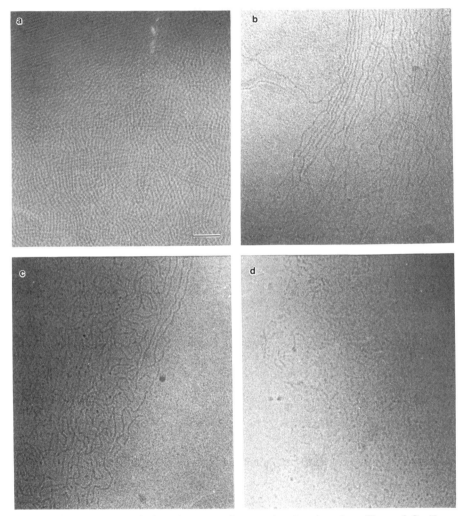

Figure 8. Cryo-transmission electron micrographs showing the effect of $C_{10}E_6$ addition to a 9 mM PS16 solution at R = $[C_{10}E_6]$/[PS16] = 0.06 (a and b), R = 0.13 (c), and R = 0.36 (d). Bar = 100 nm.

Acknowledgments

We thank Ms. B. Shdemati and Ms. J. Schmidt for their expert technical help. The work at the Technion was supported by grants from the United States-Israel Binational Science Foundation (BSF) Jerusalem and from the Fund for the Promotion of Research at the Technion.

Literature Cited

1. Israelachvili, J.; Mitchell, D.J.; Ninham, B. *J. Chem. Soc., Faraday Trans. 2* **1976**, *72*, 1525.
2. Tanford, C. *J. Phys. Chem.* **1972**, *76*, 3020.
3. Zana, R.; Talmon, Y. *Nature* **1993**, *362*, 228.
4. Zana, R.; Benrraou, M.; Rueff, M. *Langmuir* **1991**, *7*, 1072.
5. Alami, E.; Levy, H.; Zana, R.; Skoulios, A. *Langmuir* **1993**, *9*, 940.
6. Alami, E.; Marie, P.; Beinert, G.; Zana, R. *Langmuir* **1993**, *9*, 1465.
7. Kern, F.; Lequeux, F.; Candau, S.J.; Zana, R. submitted
8. Binana-Limbelé, W.; Zana, R. *Macromolecules* **1987**, *20*, 1331 and **1990**, *23*, 2731.
9. Cochin, D.; Candau, F.; Zana, R.; Talmon, Y. *Macromolecules* **1992**, *25*, 4220.
10. Zana, R.; Kaplun, A.; Talmon, Y. *Langmuir* **1993**, *9*, 1948.
11. From Tanford's (ref. 2) equation, d_s is expressed in Å units by: $d_s = 1.265$ (s+1).
12. Zana, R. in *"Surfactants Solutions : New Methods of Investigation"*; Zana, R., Ed.; Plenum Press, 1987, Chapter 5 and references therein.
13. Almgren, M. *Adv. Colloid Interface Sci.* **1992**, *41*, 9 and references therein.
14. Bellare, J.R.; Davis, H.T.; Scriven, L.E.; Talmon, Y. *J. Electron Microsc. Tech.* **1988**, *10,* 87.
15. Tachiya, M. *Chem. Phys. Lett.* **1975**, *33*, 289.
16. Almgren, M.; Grieser, F.; Thomas, J.K. *J. Am. Chem. Soc.* **1979**, *101*, 279.
17. Walter, A.; Vinson P. K.; Kaplun, A.; Talmon, Y., *Biophys. J.* **1991**, *60* , 1315.
18. Clausen, T.M.; Vinson, P.K.; Minter, J.R.; Davis, H.T.; Talmon, Y.; Miller, W.G. *J. Phys. Chem.* **1992**, *96*, 474.
19. Laughlin, R.G.; Munyon, R.L.; Burns, J.L.; Talmon, Y. *J. Phys. Chem.* **1992**, *96*, 374.
20. Lianos, P.; Lang, J.; Zana, R. *J. Colloid Interface Sci.* **1983**, *91*, 276.
21. Candau, S.J.; Hirsch, E.; Zana, R. *J. Phys., Paris* **1984**, *45*, 1263.
22. Ozeki, S.; Ikeda, S. *J. Colloid Interface Sci.* **1982**, *87*, 424.
23. Reiss-Husson, F.; Luzzati, V. *J. Phys. Chem.* **1964**, *68*, 3504.
24. Anthony, O.; Zana, R. *Macromolecules* (submitted).
25. Kamenka, N.; Kaplun, A.; Talmon, Y.; Zana, R. *Langmuir* (in press).

RECEIVED August 10, 1994

Chapter 7

Structure of Sheared Cetyltrimethylammonium Tosylate–Sodium Dodecylbenzenesulfonate Rodlike Micellar Solutions

Richard D. Koehler and Eric W. Kaler

Center for Molecular and Engineering Thermodynamics, Department of Chemical Engineering, University of Delaware, Newark, DE 19716

The viscoelastic behavior observed in aqueous solutions of the cationic/anionic surfactant mixture cetyl trimethylammonium tosylate(CTAT) and sodium dodecyl benzyl sulfonate(SDBS) provides evidence for the presence of rodlike micellar structures. In particular, highly viscous and viscoelastic gels form at low total surfactant concentrations. Small angle neutron scattering measurements reveal a strong interaction peak that becomes anisotropic with increasing shear rate. This anisotropy indicates an increase in the order of the micelles in a preferential direction. Model fits show that the rotational diffusion coefficients decrease with increasing total surfactant concentration for samples in the middle of the rodlike micellar region, but they increase on approach to the phase boundary. Thus micelles in the center of the micellar phase grow with increasing total surfactant concentration, while those near the boundary remain shorter. The shorter micelles are precursors to the vesicle and lamellar structures of nearby phases.

Rodlike micelles form in ionic surfactant solutions at high salt concentrations(1,2). They also form in solutions containing strongly binding counterions such as salicylate(3-5) or tosylate(6). These ions contain a hydrophobic part that prefers to remain in the nonpolar core of the rodlike micelle. If the counterion is replaced with an oppositely charged surfactant, there is very strong "counterion" binding, and rodlike micelles, spontaneous vesicles, and lamellar structures can form(6). In the CTAT/SDBS mixture (Figure 1) decreasing the CTAT/SDBS ratio at a constant total surfactant concentration shows a progression from a rodlike micellar phase on the CTAT rich side, to a micelle/vesicle two phase coexistence region, and through a CTAT-rich vesicle lobe to the equimolar line. The phase behavior is similar on the other side of the equimolar line. The current emphasis is on the CTAT-rich micelle phase.

Electron microscope images(6) and the viscoelastic response(1-5) of moderately concentrated rodlike micellar solutions suggests that they contain long and entangled threadlike micelles. Cates(7,8) has proposed a model for the relaxation of stress for these "living polymer" systems that accounts for the scission and recombination reactions of the micelles. The average length of the micelles and the length distribution depend on the energy of formation of two micellar endcaps during the scission reaction(7,8). However, constant shear measurements on these solutions suggest that the micelles act as rigid rods below a certain length scale(9,10). These two different points of view are not inconsistent - the threadlike micelles can be thought of as chains consisting of discrete rigid segments of a fixed persistence length (Figure 2). Since these segments are assumed to act independently, the microstructure on small length scales can be modeled as a collection of rigid rods with length equal to the persistence length.

Small angle neutron scattering (SANS) provides useful information about the structure and shapes of micelles in solution(11). However, for nonspherical particles this technique cannot provide unambiguous information about particle shape(11,12). Analysis of particle structure can be simplified by alignment of the anisotropic particles in electric(13), magnetic(13) or flow fields(9,10,13-16). The degree of alignment depends on the axial ratio of the particles, and alignment removes some ambiguity in the determination of the particle shape. Measurements of scattering in the presence of an external field also provides information about the rotational diffusion coefficient of elongated particles(17).

SANS Theory

The measured intensity from a system of rodlike colloidal particles can be written as(10,11),

$$\frac{d\Sigma}{d\Omega}(\mathbf{q}) = \frac{(\Delta\rho)^2 V_r^2}{V} \left\langle \sum_{N=1}^{N_p} \sum_{M=1}^{N_p} \exp(i\mathbf{q} \cdot \mathbf{R}_{NM}) F_N(\mathbf{q}) F_M(\mathbf{q}) \right\rangle$$

(1)

or,

$$\frac{d\Sigma}{d\Omega}(\mathbf{q}) = \frac{(\Delta\rho)^2 V_r^2}{V} \sum_{N=1}^{N_p} \left\langle F_N(\mathbf{q})^2 \right\rangle + \frac{(\Delta\rho)^2 V_r^2}{V} \left\langle \sum_{N=1}^{N_p} \sum_{M=1, M \neq N}^{N_p} \exp(i\mathbf{q} \cdot \mathbf{R}_{NM}) F_N(\mathbf{q}) F_M(\mathbf{q}) \right\rangle$$

(2)

where $\mathbf{R}_{NM}$ is a vector connecting the center of mass of particles N and M. The volume of the rod is V_r and $\Delta\rho$ is the difference between the scattering length density of the rod and the solvent. $\mathbf{q}$ is the difference between the incident and scattered intensity vectors and has magnitude $q=(4\pi/\lambda)\sin(\theta/2)$, where λ is the neutron wavelength and θ is the scattering angle. The brackets indicate averages over the orientations of the rods.

For cylindrical particles $F(\mathbf{q})$ is written(12)

$$F(\mathbf{q}) = F(q,\alpha) = \frac{\sin\left(\dfrac{qL}{2}\cos\alpha\right)}{\dfrac{qL}{2}\cos\alpha} \frac{2J_1(qR\sin\alpha)}{qR\sin\alpha}$$

(3)

where α is the angle between the rod axis, represented by a unit vector, $\mathbf{u}$, and the scattering vector, $\mathbf{q}$ (Figure 3). The amplitude function depends on the radius R and the length, L. The form factor is

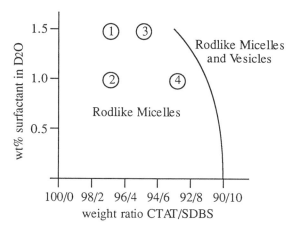

Figure 1: Schematic of the phase diagram of CTAT/SDBS showing the progression from rodlike micelle, through a micelle/vesicle two phase region, to a vesicle lobe on the CTAT rich side of the equimolar line. The circles indicate the samples studied with SANS.

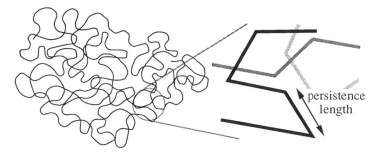

Figure 2: Schematic showing the different length scales for a threadlike micelle.

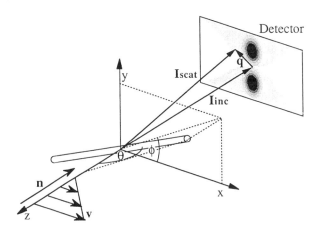

Figure 3: Scattering geometry for the shear flow experiment. Flow is in the x-direction, the shear gradient, and the neutron beam are in the z-direction. The azimuthal and polar angles are θ and ϕ.

$$\left\langle F^2(\mathbf{q}) \right\rangle = \int_{\mathbf{u}} F^2(\mathbf{q}) f(\mathbf{u}) d\mathbf{u} = \int_{-\pi/2}^{\pi/2} \int_0^{\pi} F^2(q,\alpha) f(\theta,\phi) \sin\theta \, d\theta \, d\phi$$

(4)

The normalized orientational distribution function $f(\mathbf{u})$ represents the probability that a rod is pointing between $\mathbf{u}$ and $d\mathbf{u}$ and is a function of the azimuthal, (θ) and polar, (ϕ) angles of the rod.

The second term in equation 2 can be simplified by neglecting any correlation between rod orientation and position. In this decoupling approximation we have(10,13)

$$\frac{d\Sigma}{d\Omega}(\mathbf{q}) = (\Delta\rho)^2 \Phi V_r \left\{ \left\langle F^2(\mathbf{q}) \right\rangle + n \left\langle F(\mathbf{q}) \right\rangle^2 \int_0^{\infty} \cos(\mathbf{q} \bullet \mathbf{r})[P(\mathbf{r}) - 1] d\mathbf{r} \right\}.$$

(5)

where Φ is the volume fraction of particles and $P(\mathbf{r})$ is the angularly-dependent pair distribution function. For spherical particles, $P(r)$ is isotropic and depends only on interparticle center-to-center distance. Given an interparticle potential, integral equations or Monte Carlo simulations can be used to obtain the pair distribution function(11). For rods, the interaction potential, and so $P(\mathbf{r})$, depends on both rod orientation and interparticle distance. Spherical harmonic expansions of integral equations have been used to find the angularly dependent pair distribution function for uncharged elongated particles(19). Rigid charged rods have been modeled using Monte Carlo simulations and perturbation expansions around sphericalized reference potentials(20).

If a shear field aligns the rods, the Smoluchowski equation provides the distribution function for the orientation of the rods. Doi and Edwards(21) and Larson(22,23) write the time evolution equation for the orientational distribution function as

$$\frac{\partial f}{\partial t} + \frac{\partial}{\partial \mathbf{u}} \cdot \left[(\mathbf{u} \cdot \nabla \mathbf{v} - \mathbf{uuu} : \nabla \mathbf{v}) f \right] - D_r \frac{\partial}{\partial \mathbf{u}} \cdot \left[\frac{\partial f}{\partial \mathbf{u}} + f \frac{\partial}{\partial \mathbf{u}} \left(\frac{V}{kT} \right) \right] = 0$$

(6)

where $\nabla \mathbf{v}$ is the velocity gradient tensor, D_r is the rotational diffusion coefficient, and V is the potential energy of interaction between rods. For the simple analysis given here, V is set to zero and D_r is assumed to be independent of orientation. With these assumptions the distribution function of rods in a shear flow depends on one parameter, β, which is the ratio of the shear rate Γ to the rotational diffusion coefficient. β is a measure of the ratio of strength of the flow that orients the rods to the strength of the restoring force of diffusion that randomizes them.

For calculation $f(\mathbf{u})$ can be expanded in a series of spherical harmonics(21,22)

$$f(\mathbf{u},t) = \sum_{\substack{\ell=0 \\ \ell \text{ even}}}^{\infty} \sum_{m=0}^{\ell} b_{\ell m} |\ell m) \ , \text{ where, } |\ell m) = \begin{cases} Y_\ell^m(\mathbf{u}) & \text{for } m = 0 \\ \dfrac{1}{\sqrt{2}} \left(Y_\ell^m(\mathbf{u}) + (-1)^m Y_\ell^{-m}(\mathbf{u}) \right) & \text{for } m \neq 0 \end{cases}$$

(7)

The particular form of $|\ell m)$ combines the spherical harmonic functions, $Y_\ell^m(\mathbf{u})$, so that the symmetry and boundary conditions of the problem are met. This expansion is then substituted into the Smoluchowski equation and the inner product is taken with respect to each member of the series. Using orthogonality and triple product conditions yields a set of ordinary differential equations in time for the coefficients $b_{\ell m}$.

As a result of symmetry around the flow direction the maximum of the distribution function occurs at $\phi=0$ for any value of β. The experimental geometry is not symmetric in the plane perpendicular to the vorticity, so for low values of β, when diffusion is important, the rods are pointed at 45° away from the flow direction, $\theta = \pi/4$. At higher shear rates, (large β) the particles are aligned in the flow direction with $\theta = \pi/2$. However, because this movement is out of the plane of the detector, it is not visible in the scattering experiment.

For the experiments described here the second term in equation 5 tends to zero for $q > 0.03\text{Å}$, so a fit to higher q data perpendicular and parallel to the flow direction provides information on the effective rotational diffusion coefficients of the rodlike micelles without the need for a detailed knowledge of the pair distribution function.

Experimental Section

Cetyl trimethyl ammonium tosylate (CTAT) was obtained from Sigma Chemical and recrystallized three times. Sodium dodecyl benzene sulfonate(SDBS) soft type was used as supplied from TCI. D_2O was obtained from Cambridge Isotope Laboratories and also used as supplied. The phase diagram is reported elsewhere(6).

SANS samples were prepared by completely mixing stock solutions of CTAT and SDBS in D_2O to the desired weight ratio. Two samples were in the center of the one phase rodlike micellar region: sample 1 containing 1.5wt% 97/3 CTAT/SDBS and sample 2 containing 1.0wt% 97/3 CTAT/SDBS. Two other samples were prepared closer to the rodlike micelle phase boundary, sample 3 containing 1.5wt% 95/5 CTAT/SDBS and sample 4 containing 1.0wt% 94/6 CTAT/SDBS (Figure 1). Small angle neutron scattering measurements were performed on the 30-meter SANS spectrometer at the National Institute of Standards and Technology at Gaithersburg, MD. The sample to detector distance was 5.25m and a wavelength of 6Å with $\Delta\lambda/\lambda=0.15$ yielded a q range between 0.005Å^{-1} and 0.08Å^{-1}. The samples were placed in a concentric cylinder shear cell with a 0.843mm gap width(24). Measurements were made at 25°C at steady shear rates. The samples were presheared for 300 seconds before the measurements began.

The raw 2-D data was corrected by subtracting the appropriate experimental background. Circular averages were made on shells 10° wide around a direction perpendicular to the flow direction and parallel to the flow direction, and the data was put on absolute scale with the use of a polystyrene standard. For fitting, the experimental data was divided by the prefactor in equation 5. The persistence length or effective rod length was estimated using Bragg's law, $L=2\pi/q_{max}$, where q_{max} is the q value of the peak in the scattering curve(10). The radius of the rod was found by fitting the form factor, equation 4, to the high q data at zero shear where the decay of the intensity curve only depends on the particle radius. Effective diffusion coefficients at each shear rate were obtained by fitting equation 4 to data measured for $q > 0.03\text{Å}$ both parallel and perpendicular to the flow direction.

Results and Discussion

Figure 4 shows representative scattering patterns at various shear rates from 1.0wt% 97/3 CTAT/SDBS in D_2O (sample 2). At rest, the scattering from these micellar

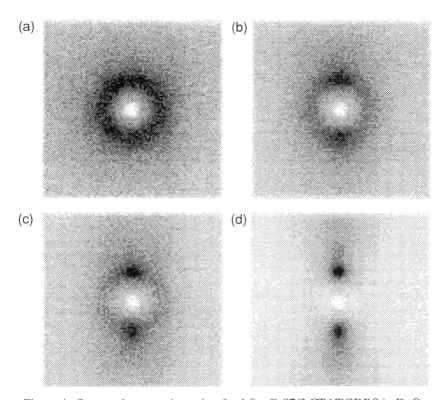

Figure 4: Scattered neutron intensity, for 1.0wt% 97/3 CTAT/SDBS in D$_2$O, showing an increase in anisotropy with shear rate. (a) shear rate 0 s^{-1}, (b) shear rate 10 s^{-1}, (c) shear rate 20 s^{-1}, (d) shear rate 40 s^{-1}.

solutions is isotropic and displays a single strong correlation peak (Figure 4a). This indicates that the micelles are charged and highly interacting. The scattering patterns become increasingly anisotropic with increasing shear rate, and show increasing intensity in the direction perpendicular to the flow direction and decreasing intensity in the direction parallel to the flow direction. This change provides clear evidence that the rodlike micelles are aligning. The increase in anisotropy implies the rods align like the rungs on a ladder, thereby reinforcing the correlation peak in the direction perpendicular to the flow and weakening it in the direction parallel to the flow.

Figure 5 shows scattering from sample 2 along lines perpendicular and parallel to the flow direction. The solid and dashed lines are the model fits of the form factor where the effective rotational diffusion coefficient is the only adjustable parameter; the radius and length are determined from the zero shear data. The good agreement between the shape of the experimental and fitted curves at high q suggests that the length estimated from the q value of the intensity peak is a good estimate of the persistence length of the micelles. Table I shows the results of this fit for the four samples studied.

Table I. Rotational Diffusion Coefficients as a Function of Shear Rate

Shear	Rotational Diffusion Coefficient D_r (s^{-1})			
Rate	wt% surfactant, ratio of CTAT/SDBS			
(s^{-1})	1.5wt%, 97/3	1.0 wt%, 97/3	1.5wt%, 95/5	1.0wt%, 94/6
5	1.7	1.0	1.7	0.7
10	1.0	1.3	1.7	1.3
20	0.4	2.0	1.7	2.7
50	0.6	2.0	1.2	6.7
100	1.0	2.0	8.3	10.0

The effective rotational diffusion coefficient decreases with increasing total surfactant concentration along the 97/3 CTAT/SDBS line, which is in the middle of the rodlike micelle region on the phase diagram. At higher SDBS ratios the effective diffusion coefficients increase with shear rate.

The rotational diffusion coefficient, D_{ro}, of a rod in dilute solution varies as L^{-3} while in a semi-dilute solution $D_r \sim D_{ro}(cL^3)^{-2}$. The rotational diffusion coefficient of the rodlike micelles depends strongly on the length of the rod. The decreasing diffusion coefficient with increasing total surfactant concentration implies an increasing micelle length or an increase in entanglement.

Near the phase boundary (samples 3 and 4), the observation of an increasing effective rotational diffusion coefficient with increasing shear rate suggests that the micelles are short or mobile enough to remain unaligned even at very high shear rates. The higher diffusion coefficient may be due to less entanglement or may be directly related to a decrease in rod length. The proximity of the phase boundary is consistent with the formation of shorter micelles and indicates that more endcaps can form near the phase boundary.

Conclusions

Rodlike micelles that can be aligned in a shear flow are present on the CTAT rich side of the CTAT/SDBS/D_2O phase diagram. Modeling the neutron scattering data provides information about the effective rotational diffusion coefficient of these

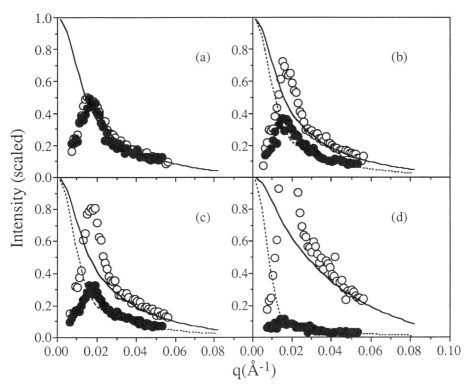

Figure 5: Scattering perpendicular, O, and parallel, ●, to the flow direction for 1.0wt% 97/3 CTAT/SDBS for a shear rate of (a) 0 s^{-1}, (b) 10 s^{-1}, (c) 20 s^{-1}, (d) 100 s^{-1}. The lines are fits described in the text.

rodlike micelles. The results indicate that longer, more entangled micelles are formed as the total surfactant concentration is increased at a constant ratio of CTAT/SDBS. Increasing the amount of SDBS at a constant total surfactant concentration leads to rodlike micelles which are much shorter and more mobile. Shorter micelles signal a change in the structures present as the phase transition is approached.

Acknowledgments

This work was supported by E.I. DuPont de Nemours & Co. The authors thank Dr. R.G.Larson for supplying the FORTRAN code used in the calculation of the orientational distribution function of rods in a shear flow. The authors also acknowledge G. Straty and Dr. J. Barker for their help with the shear cell and SANS measurements. We acknowledge the support of the National Institute of Standards and Technology, U.S. Department of Commerce, in providing the facilities used in this experiment. This material is based upon activities supported by the National Science Foundation under Agreement No. DMR-9122444.

Literature Cited

1. Kern, R.; Lemarechal, P.; Candau, S.J.; Cates, M.E. *Langmuir*, **1992**, *8*, 437.
2. Candeau, S.J.; Hirsch, E.; Zana, R.; Adam, M. *J. Coll. Interface Sci.*, **1988**, *122*, 430.
3. Hoffmann, H.; Rehage, H.; Rauscher, A. In *Structure and Dynamics of Strongly Interacting Colloids and Supramolecular Aggregates in Solution*; Kluwer Academic Publishers, Amsterdam, 1992, pp 493-510.
4. Kern F.; Zana, R.; Candau, S.J. *Langmuir*, **1991**, *7*, 1344.
5. Rehage, H.; Hoffmann, H. *J. Phys. Chem.*, **1988**, *92*, 4712.
6. Kaler, E.W.; Herrington, K.L.; Murthy, A.K.; Zadadzinski, J.A.N. J. Phys. Chem., 1992, 96, 6698.
7. Turner, M.S.; Cates, M.E. *J. Phys. II France*, **1992**, *2*, 503.
8. Turner, M.S.; Cates, M.E. *Langmuir*, **1991**, *7*, 1590.
9. Herbst, L.; Hoffmann, H.; Kalus, J.; Thurn, H.; Ibel, K.; May, R.P. *Chem. Phys.*, **1986**, *103*, 437.
10. Kalus, J.; Hoffmann, H. *J. Chem. Phys.*, **1987**, *87*, 714.
11. Kaler, E.W. *J. Appl. Cryst.*, **1988**, *21*, 729.
12. Guinier, A.; Fournet, G. *Small-Angle Scattering of X-Rays* ; John Wiley & Sons, Inc., New York, 1955, p19.
13. Kalus, J. In *Structure and Dynamics of Strongly Interacting Colloids and Supramolecular Aggregates in Solution*; Kluwer Academic Publishers, Amsterdam, 1992, pp 463-492.
14. Hayter, J.B.; Penfold, J. *J. Phys. Chem.*, **1984**, *88*, 4589.
15. Cummins, P.G.;Staples, E.; Hayter, J.B.; Penfold, J. *J. Chem. Soc., Faraday Trans. 1*, **1987**, *83*, 2773.
16. Cummins, P.G.; Hayter, J.B.; Penfold, J.; Staples, E. *Chem. Lett.*, **1987**, *138*, 436.
17. Doi, M.; Edwards, S.F. *The Theory of Polymer Dynamics*; Oxford University Press, New York, 1986; pp 295,330.
19. Perera, A.; Kusalik, P.G.; Patey, G.N. *J. Chem. Phys.*, **1987**, *87*, 1295.
20. Canessa, E.; D'Aguanno, B.; Weyerich, B.; Klein, R. *Molecular Physics*, **1991**, *73*, 175.
21. Doi, M.; Edwards S.F. *J. Chem. Soc., Faraday Trans. 2*, **1978**, *74*, 918.
22. Larson, R.G. *Recent Developments in Structured Continua*; Longman: London, 1990; Vol 2.
23. Larson, R.G.; Öttinger, R.G. *Macromolecules*, **1991**, *24*, 6270.
24. Straty, G.C. *J. Res. Natl. Inst. Stand. Technol.*, **1989**, *94*, 259.

RECEIVED July 8, 1994

Chapter 8

Rheo-optical Behavior
of Wormlike Micellar Systems

Toshiyuki Shikata[1] and Dale S. Pearson[2,3]

[1]Department of Macromolecular Science, Faculty of Science,
Osaka University, Toyonaka Osaka 560, Japan
[2]Department of Chemical and Nuclear Engineering and Materials
Department, University of California, Santa Barbara, CA 93106

Cetyltrimethylammonium bromide, CTAB, forms long wormlike micelles in aqueous solutions with sodium salicylate, NaSal. The system makes an entanglement network similar to concentrated polymer systems, and shows profound viscoelastic behavior highly dependent on concentration of CTAB, C_D, and NaSal, C_S. The system also is strongly flow birefringent. The refractive index tensor and the stress tensor are linearly related by a stress-optical coefficient $C = -3.1 \times 10^{-8} \ cm^2 dyn^{-1}$ essentially independent of C_D and C_S. This linearity suggests that the origin of elasticity of the wormlike micellar system is orientation of micellar portions between entanglement points. The sign and magnitude of C is consistent with the proposed structure in which most of CTA^+ ions are arranged in a radial pattern with their molecular axis mostly perpendicular to the axis of the wormlike micelle.

Various kinds of surfactant molecules form very long wormlike or threadlike micelles in aqueous systems with some additives (1-8). Especially systems consisting of a cationic surfactant such as cetyltrimethylammonium bromide, CTAB, (4-8) or cetylpyridinium bromide, CPB (3), and some salts such as sodium salicylate, NaSal, show very unique viscoelastic behavior. In those systems the wormlike micelles sometimes behave like polymer systems; for example, they make very dense entanglement networks (2-8). The fact (2,3,5-8) that the high frequency plateau modulus of entangling micellar systems is approximately proportional to the square of the concentration of surfactant molecules is essentially identical to that observed in polymer systems (9). On the other hand, the relaxation spectrum in the terminal zone

[3]Deceased

0097-6156/94/0578-0129$08.00/0
© 1994 American Chemical Society

of the entangling micellar system could be well expressed by only one spike like spectrum (2,3,5-8), but the entangling polymer system has a broad box type spectrum (9). This similarity and difference mean that the entanglement among the wormlike micelles is the essential reason for the elastic properties as it is for the polymer systems, whereas the relaxation mechanism in the wormlike micellar system is not controlled by diffusion of micelles along themselves, which happens in the polymeric systems (9,10), but is controlled by a completely different mechanism that reflects the unique characteristics of the wormlike micelle (5,7,11,12).

The physical origin of elasticity in the micellar system is not fully understood so far. The first possibility is that the wormlike micelle is rather flexible and portions between entanglement points have enough length to behave like Gaussian chains without changing their contour length, just like polymer systems(10,11). In this case, the refractive index tensor must be a function of orientation of the portions and proportional to the stress tensor when the applied strain is not so large. Then, the so called *stress-optical law* would hold as in polymer systems. The other possibility is that the wormlike micelle could be really extended, changing its contour length, and/or bent with certain elastic constants. If this is the main contribution to the stress of the wormlike micellar system, the stress-optical law will not hold, because the structure of the micelle would be altered due to elongation and the refractive index tensor would not be a linear function of the stress tensor.

Flow birefringence measurement should be the most powerful and useful technique to examine the relationship between the stress and refractive index tensors of the micellar system. In this paper, viscoelastic and flow birefringence behavior of an aqueous wormlike micellar system, CTAB:NaSal/W, consisting of CTAB and NaSal will be reported, and the applicability of the stress-optical law to the system will be discussed. Furthermore, the persistence length of the wormlike micelle will be estimated from the birefringence data.

Experimental Section

Materials. CTAB and NaSal, were purchased from Fisher Scientific. Highly deionized (resistance > 16MΩ) dust free water was used as the solvent. The concentrations of CTAB, C_D, were 0.01, 0.03, and 0.1 M, and concentration of NaSal, C_S, were changed from 0.15 to 0.4 M. Prepared solutions, CTAB:NaSal/W, were rested for at least two days at room temperature before measurements for complete equilibration.

Viscoelastic and Flow Birefringence Measurements. A conventional rheometer (Rheometrics, RMS800) was operated at 25°C (room temperature) with a cone and plate geometry of 25 mm in radii and of 0.02 rad in cone angle. The shear stress, σ_{xy}, and the first normal force difference, $\sigma_{xx}-\sigma_{yy}$, were measured as functions of time under given constant shear rates, $\dot{\gamma}$, from 0.01 to 100 s^{-1}.

We operated a home made flow birefringence instrument with a photo-elastic modulator and determined birefringence, Δn, and extinction or orientational angle, χ, simultaneously as functions of time, t, under a constant given shear rate. A precise description of this equipment was published elsewhere (15). A Couette geometry with inner and outer cylinders of 15 and 16 *mm* in radii was employed. The tested $\dot{\gamma}$ was 0.2 to 10 s^{-1} and the temperature was also room temperature.

Results and Discussion

Behavior under Weak Flow. Rheo-optical behavior of the CTAB:NaSal/W micellar system was roughly classified into two kinds of conditions depending on $\dot{\gamma}$. We call the condition in which the product of the applied $\dot{\gamma}$ and the longest relaxation time of the system, $\tau\dot{\gamma}$, is less than a certain critical value (depending on C_D) *weak flow*. On the other hand, the condition of $\tau\dot{\gamma}$ greater than that value *strong flow*. A typical result for the time dependent behavior of the orientational angle, χ , and birefringence, Δn, for a micellar system with $C_D = 0.03$ and $C_S = 0.23$ *M* under weak flow is shown in Figure 1. The critical $\tau\dot{\gamma}$ value for this system was a decreasing function of C_D and was about 6 for this $C_D = 0.03$ *M* solution. χ starts from 45° and decreases with time until finally reaching a steady value. The steady state value of χ depended on $\dot{\gamma}$, and χ always started from 45° and never showed a minimum even at high $\dot{\gamma}$ in every solution we tested. This behavior is similar to that usually observed in polymer systems (15). A striking fact in Δn is that the sign of Δn is minus and the magnitude observed is much larger than ordinary polymers (15). $-\Delta n$ starts from zero and showed a small maximum around t $\approx$ 5 s, which is called *over-shoot*, and decreases slightly, then reaches a certain steady state value. In the linear regime $\dot{\gamma}t <$ 1, none of the solutions showed a maximum in $-\Delta n$, while in the nonlinear regime $1 < \dot{\gamma}t$ maxima were always observed and the difference between the maximum and the steady value in $-\Delta n$ strongly depended on $\dot{\gamma}$. This behavior in $-\Delta n$ is also very similar to that observed in polymer systems (15).

Time dependence of the xy component of the refractive index tensor, $n_{xy} = -\Delta n\sin(2\chi)/2$, and the difference between xx and yy component, $n_{xx}-n_{yy} = -\Delta n\cos(2\chi)$, obtained from Figure 1 were compared with the shear stress, σ_{xy}, and the normal stress difference, $\sigma_{xx}-\sigma_{yy}$, of the same solution. Corresponding pairs of components had quite similar time dependence, and both pairs of curves could be superposed well by shifting only in the vertical direction. Thus, it is obvious that equations 1 and 2 shown below hold well in the entire time region with a constant of proportionality.

$$\sigma_{xy} = \frac{\Delta n \sin(2\chi)}{2\ C} \tag{1}$$

$$\sigma_{xx} - \sigma_{yy} = \frac{\Delta n \cos(2\chi)}{C} \tag{2}$$

This proportionality between the stress and refractive index tensors is called *the stress-optical law* and the proportionality constant, C, is called *the stress-optical constant*(13,14) and was evaluated as $-3.1 \times 10^{-8}\ cm^2 dyn^{-1}$ for this solution. All of the tested solutions showed the validity of the stress-optical law not only in the steady state behavior but also in transient behavior under the weak flow condition.

It has been reported that the rheological properties of CTAB:NaSal/W system are very sensitive to C_D and also C_S (or concentration of free salicylic anions, C_S^*, which could be estimated as $C_S^* = C_D - C_S$ (5,7,8)). Particularly, the longest relaxation time of the system is a function of only C_S^*, whereas the high frequency plateau modulus, G_N^0, is a function of only C_D. Therefore, one can easily control all of the rheological parameters by changing C_D and C_S^*, because the zero shear viscosity can also be expressed as $\eta_0 \approx G_N^0 \tau$ in this system. Therefore, we tested the applicability of the stress-optical law with the steady state value in solutions with different τ and G_N^0.

Dependence of the steady state values of σ_{xy} and $\sigma_{xx} - \sigma_{yy}$ on $\dot{\gamma}$ of the same solution as (with $\tau \approx 2$ s) in Figure 1 is plotted in Figure 2. Optically obtained $\Delta n \sin(2\chi)/(2C)$ and $\Delta n \cos(2\chi)/C$ are also plotted in the same figure with the same $C = -3.1 \times 10^{-8}\ cm^2 dyn^{-1}$. The agreement between rheological and optical data is perfect.

Such agreement is also observed in a solution with $C_D = 0.1\ M$ and the same $C_S^* = 0.2\ M$ (or the same τ). We also confirmed applicability of the stress-optical law in a solution with much lower $C_D = 0.01\ M$ and the same $C_S^* = 0.2\ M$ by using the same C value again. Furthermore, we checked the stress-optical law in solutions with $C_D = 0.1\ M$ and different $C_S^* = 0.05$ and $0.3\ M$, too. The solution with $C_S^* = 0.3\ M$ ($\tau \approx 0.3$ s) had the same C value as former solutions. The other solution with much lower $C_S^* = 0.05\ M$ ($\tau \approx 2$ s) showed the stress-optical law with the same C value. However, the shear rate range in which the law held was smaller than that of other solutions and was restricted to the linear region of $\tau \dot{\gamma} < 1$. This difference will be discussed later.

Behavior under Strong Flow. Typical time evolution of χ and Δn of the solution with $C_D = 0.1$ and $C_S^* = 0.2\ M$ ($\tau \approx 2$ s) under the strong flow condition at $\dot{\gamma}$ = 6.3 s^{-1} is plotted in Figure 3A. Until $t \approx 4$ s the time dependence of χ and Δn looks normal; however, after $t \approx 4$ s instability occurs in both χ and Δn; therefore, we

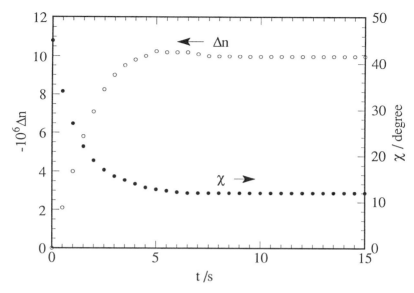

Figure 1.   A typical time dependence of birefringence, Δn, and orientational angle, χ, for a system with $C_D = 0.03$ M and $C_S = 0.23$ M under the weak flow condition at shear rate $\dot{\gamma} = 1.6$ s^{-1}.

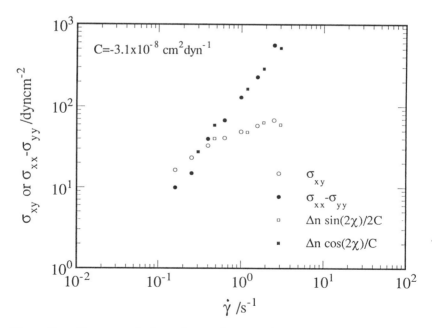

Figure 2.   Shear rate, $\dot{\gamma}$, dependence of the steady values of the shear stress, σ_{xy}, and the first normal stress difference, $\sigma_{xx}-\sigma_{yy}$, and also of optically obtained $\Delta n \sin(2\chi)/(2C)$ and $\Delta n \cos(2\chi)/C$ of the solution with $C_D = 0.03$ M and $C_S = 0.23$ M ($C_S^* = 0.2$ M and $\tau \approx 2$ s).

could not continue measurment. In the unstable region, χ and Δn are badly scattered, and the steady state value of them could not determined. Also, the time dependence of χ and Δn in the instability was not reproducible. This instability was always observed after χ approached a very small value, close to $0°$, in all the solution under the strong flow condition.

Supposing the same $C = -3.1 \times 10^{-8}$ $cm^2 dyn^{-1}$, agreement between optical data and rheological ones was recognized and the stress-optical law held until $t \approx 1$ s or $\gamma \approx 6$ (Fig 3B). However, increases in σ_{xy} and $\sigma_{xx}-\sigma_{yy}$ with time were much greater than for the optical ones, and the stress-optical law did not hold any more above $\gamma \approx 6$. According to the theoretical interpretation of the stress-optical law (10,13,14) for polymer systems, the validity of the stress-optical law is based on the Gaussian behavior of component chains; therefore, this break down of the stress-optical law around $\gamma \approx 6$ would mean that the behavior of the wormlike micelle in this system is no longer approximated by the Gaussian chain statistics above $\gamma \approx 6$. The critical strain of validity of stress-optical law at the strong flow condition was a decreasing function of C_D , and is related to the size of the mesh or the distance between entanglement points in the micellar network.

The instability in the birefringence data occured around the time (or strain) at which σ_{xy} and $\sigma_{xx}-\sigma_{yy}$ show the first overshoot. The σ_{xy} value showed a couple of damping oscillation after the first overshoot, and then reached a certain steady state value (7). The instability in the optical data should mean some dramatic structural change of the entanglement network and also of the micelle; for example break down of the micelle. Therefore, during the damping oscillation in σ_{xy} and $\sigma_{xx}-\sigma_{yy}$, the inner structure of the entanglement network of the micelles must be destroyed or altered by very strong shear deformation. The possibility of structural change of the wormlike micellar system by shearing (16,17), and also phase transition from isotropic liquid to a liquid crystalline phase by strong shearing has been pointed out (18).

In the solution with $C_S^* = 0.05$ M, the instability began at a shear rate belonging to the nonlinear region ($\dot{\gamma}\tau > 1$), and the critical strain of validity of the stress-optical law was also much smaller than that of the system with $C_S^* = 0.2$ and 0.3 M. The remarkable difference in response to a stress relaxation experiment under large step strains inducing nonlinear behavior was observed in the CTAB:NaSal/W system around $C_S^* \approx 0.1$ M. At $C_S^* < 0.1$ M the system showed increases in instantaneous relaxation modulus with increasing magnitude of strain, whereas at $C_S^* > 0.1$ M decreases in the modulus with strain were observed (7). However, no C_S^* dependence of the modulus in the linear response regime under a small amplitude stress relaxation experiment and a dynamic test was identified (7). From these it is likely that mechanical and/or structural features of the wormlike micelles at $C_S^* > 0.1$ M is slightly different from those at $C_S^* < 0.1$ M, and the difference could not be detected by the viscoelastic behavior and flow birefringence in the linear region.

The Stress-Optical Coefficient. The stress-optical law holds well in the wormlike micellar system with $C = -3.1 \times 10^{-8}$ $cm^2 dyn^{-1}$ at least in the linear region. Rehage and Hoffmann also reported a quite similar C value for several kinds of cationic surfactant systems (3). From this and the fact that the plateau modulus of the wormlike micellar system is approximately proportional to C_D^2, like the polymer system, we could conclude that the essential origin of the elasticity of the entangling wormlike micellar system is excess orientation of micellar portions between the entanglement points and their Gaussian behavior, similar to polymeric systems. Therefore, we could estimate the stress-optical coefficient of the wormlike micelles with a basic model for the polymer system or rubber.

The stress-optical coefficient for the linear flexible free jointed (Gaussian) chain system had been calculated theoretically as below (10,13,14)

$$C = \frac{2\pi(n^2+2)^2\Delta\alpha}{45nk_BT} \tag{3}$$

where n, $\Delta\alpha$, and k_B are the mean values of refractive index, anisotropy of polarizability of an equivalent random walk (Kuhn) segment, and Boltzmann's constant, respectively: $\Delta\alpha = \alpha_1 - \alpha_2$, and α_1 is the polarizability in the direction along the backbone of the Kuhn segment and α_2 is that in the direction perpendicular to its backbone. Supposing equation 3 is valid in the wormlike micellar system, the optical anisotropy of the Kuhn segment of the micelle could be estimated as $\Delta\alpha = -8.5 \times 10^{-22}$ cm^3 with $C = -3.1 \times 10^{-8}$ $cm^2 dyn^{-1}$. This value is exactly the same as that for the wormlike micelles of CPSal/W reported by Schorr(3,19), using the technique of electric birefringence.

In polymer systems, the optical anisotropy of Kuhn segments could be determined from both the optical anisotropy of the monomer of the flexible chain, $\Delta\alpha^o$, and the equivalent Kuhn length of the chain, $<r^2>/L$, or persistence length, $q = <r^2>/(2L)$, as shown in equation 4 below (13):   L is the contour length of the chain and $<r^2>^{1/2}$ is the mean square end-to-end distance of the chain in the unperturbed state.

$$\Delta\alpha = \Delta\alpha^o \frac{<r^2>}{L\lambda} \quad \text{or} \quad = 2\Delta\alpha^o \frac{q}{\lambda} \tag{4}$$

where λ is the length of the monomer of the chain. In the entangling wormlike micellar system, the discussion above would be satisfied by replacing L and $<r^2>^{1/2}$

with the contour length, L_e, between entanglement points and the mean square distance between entanglement points, $<r_e^2>^{1/2}$, respectively. Thus, if one could estimate both values of $\Delta\alpha^\circ$ and λ, the persistence length, q, of the wormlike micelle could be determined.

Monomers and the Persistence Length of the Micelle. Δn of the wormlike micellar system could be expressed as the product of the order parameter of micellar monomers, S, and the characteristic birefringence of a perfectly aligned system, Δn°. Δn° is proportional to the optical anisotropy of the monomer as below (20)

$$\Delta n = S\Delta n^\circ \tag{5}$$

$$\Delta n^\circ = \frac{2\pi(n^2+2)^2\rho_m\Delta\alpha^\circ}{9n} \tag{6}$$

where ρ_m is number density of the monomer.

Here, we propose a simple structural model for the wormlike micelle. The wormlike micelle is constructed by disc shape monomers consisting of m of CTA^+Sal^- complexes (5). CTA^+ is arranged in a radial pattern keeping its methylene chain perpendicular to the axis of the micelle. Sal^- is located beside an ammonium group of CTA^+, but its motion is rather fast (5). Since the polarizability of methylene chain parallel to the molecular axis is larger than that perpendicular to the axis, this model could successfully explain the large negative $\Delta\alpha$ (3,21). With this model we could obtain the relation between Δn° and $\Delta\alpha^\circ$ as

$$\Delta n^\circ = \frac{2\pi(n^2+2)^2 N_A C_D \Delta\alpha^\circ}{9000nm} \tag{7}$$

where N_A is Avogadro's number.   Then, we are concerned with a critical condition just before instability under strong flow. Because the system shows χ very close to zero at the critical condition, the order parameter, S, of the micelle should be approximately equal to unity, so that the extreme value of Δn at the critical condition could be considered as Δn°.   Figure 3A is a typical example showing the critical condition, and one could estimate $\Delta n^\circ \approx -11 \times 10^{-5}$ around t $\approx$ 4 s for a solution with $C_D = 0.03$ M. Taking the slope of a plot between Δn° and C_D at the critical condition, we could evaluate $\Delta\alpha^\circ/m \approx -8.0 \times 10^{-25}$ cm^3 by means of equation 7. On the other hand, m could be roughly estimated as follows. Suppose the radius of the micelle is 2.4 nm (22-24), circumference of the monomer disc could be estimated as 15 nm. And the size of a head of a CTA^+Sal^- complex could be evaluated as about 0.85 nm

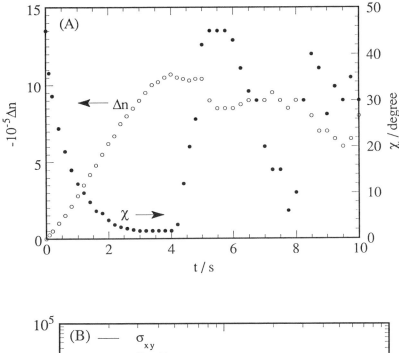

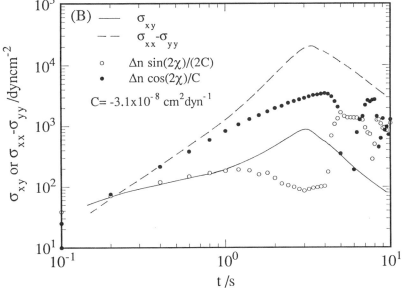

Figure 3. (A): A typical time dependence of birefringence, Δn, and orientational angle, χ, for a system with C_D = 0.03 *M* and C_S = 0.23 *M* under the strong flow condition at shear rate $\dot{\gamma}$ = 6.3 s^{-1}. (B): Comparisom between optical data and rheological ones as a functions of elapsed time supposing C = 3.1 x 10^{-8} *cm²dyn^{-1}*.

by means of Imae's data (23) for spherical micelles of a CTASal:NaSal/W system. Consequently m could be obtained as about 18, then $\Delta\alpha^o$ = -1.4 x 10^{-23} cm^3 could be estimated. We also estimate $\lambda \approx 0.85nm$, because λ is equal to the width of the CTA^+Sal^- complex.

Finally, we could determine the persistence length as $q \approx 26$ nm by means of equation 4 with m $\approx$ 18 and $\lambda \approx 0.85$ nm; therefore, the Kuhn length of the micelle could be also estimated as about 52 nm. Imae also estimated q as 110~150 nm for a CTASal:NaSal/W system (23) with a light scattering technique at a C_D condition much lower than the entanglement concentration. But her q value is considerably larger than ours. The reason for this discrepancy is not clear at this stage. However, the C_D range she studied was much lower than that of this study, and it could be possible that q has C_D dependence (21). On the other hand, Appell et al (25) also estimated q of a CPB:NaBr/W system with a simple salt NaBr by use of light scattering and magnetic birefringence, and they obtained q = 20 $\pm$ 5 nm under a C_D condition lower than the entanglement concentration. Interestingly, their q value is much closer to ours than Imae's q. It is well known that the slowest relaxation mechanism of CTAB:NaSal/W is quite different from that of systems with simple salts such as CTAB:NaBr/W or CPB:NaBr/W. Although Cates' model (11,12) for the slow relaxation mechnism has been successfully applied to systems with simple salts, his model could not explain the C_S^* dependent relaxation time in the CTAB:NaSal/W and CPB:NaSal/W systems (2,8). It is not clear whether there is considerable difference in the short time response and structural features of micellar portions between entanglement points and also in the value of q of these systems.

The rheo-optical technique, which we proposed here, must be one of the useful and powerful methods to evaluate the persistence length, q, of the wormlike micelles.

Conclusions

The stress-optical law held in the entangling CTAB:NaSal/W wormlike micellar system. This suggests that the origin of elasticity of the micellar system is excess orientation or Gaussian behavior of micellar portions between entanglement points similar to the rubber elasticity in ordinary flexible polymer systems.

The sign and magnitude of the stress-optical coefficient was consistent with the proposed micellar structure.

We proposed a new method to estimate the persistence length of the wormlike micelle through the flow birefringence data, and the persistence length of the micelle was evaluated as q = 26 nm.

Literature Cited

1. Gravsholt, S. *J. Colloid Interface Sci.* **1976**, *57*, 575.
2. Candau, S. J. ; Hirsh, E. ; Zana, R. ; Delsani, M. *Langmuir* **1989**, *5*, 1225.
3. Rehage, H. ; Hoffmann, H. *Mol. Phys.* **1989**, *5*, 1225.
4. Shikata, T. ; Hirata, H. ; Kotaka, T. *Langmuir* **1987**, *3*, 1081.
5. Shikata, T. ; Hirata, H. ; Kotaka, T. *Langmuir* **1988**, *4*, 354.
6. Shikata, T. ; Hirata, H. ; Kotaka, T. *Langmuir* **1989**, *5*, 398.
7. Shikata, T. ; Hirata, H. ; Takatori, E. ; Osaki, K. *J. Non-Newtonian Fluid Mech.* **1988**, *28*, 171.
8. Shikata, T. ; Kotaka, T. *J. Non-cryst. Solids* **1991**, *131-133*, 831.
9. Ferry, J. D.*Viscoelastic Properties of Polymers, 3rd Ed.*, Wiley, New York, 1980.
10. Doi, M. ; Edwards, S. F. *The Theory of Polymer Dynamics*, Oxford University Press, Oxford, 1986.
11. Cates, M. E. *Macromolecules* **1987**, *20*, 2289.
12. Cates, M. E. *Macromolecules* **1988**, *21*, 256.
13. Flory, P. J. *Statistical Mechanics of Chain Molecules*, Interscience Publishers, New York, 1969.
14. Treloar, L. R. G. *The Physics of Rubber Elasticity*, Clarendon Press, Oxford, 1975.
15. Pearson, D. S. ; Kiss, A. ; Fetters, L. *J. Rheol.* **1989**, *33*, 517.
16. Rehage, H. ; Wunderlich, I. ;Hoffmann, H. *Progr. Colloid Polym. Sci.* **1986**, *72*, 51.
17. Hu, Y. ; Wang, S. Q. ; Jamieso, A. M. *J. Rheol.* **1993**, *37*, 531.
18. Kalus, J. ; Hoffmann, H. ; Lindner, P. *Progr. Colloid Polym. Sci.* **1989**, *79*, 233.
19. Schorr, W. Ph D. Thesis, University of Bayreuth, 1982.
20. Stein, R. S. ; Tobolsky, A. V. *Textile Res. J.* **1948**, *18*, 201.
21. Shikata, T. ; Pearson, D. S., submitted to *Langmuir*, 1994.
22. Vinson, P. K. ; Talmon, Y. *J. Colloid Interface Sci.* **1989**, *133*, 288.
23. Imae, T. *J. Phys. Chem.* **1990**, *94*, 5953.
24. Herbst, L. ; Kalus, J. ; Schmelzer, U. *J. Phys. Chem.* **1993**, *97*, 7774.
25. Appell, J. ; Porte, G. *J. Colloid Interface Sci.* **1982**, *87*, 492.

RECEIVED July 8, 1994

Chapter 9

Spinnability of Viscoelastic Surfactant Solutions and Molecular Assembly Formation

Toyoko Imae

Department of Chemistry, Faculty of Science, Nagoya University, Chikusa, Nagoya 464, Japan

The spinnability was measured for aqueous viscoelastic solutions of cationic surfactants with aromatic salicylate counterion in the absence and presence of sodium salicylate. Aqueous solutions of alkyl- and oleyldimethylamine oxides also exhibited the characteristic spinnable behavior. The spinnability of these surfactant solutions was classified into the ductile failure type and the cohesive fracture failure type. The behavior was compared with that of spinnable polymer solutions. The spinnability is closely related to the viscoelasticity, that is, the plateau modulus or the dynamic (structural) relaxation. The assembly formation by surfactant molecules is discussed in connection with rheological properties of solutions.

When a rod in a liquid is pulled up at a constant rate, the liquid is stretched to form a thread. This rheological phenomenon is called the "spinnability", which is related to the Weissenberg effect. The spinnable liquids are generally non-Newtonian and viscoelastic, and the thread-forming property is characterized by the length of the liquid thread. The spinnability was reported for native and synthetic polymer solutions (*1-5*). It was suggested that the thread-forming state was described by the elastic deformation superposed on the viscous flow and that the spinnability was most remarkable when the drawing velocity coincided with the mechanical relaxation time in Maxwell's flow.

Similar characteristics were also observed for aqueous solutions of cationic surfactants with certain kinds of aromatic counterions such as salicylate (*6*). Such surfactants exhibited the strong viscoelasticity even in very dilute aqueous solutions. The micellar structure in the viscoelastic solutions and the mechanism for the inducement of the viscoelasticity are discussed in several works. Some investigators (*7-9*) confirm the formation of rodlike micelles in the viscoelastic solutions. It has been assumed that rodlike micelles build the three-dimensional network, where the contacts between micelles are constantly formed and broken (*7*). Later work indicates that the effect of micellar kinetics has to be taken into consideration at the same time

0097–6156/94/0578–0140$08.00/0

(*10*). The experimental data have also been explained by the formation of an entangled network of threadlike micelles and the breakdown-reformation of the network (*11*).

Since salicylate ions exhibit specific adsorption and penetration ability (*12*), the electric structure such as surface potential and electrical double layer, and as a result the micelle structure and their properties in a solution, may change remarkably. Therefore, the role of salicylate ions on micelle formation must be elucidated. It has been emphasized that the unique orientation of salicylate ions on the micellar surface is the crucial factor for inducing viscoelasticity, and two mechanisms have been proposed (*13*).

Dilute aqueous solutions of hexadecyl- and octadecyldimethylamine oxide ($C_{16}DAO$, $C_{18}DAO$) present iridescent colors at lower temperatures and are transparent at higher temperatures. It has been confirmed from microscopic observations that platelike assemblies and lamellar layers are constructed in the iridescent solutions (*14,15*). Rodlike micelles are formed in transparent solutions of $C_{16}DAO$ as well as of oleyldimethylamine oxide (ODAO) (*16,17*). Aqueous solutions of $C_{16}DAO$ and ODAO are viscoelastic (*18*), and the spinnability of aqueous ODAO solutions is especially remarkable.

Extract from the root of *Abelmoschus Manihot* is used to disperse paper fibers when Japanese paper is manufactured. It exhibits the remarkable viscoelasticity as well as sodium alginate, poly(ethylene oxide) (PEO), and sodium carboxymethylcellulose (NaCMC) which are also useful to disperse fibers (*19*). Although the spinnability phenomenon is known for these polymers, it is not necessarily discussed in relation to the viscoelasticity and the polymer structure in solutions.

In this work, the spinnability property is quantitatively investigated for aqueous solutions of tetradecyl- and hexadecyltrimethylammonium salicylates ($C_{14}TASal$, $C_{16}TASal$) in the absence and presence of sodium salicylate (NaSal) and is compared with that of C_nDAO, ODAO and polymers described above. Moreover, the spinnability mechanism in dilute surfactant solutions is discussed. Such a mechanism has never clearly been identified and reported.

It may be noticed that aqueous C_nTASal solutions investigated here do not include any additive counterions like halides, since samples of C_nTASal were synthesized by counterions-substituting alkyltrimethylammonium halides with NaSal (*20*). Then the solution behavior of C_nTASal is partly different from that of alkyltrimethylammonium halides mixed with NaSal (*20,21*), because the specific adsorption and the penetration ability of salicylate ions are hindered by the adsorption of halide ions.

Spinnability

Spinnability was measured on an apparatus constructed as previously reported (*2*). A rod in a test solution was driven up at a constant rate (*20*). The maximum thread length, namely, the drawing length was measured as a length from the rod to the surface of the solution. The drawing lengths at various immersion depths were extrapolated to the zero depth to obtain the intrinsic drawing length.

Figure 1 gives the schematic representation of the intrinsic drawing length L_0 as a function of drawing velocity v. The L_0 values increase linearly with v in some cases,

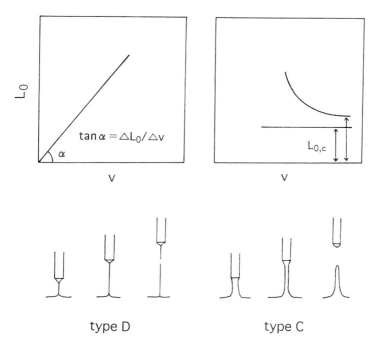

Figure 1. Schematic representation of the variation of drawing length against drawing velocity (upper) and the drawing process (lower) for the type D and C spinnability. (Reproduced with permission from ref. 20. Copyright 1990 Verlag GmbH.)

while others decrease with v or are almost independent of v. Both cases are respectively characterized by a slope $\triangle L_0/\triangle v$ of the straight line and by a constant intrinsic drawing length $L_{0,c}$ at the finite drawing velocity.

The former (type D) spinnability is a kind of ductile failure (or capillary-ductile failure), which is observed for viscoelastic solutions with low viscosity (22). In this case, the liquid thread becomes thin with drawing until it breaks, as seen in the schematic representation in Figure 1. In the latter (type C) spinnability, the liquid thread keeps almost constant thickness during drawing and behaves like cohesive fracture failure, which occurs in rubber or gel elasticity (22).

A liquid thread in the type C spinnability is pulled back into the liquid reservoir, when it is broken, since the strong elasticity acts against the deformation caused by drawing up the rod. The liquid thread snaps at a position near the rod when the tension (retracting force) of a solution overcomes the adhesive force of the liquid thread to the rod. The tension may be approximately inversely proportional to the length of the liquid thread. If the liquid thread deforms more quickly than its maximum relaxation time, the thread is broken at a certain drawing length, regardless of the drawing velocity of the rod. Therefore, the constant drawing length can be related to the elasticity of a solution. On the other hand, it is reported that the shear modulus increases as the solute concentration increases (18). Hence the drawing length may become shorter with an increase in solute concentration. Such a relationship for the type C spinnability was confirmed by the present work (19,20).

Aqueous Solutions of Tetradecyl- and Hexadecyltrimethylammonium Salicylates

In aqueous solutions without NaSal, $C_{14}TASal$ molecules associate into elliptic or short rodlike micelles at temperatures below 25 °C (23). Since the external interference effect indicating the intermicellar correlation is strong at micelle concentrations above 10^{-3} g cm^{-3}, short rodlike micelles may interact and pseudolink with each other. The linked micelles form the pseudonetwork structure, as shown in Figure 2a.

The absolute complex viscosity $|\eta*|$, the storage modulus G', and the loss modulus G" were measured as a function of angular frequency, and the zero shear viscosity η_0, the plateau modulus G_N, and the relaxation time τ were evaluated from experimental data (21). While the relaxation time is almost independent of micelle concentration, the zero shear viscosity and the plateau modulus increase at micelle concentrations above 10^{-3} g cm^{-3}.

Type D spinnability is observed at 20-40 °C, where the $\triangle L_0/\triangle v$ values increase with increasing micelle concentration (20). The surfactant solutions at higher micelle concentrations provide type C behavior, and the $L_{0,c}$ values decrease gradually as the micelle concentrations are increased. The spinnability changes from type D to C at 4 x 10^{-2} g cm^{-3} micelle concentration and is strongest in the transitional region of type D to C. At the transition region, type D is observed at low drawing velocities, and type C is observed at high velocities.

When the micelle concentration is diluted, $C_{14}TASal$ micelles are short and few, and the intermicellar correlation is rather weak. Then the loose pseudonetwork composed of short rodlike micelles is formed in micellar solutions. Such structure

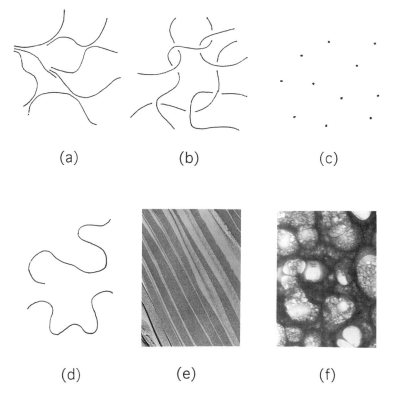

Figure 2. Structures of various assemblies. (a) Linked short rodlike micelles; (b) entangled rodlike micelles; (c) small micelles; (d) rodlike micelles; (e) lamellar layers; (f) super-network.

induces the type D spinnability. The type D spinnability is observed for solutions where the viscous flow overcomes the elastic deformation. The spinnability strengthens with increasing micelle concentration, owing to the increase in the amount of micelles and/or to the micellar growth. Then the pseudolinkages between micelles increase and tighten. Therefore, the elasticity becomes more operative. In solutions having a drawing length too long to measure by the apparatus in this work, the elasticity and the viscosity must be favorably balanced.

At higher micelle concentrations, the contribution of the strong elasticity from the sufficiently developed pseudonetwork of micelles is superior to that of the viscosity and, therefore, type C spinnability appears. In this case the effective tension of the solution strengthens enough to pull the thread back into the solution and to break the thread. The drawing length thus becomes short, and the spinnability converts from the ductile failure to the cohesive fracture; that is, from the viscous response to the elastic deformation.

The pseudonetwork composed of pseudolinkages between short rods in which salicylate ions participate is formed even in aqueous C_{16}TASal solutions without NaSal (*24*). As the alkyl chain length of surfactant becomes longer, larger or longer rodlike micelles are formed. Hence the pseudonetwork is formed at lower micelle concentrations and will tighten. Such structure may induce the type C spinnability. In fact, type C behavior is predominant for aqueous C_{16}TASal solutions, when the aspect is compared with that of aqueous C_{14}TASal solutions (*20*).

The pseudonetwork in aqueous CnTASal solutions without NaSal is constituted by the pseudolinkages but not by the entanglement. These kinds of assembly structures are distinguished by the contrary angular dependence of static light scattering (*23,24*). In the pseudolinkages, salicylate counterions penetrate into rodlike micelles and hydrogen-bond between counterions through water. This kind of network is indispensable for inducing strong spinnability and viscoelasticity, because such properties are not induced by only the entanglement. In this connection, semidilute solutions of entangled rodlike micelles of alkyltrimethylammonium halides and oligooxyethylene alkyl ethers exhibit strong viscosity but very low viscoelasticity (*25,26*).

Aqueous NaSal Solutions of Tetradecyl- and Hexadecyltrimethylammonium Salicylates

When 0-0.4 M NaSal is added into aqueous C_{14}TASal solutions at a surfactant concentration of 1.6 x 10^{-2} g cm^{-3}, the type D behavior is observed at 15-35 ℃ (*20*). In Figure 3, rheological and light scattering results are compared as a function of ionic strength $C_0 + C_s$ for aqueous C_{14}TASal solutions at 25 ℃ (*21,23,24*), where C_0 and C_s are critical micelle concentration and salt concentration, respectively. M_{app} is the apparent molecular weight (*24*) and U is the electrophoretic mobility. While the plateau modulus is independent of ionic strength, the zero shear viscosity, the relaxation time, and the ductile failure spinnability increase with addition of a small amount of NaSal, since the intermicellar correlation diminishes above $\sim$ 0.001 M ionic strength and rodlike micelles lengthen. On further addition of NaSal, the zero shear viscosity and the spinnability decrease through a second maximum and a shoulder, respectively, at

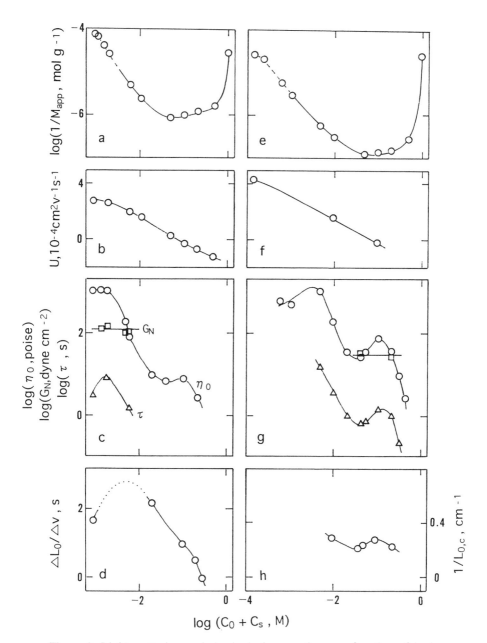

Figure 3. Light scattering and rheological properties as a function of ionic strength for aqueous CnTASal solutions at 25 °C. Surfactant: a-d, $C_{14}TASal$; e-h, $C_{16}TASal$. Surfactant concentration (10^{-2} g cm^{-3}): a, 0.8; b, 0.8; c, 1.6; d, 1.6; e, 0.2; f, 0.055; g, 1; h, 0.2. Broken lines in a and e represent the contribution of external interference. (Reproduced with permission from ref. 23. Copyright 1992 American Chemical Society.)

$\sim$ 0.1 M NaSal, where the micelle size is maximum. The spinnability disappears at higher ionic strength.

As seen in Figure 3, viscoelasticity and light scattering of aqueous $C_{16}TASal$ solutions behave as well as those of aqueous $C_{14}TASal$ solutions (20,21,24). The zero shear viscosity has two maxima, and the corresponding maximum around 0.1 M NaSal is also obtained for the relaxation time. The type C spinnability is observed at 25 and 35 °C, and the reciprocal $L_{0,c}$ values and the micelle size have a maximum at $\sim$ 0.1 M NaSal.

Electrophoretic light scattering is a useful method for examining the surface potential and the charge of colloidal particles, which are essential for the stability of particles in a medium. Electrophoretic mobility examination indicates that CnTASal micelles in water exhibit a positive, low net surface charge, as seen in Figure 3, suggesting the specific adsorption and penetration of salicylate ions to micelles (23, 24). With addition of NaSal, the net charge of micelles is converted from positive to negative through neutral at $\sim$ 0.1 M NaSal. This indicates that the specific adsorption and the penetration of salicylate ions dominate the micelle size and the solution behavior such as spinnability and viscoelasticity. The added NaSal not only promotes the specific adsorption and penetration but also decreases the electric double layer. Thus, 0.1 M NaSal must be enough to reach to the zero thickness double layer.

Since addition of a small amount of NaSal stimulates the growth of rodlike micelles by the salting-out effect, the rheological character, spinnability and viscoelasticity, is slightly emphasized. Accompaning the increase of viscosity, the spinnability tends to be type D rather than type C and becomes stronger. The further addition of NaSal partly destroys the intermicellar correlation and the pseudolinkages in the pseudonetwork, owing to the electric shielding effect. As a result, the rheological character diminishes at NaSal concentrations above the first extreme.

The addition of excess NaSal promotes the specific adsorption of salicylate counterion at the Stern layer of the micelle surface and their penetration into micelles. This results in rodlike micelles with zero net surface charge at $\sim$ 0.1 M NaSal and, therefore, the elongation and the entanglement of rodlike micelles in the semidilute region progress because of the diminution of electrostatic repulsion in the micelles. Then the rheological behavior again increases slightly. The pseudonetwork in semidilute solutions is constituted by the entanglement and a small fraction of pseudolinkages. While the pseudonetwork composed of the pseudolinkages presents rather elastic rheological character, the pseudonetwork by the entanglement has more viscous character.

Above 0.1M NaSal, the negative net charge of a micelle increases due to the excess adsorption and the penetration of salicylate ions. Therefore, the micelle size diminishes because of the electrostatic repulsion in "anionic" micelles, until small micelles with a negative charge are formed around 1 M NaSal. Simultaneously, the rheological behavior decreases and disappears with the diminution and the disappearance of entanglement and pseudolinkages. Since simple salts such as sodium halides generally tend only to increase micelle size, the destruction of the micelles by 1 M NaSal is unusual. The remarkable effect of NaSal may be attributed to the superior ability of NaSal to change water structure. Schematic models of typical micellar structures are illustrated in Figures 2a-2d.

Aqueous NaCl Solutions of Alkyl- and Oleyldimethylamine Oxides

While aqueous $C_{14}DAO$ solutions at 25 ℃ present no spinnability even in the presence of 0.2 M NaCl, the type D spinnability is observed on addition of HCl for solutions with 0.2 M NaCl (18). The $\triangle L_0/\triangle v$ values have a maximum at Ca/C ratio of $\sim$ 0.3, where Ca and C are mole concentrations of HCl and $C_{14}DAO$, respectively. For aqueous $C_{16}DAO$ solutions without salt, the type C spinnability is observed at 20 and 25 ℃, and the type D is at 35 - 45 ℃. The $L_{0,c}$ and $\triangle L_0/\triangle v$ values increase with the initial increase of surfactant concentration. Aqueous ODAO solutions with and without NaCl present only the type D spinnability. Whereas the $\triangle L_0/\triangle v$ values are independent of low NaCl concentrations, they increase above 0.1 M NaCl.

When a moderate amount of HCl is added, C_nDAO molecules with charge make pairs with noncharged molecules, and rodlike micelles grow and entangle (16). This strengthens the type D spinnability. At higher Ca/C ratios, the micelle size decreases by the electrostatic repulsion between charged surfactants in a micelle, and the spinnability weakens.

The viscoelastic behavior of dilute solutions of $C_{16}DAO$ without salt at 35 ℃ and ODAO at 25 ℃ is similar to that of dilute nonlinked polymer solutions; and that of concentrated solutions is like that of entangled polymer solutions (18). The G' and G" values increase with surfactant concentration. Moreover, the values increase sharply with angular frequency ω and there is even a small peak in a G"- ω plot for a concentrated solution. This aspect is consistent with rodlike micelles being formed in dilute surfactant solutions and entangling each other at high surfactant concentrations. Then the spinnability is affected by the surfactant concentration.

The addition of NaCl enhances the growth and the entanglement of rodlike ODAO micelles but shields the intermicellar interaction (17). The spinnability and the viscoelasticity are not strongly affected by the addition of NaCl below 0.1 M as a result of the compensation between these two opposite effects. On the other hand, the growth and the entanglement of micelles are remarkable above 0.1 M NaCl, and, therefore, the spinnability and the viscoelasticity increase.

Aqueous solutions of $C_{16}DAO$ and $C_{18}DAO$ without salt at room temperature present iridescence at 0.3 - 2 wt % surfactant concentrations. Optical and electron microscopic observation confirmed the formation of platelike assemblies and lamellar layers (14,15), as shown in Figure 2e. Such solutions exhibit rheological behavior different from that of rodlike micelles (18). For aqueous $C_{16}DAO$ solutions at 20 ℃, which exhibit type C spinnability, the frequency dependence of G' and G" is small, and the G" values are larger than the G' values. The viscoelastic behavior of aqueous $C_{18}DAO$ solutions without salt at 25 ℃ is very similar to that of $C_{16}DAO$ at 20 ℃. Long-range structure exists in the iridescent solutions, and such structure has a relaxation time longer than the experimental time scale. Therefore, the stretched solutions produce the large stress by the elastic force, but the relaxation of the elasticity by flow may not act there.

Polymers

Extract from the root of *Abelmoschus manihot* is a polysaccharide that is composed of rhammnose and galacturonate (*19*). The extract in water is viscoelastic, since polymer chains form junctions. The major component of a brown seaweed, sodium alginate, is a linear copolymer of L-guluronate and β-D-mannuronate. Sodium alginate forms intermolecular linkages mediated by cations in aqueous medium. PEO chains form helical structures in water and interact through hydrogen bonding between polymer and water. Moderately substituted NaCMC molecules are joined by noncovalent linkages in a relatively concentrated solution and result in chain-bundles. Such junction formation between polymers should be related to their viscoelastic properties.

Aqueous solutions of extract from the root of *Abelmoschus manihot* and PEO exhibit type D spinnability as well as those of sodium alginate at low concentrations (*19*). On the other hand, type C spinnability is observed for aqueous solutions of sodium alginate at high concentrations and for NaCMC. The $\triangle L_0 / \triangle v$ or reciprocal $L_{0,c}$ values always increase with increasing polymer concentration. The $\triangle L_0 / \triangle v$ is the reciprocal deformation rate of a thread when the thread is broken, and it corresponds to the relaxation time. It can be interpreted that the relaxation time of aqueous PEO solutions increases with concentration (*27*). This is consistent with the concentration dependence of the $\triangle L_0 / \triangle v$ values.

The $L_{0,c}$ value reflects the elastic modulus of the solution and may decrease with increasing modulus. The reciprocal $L_{0,c}$ and the dynamic modulus for aqueous sodium alginate solutions at high concentrations increase with concentration (*18,19*). It is found that the dynamic modulus of aqueous NaCMC solutions increases with increasing concentration (*28*). These results support the explanation described above.

Cryo transmission electron microscopic observation was carried out (*29*). Network images were observed for both sodium alginate and PEO solutions. The network is not homogeneous and the thick network domains segregate from the thin network domains, as seen in Figure 2f. The segregation produces a "super-network" structure. Then the characteristics of viscoelastic and spinnable solutions are due to the super-network structure based on the pseudonetwork formation.

It is suggested in many investigations that the high-order structure constructed by intermolecular interactions participates in the viscous property and the elastic response of polymer solutions. Parts of polymer chains form the linkages to constitute a junction zone. The network consists of junction zones and free residues of polymer chains. The junction-rich and free-residue-rich domains result in thick and thin networks, respectively. The segregated network domains form a supernetwork. Such three- dimensional network structures leads to viscoelastic behavior with long lifetime processes, where motions of networks relax with long relaxation times.

The difference between the type C and D spinnability may depend on life time, and the amount and strength of junctions among polymers. The type D spinnability appears for polymers with few or weak junction linkages, because the solution can easily deform and flow, allowing the stress to be dissipated. The type C spinnability is superior to polymers largely linked through the interaction, since the structure brings the elastic response. Thus the solution does not flow, and the stress is maintained.

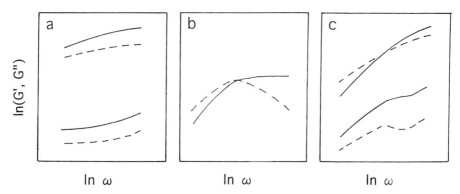

Figure 4. Schematic representation of the variation of dynamic modulus against angular frequency. (a) Gel-like; (b)Maxwell-like, (c)polymer-like viscoelasticity. ———— G', - - - G".

Conclusions

Some types of molecular assembly structures composed of surfactant molecules are formed in aqueous solutions of CnTASal and CnDAO. Such structures confer viscoelastic and spinnable properties on aqueous surfactant solutions. Then the spinnability depends on the strength of such structures or on the balance between the elasticity and the viscosity in which the structures result, with respect to the analogy of viscoelastic polymer solutions and of concentrated dispersions.

Three types of viscoelasticity illustrated in Figure 4 are observed in aqueous solutions of surfactants and polymers examined here. The storage modulus G' and the loss modulus G" in the gel-like viscoelasticity are almost independent of or slightly dependent on angular frequency ω. In the Maxwell-like viscoelasticity, the G' values increase at low frequencies and reach the plateau at high frequencies, while the G" values have a maximum. Both the G' and G" values in the polymer-like viscoelasticity increase with the initial slopes of 2 and 1, respectively, in a double logarithmic plot.

The types of viscoelasticity are listed in Table I for aqueous solutions of surfactants and polymers and compared with the types of spinnability and the assembly structures, which are drawn in Figures 1 and 2. The ductile failure spinnability can be connected with the Maxwell-like and the polymer-like viscoelasticity, and the cohesive fracture failure spinnability is connected with the gel-like viscoelasticity. The former is observed for entangled rodlike micelles in semidilute solutions and polymers with the super-network structure, and the latter is observed for linked rodlike micelles and lamellar layers.

Nakagawa (4) concluded that the relaxation time was comparable with the spinning time. This means that the spinnability depends on the dynamic relaxation of a solution. In fact, the spinnability for aqueous CnTASal solutions at 25 ℃ behaves similarly to the relaxation time, as seen in Figure 3. The relation of uniaxial elongation to the spinning process of the fiber was reported. When the Maxwell liquid is elongated at a constant velocity, the stress σ at the time t is described by

$$\sigma \sim (Bt+C) \; exp \; (-t/\tau)$$

Table I. Rheological Characteristics and Assembly Structure of Aqueous Solutions of Surfactants and Polymers (*14-21,23,24,29*)

Solute	*solvent*	*T, $\,^{\circ}C$*	*c,* $10^{-2}gcm^{-3}$	*spin-nability*	*visco-elasticity*	*assembly structure*
$C_{14}TASal$	water	25	0.8-3.2	D	Maxwell+gel	linked short rodlike micelle
	water	25	10	C	gel	linked short rodlike micelle
	1-4mMNaSal	25	1.6	D	Maxwell	
	0.02-0.2MNaSal	25	1.6	D	polymer	entangled rodlike micelle
	>0.3MNaSal	25	1.6	no	no	small micelle
$C_{14}DAO$	0.2MNaCl,Ca/C=0-1	25	5	D	polymer	rodlike micelle
$C_{16}DAO$	water	20	1-3	C	gel	platelike assembly, lamellar layer
	water	35	1-3	D	polymer	rodlike micelle
$C_{18}DAO$	water	25	0.5-3	C	gel	platelike assembly, lamellar layer
ODAO	water	25	0.45-6	D	polymer	rodlike micelle
	0-2MNaCl	25	0.9	D	polymer	rodlike micelle
Abelmoschus Manihot	water	15-35	0.03-0.08	D		
PEO	water	15-25	6-12	D	polymer	super-network
sodium alginate	water	25	1	D	polymer	
	water	25	2-5	D	polymer+gel	super-network
	water	25	7,10	C	gel	
NaCMC	water	15-25	2-3	C	gel	

where B and C are constants (30). It is recognized that the stress reaches a maximum at the relaxation time τ. Therefore, with solutions exhibiting type D spinnability, the elongated thread is broken at the relaxation time τ. In the type C spinnability, where the elastic modulus is very strong, the maximum value of the stress is so high that the stress by the elongation is larger than the cohesion force of the thread, and the thread is pulled back elastically into the solution reservoir.

Acknowledgments

This article is dedicated to the memory of Yasuo Imae who died in July, 1993.

Literature Cited

1. Jochims, J. *Koll. Z.* **1927**, 43, 361.
2. Erbring, H. *Koll. Beih.* **1936**, 44, 171.
3. Thiele, H.; Lamp. H. *Koll. Z.* **1952**, 129, 25.
4. Nakagawa, T. *Bull. Chem. Soc. Japan* **1952**, 25, 88; 25, 93.
5. Inagaki, T. *J. Colloid Sci.* **1956**, 11, 226.
6. Gravsholt, S. *J. Colloid Int. Sci.* **1976**, 57, 575.
7. Rehage, H.; Hoffmann, H. *Faraday Discuss. Chem. Soc.* **1983**, 76, 363.
8. Angel, M.; Hoffmann, H.; Lobl, M.; Reizlein, K.; Thurn, H.; Wunderlich, I. *Progr. Colloid Polym. Sci.* **1984**, 69, 12.
9. Shikata, T.; Sakaiguchi, Y.; Uragami, H.; Tamura, A.; Hirata, H. *J. Colloid Inteface Sci.* **1987**, 119, 291.
10. Rehage, H.; Hoffmann, H. *J. Phys. Chem.* **1988**, 92, 4712.
11. Shikata, T.; Hirata, H.; Kotaka, T. *Langmuir* **1987**, 3, 1081; **1988**, 4, 354.
12. Olsson, U.; Soderman, O.; Guering, P. *J. Phys. Chem.* **1986**, 90, 5223.
13. Rao, U. R. K.; Manohar, C.; Valaulikar, B. S.; Lyer, R. M. *J. Phys. Chem.* **1987**, 91, 3286
14. Imae, T.; Trend, B. *J. Colloid Interface Sci.* **1991**, 145, 207.
15. Imae, T.; Iwamoto, T. *J. Colloid Interface Sci.* **1992**, 152, 289.
16. Imae, T. *J. Jpn. Oil Chem. Soc.* **1992**, 41, 616.
17. Imae, T.; Ikeda, S. *J. Colloid interface Sci.* **1984**, 98, 363; *Colloid Polym. Sci.* **1985**, 263, 756.
18. Hashimoto, K.; Imae, T. *Langmuir* **1991**, 7, 1734.
19. Hashimoto, K.; Imae, T. *Polym. J.* **1990**, 22, 331.
20. Imae, T.; Hashimoto, K.; Ikeda, S. *Colloid Polym. Sci.* **1990**, 268, 460.
21. Hashimoto, K.; Imae, T.; Nakagawa, K. *Colloid Polym. Sci.* **1992**, 270, 249.
22. Ide, Y.; While, J. L. *J. App. Polym. Sci.* **1976**, 20, 2511.
23. Imae, T.; Kohsaka, T. *J. Phys. Chem.* **1992**, 96, 10030.
24. Imae, T. *J. Phys. Chem.* **1990**, 94, 5953.
25. Imae, T.; Sasaki, M.; Ikeda, S. *J. Colloid Interface Sci.* **1989**, 127, 511.
26. Sasaki, M.; Imae, T.; Ikeda, S. *Langmuir* **1989**, 5, 211.
27. Lance-Gomez, E. T.; Ward, T. C. *J. Appl. Polym. Sci.* **1986**, 31 , 333.
28. Thurston, G. B.; Martin, A. *J. Pharm. Sci.* **1978**, 67, 1499.
29. Imae, T. *Polym. J.* **1993**, 25, 201.
30. Fisher, R. J.; Denn, M. M. *ALChE J.* **1976**, 22, 236.

RECEIVED August 10, 1994

Chapter 10

Correlation of Viscoelastic Properties with Critical Packing Parameter for Mixed Surfactant Solutions in the L_1 Region

Craig A. Herb, Liang Bin Chen, and Wei Mei Sun

Helene Curtis, Inc., Surface Science Laboratory,
4401 West North Avenue, Chicago, IL 60639–4769

We have investigated the steady shear and dynamic rheological properties of aqueous solutions containing sodium dodecylsulfate and n,n-bis(2-hydroxyethyl) dodecanamide over a large part of the normal micelle (L_1) region of the phase diagram. The solutions are found to exhibit viscoelastic behavior typical of systems that are composed of elongated, rod-like micelles. We have mapped lines of constant zero-shear viscosity, η_0, and relaxation time, τ_R, onto the phase diagram and show that there is a relationship between the shape of the L_1 phase boundary and the location and values of these lines. Several other anionic/nonionic and cationic/anionic systems are compared and found to follow the same relationship. We also show that lines of constant τ_R for our system follow the same paths as our estimated lines of constant *critical packing parameter*, suggesting the possibility of being able to predict the rheology of these surfactant solutions from more fundamental structure/property considerations for the micelles.

Mixed anionic/nonionic surfactant solutions are of great importance in a wide range of industrial and consumer products. The flow properties of these systems affect the manufacturing process, package selection, and consumer perception of the products. It is of interest, therefore, to understand the underlying mechanisms well enough to be able to predict and control the rheology. The hair care industry provides an obvious example of the practical need for this understanding. The consumer equates high viscosity (5 to 10 Pa s) with efficacy, ease of use, and value of a shampoo. This perception, which is restricted to the flow behavior in the bottle and the hand, falls into the shear rate range of about 5 to 20 $\sec^{-1}$, and is usually on the zero-shear viscosity plateau, η_0, for a simple shampoo. However, it is difficult or impossible for the consumer to pump or squeeze 5 to 10 Pa s material out of a bottle. Clearly, the viscosity at the conditions encountered during dispensing must be considerably lower than this.

0097–6156/94/0578–0153$08.00/0

Simple calculations of the flow rate required and the shear stress produced by the dispenser show that the apparent viscosity must be no higher than 0.5 Pa s and should be even lower. Thus, the viscosity must drop off by 1 or 2 orders of magnitude to be acceptable and must do it at an applied shear stress less than or equal to that supplied by the delivery system. Successful shampoo formulations combine the required high η_0 value with this rapid drop off.

It turns out that such shampoo formulations are viscoelastic surfactant solutions found in the normal micelle (L_1) region of the phase diagram. Over the last decade it has been established that these systems are composed of elongated, rod-like micelles that, because of their length and flexibility, exhibit entangled polymer-like flow behavior. This work has been reviewed by Hoffmann, Rehage, and co-workers at several stages (*1-5*), the first chapter of this book being the most recent. These "worm-like" micelles differ from polymers in that they can break and re-form fairly rapidly. By modifying reptation theory (*6,7*) to accommodate chains that are capable of such "reversible scission" reactions, Cates (*8-11*) has developed a model suitable for describing the rheology of these surfactant solutions. This model predicts a single exponential stress relaxation for systems in which micelle breaking and reforming are rapid compared to the reptation process for an average micelle. This single relaxation time is not observed for systems in which the relaxation time for micelle breaking, τ_{break}, is larger than the relaxation time for chain reptation, τ_{rep}. Many surfactant solutions in the L_1 phase can, indeed, be modeled as single relaxation Maxwell materials (*5*). This is also true of the anionic/nonionic system to be discussed in this paper, suggesting that we are in the regime where $\tau_{break} \leq \tau_{rep}$. Under these conditions Cates theory (*8*) predicts that the rheologically measured relaxation time, τ_R, is approximately equal to $(\tau_{break} \tau_{rep})^{\frac{1}{2}}$, suggesting that the relaxation process still involves cooperation between breaking and reptation. See chapters 2 and 3 of this book for further details.

In addition, the shear modulus, G_0, scales with c/c* to the 2.3 power (*3*), where c is the concentration of the surfactant and c* is the overlap concentration. The value of G_0 appears to be dependent only on the reduced concentration (c/c*) of surfactant. Thus, we should be able to predict G_0 from determinations of c* as a function of micelle composition. As there is evidence that the power law slope changes from 2.3 to about 1.7 at high salt or surfactant concentration in some systems (*12*), caution should be used to ensure that the 2.3 scaling holds over the region of interest. Since we now have an estimate of G_0, and η_0 is simply the product of τ_R and G_0, we can estimate η_0 for much of the L_1 region if we can find a way to predict τ_R as a function of solution composition.

In this paper we examine the phase behavior and rheology of aqueous solutions containing sodium dodecyl sulfate (SDS) and n,n-bis(2-hydroxyethyl) dodecanamide over a large part of the L_1 region of the phase diagram. The compound n,n-bis(2-hydroxyethyl) dodecanamide is usually referred to as "lauric diethanolamide," "lauramide DEA" or simply "LDEA" by the manufacturers and formulators who use it. For convenience, we will refer to it as LDEA in this paper. Values of the *critical packing parameter* (*13*) for this system are also calculated based on interfacial tension measurements and counter ion binding estimates. The measured values of τ_R and η_0

from the rheological data will be mapped onto the phase diagram, and we will show that the lines of constant relaxation time over the L_1 region qualitatively follow the predicted lines of constant packing parameter for our anionic/nonionic system. This suggests a possible path from solution composition, through micelle formation and morphology, to flow behavior prediction. We will also show data for other systems to support the hypothesis that the presence of an indentation in the L_1 phase boundary and the rheological maxima are linked and observed only in systems for which the critical packing parameter goes through a maximum as the composition of the micelle changes. We will provide results suggesting that this behavior is general and explains the rheological maxima observed when the surfactant concentration is increased along lines of constant surfactant composition.

Micelle Relaxation Times

There are two fundamental chemical relaxation processes that are reported for micelles. The first, referred to as the fast process, involves the rapid exchange of surfactant monomers with the micelles. This process is often diffusion controlled. The second, or slow process, involves the break up of the entire micelle. These two processes can be investigated by a number of experiments including temperature or pressure jump measurements (*14,15*). At high concentrations, the slow relaxation process is assumed to involve mainly micelle breaking and coalescence (*16*). Thus, τ_{break} is essentially the slow chemical relaxation time. The greater the free energy change required to create two smaller micelles from a large micelle, the less likely it is that a break will occur in a given time period. That is, we should observe a longer relaxation time for these more stable micelles. In general, factors that decrease the free energy of a surfactant molecule in a micelle should promote larger values of τ_{break}. Thus, closer packing of the head groups will lead to a reduction of the free energy per molecule along the cylindrical portion of the micelle by reducing the contact between the water and hydrocarbon tails. Also, the difference between the surface area per head group required by the geometry of the hemispherical end caps and the optimal surface area desired by the molecules, a_0, increases as a_0 decreases (*17*). This suggests that cylinder ends become more and more unfavorable energetically as a_0 decreases. Although this is, admittedly, an oversimplification of the situation, the combination of these two effects should lead to larger, more stable micelles that will exhibit larger τ_{break} and τ_{rep} values. Thus, as discussed below, it is reasonable to expect a positive correlation between τ_R and the critical packing parameter.

The Critical Packing Parameter

For a given optimal area per head group a_0, volume of the hydrocarbon tail v, and critical chain length l_c, a dimensionless packing parameter $v/a_0 l_c$ will determine the micelle morphology (*18*). Because of the impact of this packing parameter on the expected size and stability of cylindrical micelles, it is reasonable to look for a link between it and the rheological behavior of viscoelastic surfactant solutions. For saturated n-alkanes the quantity v/l_c is essentially constant at 0.21 nm² for pentane and above (*19*). Thus, the packing parameter, $P*$, can be evaluated from measured values of a_0. We determined a_0 from interfacial tension measurements against hexadecane using the multicomponent Gibbs adsorption equation (*20,21*). Making all the usual dilute solution assumptions, we can write the Gibbs equation for our multicomponent system as

$$d\gamma = -RT \left[\Gamma_{LDEA} \, d \ln c_{LDEA} + \Gamma_{DS^-} \, d \ln c_{DS^-} + \Gamma_{Na^+} \, d \ln c_{Na^+} \right] \tag{1}$$

where γ is the interfacial tension, R is the gas constant, T is the absolute temperature, c_{LDEA} and c_{DS^-} are the bulk molar concentrations of LDEA and dodecyl sulfate respectively, and Γ_{LDEA} and Γ_{DS^-} are the surface excess concentrations of LDEA and dodecyl sulfate respectively. We can write

$$\Gamma_{LDEA} = \frac{-1}{RT} \left(\frac{\partial \gamma}{\partial \ln c_{LDEA}} \right)_{T, c_{DS^-}, c_{Na^+}} \quad \text{and} \quad \Gamma_{DS^-} = \frac{-1}{RT} \left(\frac{\partial \gamma}{\partial \ln c_{DS^-}} \right)_{T, c_{LDEA}, c_{Na^+}} \tag{2}$$

Thus, by measuring γ for solutions in which c_{DS^-}, c_{Na^+}, and T are held constant while c_{LDEA} is varied, we can obtain Γ_{LDEA} from the slope of γ versus $\ln c_{LDEA}$. Likewise, values of Γ_{DS^-} are obtained by holding c_{LDEA}, c_{Na^+}, and T constant while c_{DS^-} is varied. The limiting slope as the CMC is approached provides the maximum Γ_{LDEA} and Γ_{DS^-}. The average area per head group is then given by

$$a_0 = \frac{1}{N_0 \left(\Gamma_{LDEA} + \Gamma_{DS^-} \right)} \tag{3}$$

where N_0 is Avogadro's number. Note that this is the value of a_0 for the conditions found at this CMC, at which the mole fraction of LDEA in the micelle is given by $\Gamma_{LDEA} / (\Gamma_{LDEA} + \Gamma_{DS^-})$. With these data we can calculate the critical packing parameter as a function of micelle composition and sodium activity in the solution using

$$P* = \frac{v}{a_0 l_c} = \frac{v N_0}{l_c} \left(\Gamma_{LDEA} + \Gamma_{DS^-} \right) \tag{4}$$

Note that the value of $P*$ calculated in this way is essentially just a constant divided by a_0. More sophisticated approaches consider the fact that l_c varies to achieve the optimal micelle structure (*22,23*).

Materials and Methods

Materials. The nonionic surfactant used for the majority of this study is n,n-bis(2-hydroxyethyl) dodecanamide or LDEA as it is referred to in this paper. The commercial version of this material used for the rheological study was Witcamide 6511, which contained 93.5% of the desired amide, 4.8% free diethanolamine, 0.5% of the mono- and di-esters formed by the reaction of the acid with the hydroxyl groups on the amine, 0.5% water, and <1% other impurities. Also note that the hydrophobic moiety was only 96.1% C_{12} with 3.3% C_{14} and 0.6% C_{10} chains. The LDEA used for the interfacial tension measurements and critical packing parameter calculations was kindly prepared for us by Dr. Herman Chen. As supplied, it was 99% pure, as determined by HPLC. The C$_{14}$ impurity was removed by multiple recrystallizations from hexane. Because of the difficulty and potential expense of obtaining pure LDEA in large quantities, the initial rheological work reported in this paper was carried out with the commercial material. The interfacial tension work necessary for the packing parameter estimates demanded the use of pure components. Thus, it must be noted that the packing parameter and rheological results are compared for slightly different systems. These differences and the changes in the phase diagrams will be noted again in a later section when the comparisons are being made.

The high purity SDS (99%) obtained from BDH Chemicals was used as received for the rheological work and was recrystallized twice from absolute ethanol for the packing parameter work. No minimum was observed in the surface tension data versus concentration. The tetra- and hexa-ethyleneglycol nonylphenol surfactants, Igepal 430 and Igepal 530 respectively, were used as received from Rhône-Poulenc. The tetradecyltrimethylammonium bromide (TTAB) obtained from Sigma was 99% pure, the cetylpyridinium chloride monohydrate (CPyCl) obtained from Aldrich was 98% pure, and the sodium salicylate from Fluka was >99.5% pure. All were used as received. The hexadecane used for the interfacial tension measurements was obtained from Aldrich (99+%, anhydrous). Water for all samples was from a Milli-Q Plus system (filtration, activated carbon, deionization, and an additional organic scavenger unit), which exceeds Type 1 standards for Reagent Grade Water.

Methods. The general features of the ternary phase diagrams at 25°C were determined by titration to the clarification and turbidity points of the system. The exact locations of the boundaries of the one-phase regions were checked by long-term, thermostatted storage of samples on both sides of each boundary. The identities of the one-phase regions were determined by visual inspection and by microscopic observation under polarized light.

Interfacial tension measurements were made with a Krüss K-12 Process Tensiometer with a 9.545 mm radius ring having an R/r ratio of 51.6. Corrections were made automatically using Zuidema and Waters fit (*24*) to the Harkins and Jordan tables (*25*). All measurements were made at 25°C. The hexadecane was saturated with water prior to contact with the surfactant solution. The Krüss K12 instrument stops each measurement as soon as the force maximum is passed, and moves the ring back to the starting position.   No detachment occurs and measurements can be automatically repeated. The measurements were continued until the readings stopped changing with time.

Most rheological measurements were made on a Bohlin VOR rheometer with the C25 cup and bob geometry. The cup diameter is 27.5 mm, the bob diameter is 25 mm, and the bob height is 37.5 mm. Evaporation was controlled by floating several millimeters of 10 mPa s silicone oil on top of the sample. The oil is not solubilized in the surfactant solution over the time frame of the experiments. Even when the oil and sample were left in the rheometer for several days, there was no detectable effect on the rheological results. All tests were run at 25°C. All oscillatory measurements were carried out in the linear viscoelastic region as determined by strain sweep tests. The VOR is capable of covering the frequency range of 0.01 to 125 rad/sec. The relaxation time, τ_R, and the shear modulus, G_0, were obtained from Maxwell model fits to the frequency sweep data when appropriate. For samples with relaxation times shorter than about 10 msec, extrapolation to the G' plateau to obtain G_0 becomes questionable. Thus, a Rank Pulse Shearometer (26) was used to obtain values of G_0 for the samples as well.   The instrument provides only a single frequency measurement at approximately 1300 rad/sec and no measure of G". One must, therefore, assume that 1300 rad/sec is on the plateau and that at that frequency G" is much smaller than G'. To ensure that our values of G_0 were reasonable, several of the samples were also run on a multiple lumped resonator (MLR) rheometer (27,28) by Eric Amis in his laboratory at the University of Southern California. The MLR tests covered the range of 630 to 38,600 rad/sec. The four samples run verified our Pulse Shearometer data.

Results and Discussion

Phase Behavior. The phase behavior of the SDS/LDEA/Water system is represented in Figure 1. Concentrations are in weight percent. This is the phase diagram for the commercial grade LDEA and "as received" SDS system that was used in the rheological study. The normal micelle (L_1), hexagonal liquid crystal (E), and lamellar liquid crystal (D) phases are shown. In addition to these one-phase regions, the two-and three-phase regions are also labeled, with the measured tie lines shown in the L_1 + E area. Note that there is an L_1 + solid LDEA region above the L_1 phase due to the limited aqueous solubility of LDEA. Note also that the boundary between the L_1 and the

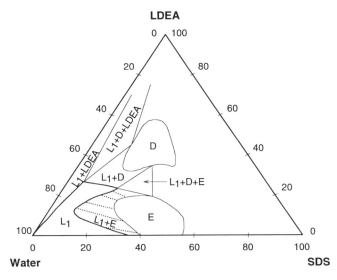

Figure 1. Phase diagram for the SDS/LDEA/Water system at 25°C. This phase diagram is for the commercial LDEA and "as received" SDS as described in the text.

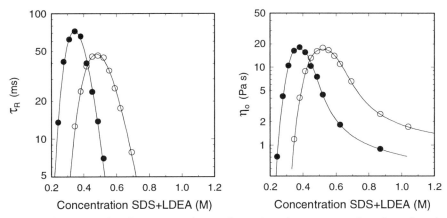

Figure 2. Relaxation time, τ_R , and zero-shear viscosity, η_0 , as a function of total surfactant concentration for two SDS/LDEA ratios: (●) 25/75 and (○) 33/67.

L_1 + E regions exhibits a large indentation that splits the L_1 phase into two legs. We believe, as discussed below, that the existence of this indentation is an indicator of a system that will exhibit maxima in η_0 and τ_R with increasing surfactant concentration.

Rheology. The rheology of solutions in the L_1 region was determined either (a) as a function of surfactant concentration (0.2 M to 1.0 M) with the ratio of SDS to LDEA held constant or (b) as a function of SDS to LDEA ratio with the total surfactant concentration held constant. The surfactant concentration series were run at SDS/LDEA weight ratios of 25/75, 33/67, 45/55, and 60/40. As the molecular weights of SDS and LDEA are 288.4 and 289.5 respectively, these ratios do not differ significantly from the molar ratios. As mentioned earlier we find that our samples can be modeled as single relaxation Maxwell materials, suggesting that we are in the regime where $\tau_{break} \leq \tau_{rep}$. Plots of log G_0 versus log concentration are linear with a slope of 2.3 for all four of the surfactant concentration series. There was no indication of a change of slope at the higher concentrations for this system. Figure 2 shows the zero-shear viscosity and relaxation times for the 25/75 and 33/67 ratio series. The maxima in these curves and the steady increase of G_0 with increasing concentration are consistent with the behavior reported for other viscoelastic surfactant systems in the L_1 region. When the surfactant concentration is held constant and the ratio of SDS to LDEA is varied, we again find maxima. Figure 3 plots η_0 and τ_R against surfactant composition at 10%, 15%, and 20% total surfactant by weight. To see how these two quantities vary as a function of composition, it is interesting to look at lines of constant η_0 and τ_R plotted directly onto the phase diagram. We have enlarged the water-rich corner of the phase diagram in Figures 4 and 5 in order to present the results for the L_1 region. Note that the lines do not radiate from the pure water corner, but appear to start at the boundary with the L_1 + solid LDEA two-phase region. Thus, if the experimentalist follows the logical path along lines of constant surfactant composition, as we did in Figure 2, the data will pass over the ridge of maximum η_0 and τ_R . Note that the maximum line appears to be

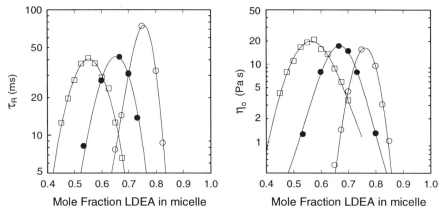

Figure 3. Relaxation time, τ_R, and zero-shear viscosity, η_0, as a function of mole fraction of LDEA in the micelle at three different levels of total surfactant concentration: (○) 10%, (●) 15% and (□) 20%.

in line with the indentation in the phase boundary between the L_1 and the L_1 + E regions. The values of η_0 and τ_R drop off rapidly on either side of the maximum. From the point of view of prediction, it is of interest to determine if the lines of constant rheological properties correlate with a more fundamental property of the micelles.

Critical Packing Parameter, P^*. When we choose lines of constant surfactant ratio, we are trying to hold the intramicellar properties constant as we vary the surfactant concentration. Unfortunately, a constant surfactant ratio does not necessarily provide the constant properties expected. Lines of constant critical packing parameter, on the other hand, should map out paths over which the surfactants create nearly identical structures. As discussed earlier, interfacial tension data were used to calculate values of P^* as a function of micelle composition and sodium activity for the SDS/LDEA/Water system. (Pure components were required for this part of the work, so the phase behavior differs slightly from the commercial system used for the rheology study.) Figure 6 shows the critical packing parameter versus mole fraction LDEA in the micelle at four levels of sodium activity. With increasing LDEA content, each of the curves increases slowly at first, then goes through a sharp maximum, followed by a rapid decrease in P^*. An unexpected feature is the precipitous drop of P^* at LDEA mole fractions of 0.87 for the 0.017 M Na^+ curve and 0.74 for the three higher Na^+ curves. Interestingly, these points correspond to points on the phase boundary between the L_1 and the L_1 + E regions. The next step is to transform this information into lines of constant P^* on the ternary phase diagram. To do this, we need to know the sodium activity throughout the L_1 region. Ideally, we would obtain this information experimentally. For this paper, we assume that counter ion binding changes only with micelle composition, not with total surfactant concentration, and use Scamehorn's (29) data of counter ion binding for SDS/nonionic systems. Given values of the sodium activity and mole fraction LDEA as a function of position in the L_1 region, the data from Figure 6 are mapped onto the phase diagram (Figure 7). (Note that the shape of the L_1 region is

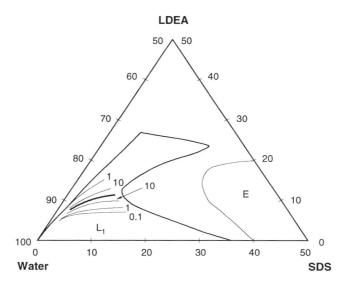

Figure 4. Lines of constant zero shear viscosity in the L_1 region of the ternary phase diagram for the SDS/LDEA/Water system at 25°C. The values for the lines are given in Pa s. The heavy line connects the maximum η_0 values along the ridge.

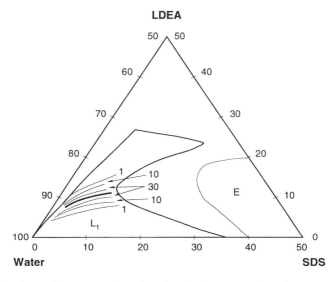

Figure 5. Lines of constant relaxation time in the L_1 region of the ternary phase diagram for the SDS/LDEA/Water system at 25°C. The values for the lines are given in milliseconds. The heavy line connects the maximum τ_R values along the ridge.

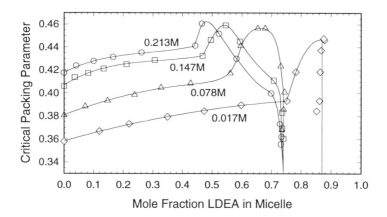

Figure 6. Calculated values of the *critical packing parameter*, P^*, as a function of micelle composition at four levels of sodium ion activity as labeled.

slightly different for this purified system, but retains most of the same features seen in the earlier diagram of Figure 1.) The path of maximum P^* is indicated by the bold line, and its location coincides reasonably closely with the center of the indentation. The values of P^* drop off on either side of this ridge. The constant P^* lines at and above the maximum do not originate in the water corner, but from points on the phase boundary. If this diagram is compared with the lines of constant relaxation time in Figure 5, it is seen that there is good qualitative agreement between the shape and location of the maximum curves and the way in which the values drop off as we move to higher and lower LDEA contents. Another interesting and unexpected feature that emerges is the way in which the lines of constant P^* mimic the shape of the phase boundary in the low LDEA area. Remember that all our data for P^* was obtained below the CMC, that our values of sodium activity are only estimates, and that the systems for P^* and the rheological work are slightly different. With this in mind the qualitative correlation of P^* with the phase boundary features and the relaxation time data is quite encouraging.

Generalization to Other Systems. The estimated contours of P^* shown in Figure 7 provide a possible explanation for both the indentation in the phase boundary and the rheological behavior of systems that exhibit such an indentation. Both features are characteristic of surfactant mixtures that pass through a P^* maximum as the surfactant ratio is changed. Because SDS and LDEA have equal chain lengths, the maximum in P^* for our system corresponds to a minimum in a_0. The initial spacing of the sulfate head groups is large because of electrostatic repulsion. The nonionic surfactant molecules can at first be inserted into this surface without requiring the sulfate head groups to move apart as much as would another sulfate headgroup. Initially, then, the average area per head group decreases on addition of the nonionic. If the area required by the nonionic head group is larger than the hydrated sulfate ion, however, a point will be reached at which a_0 will begin to increase again as the nonionic begins to dominate the surface. This minimum in a_0 would not be expected if the nonionic head group

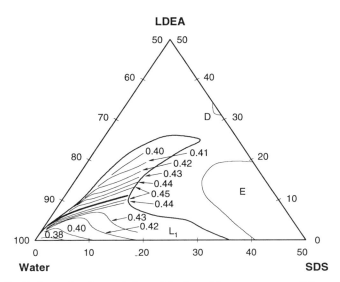

Figure 7. Estimated values of the *critical packing parameter*, P^*, mapped into the L_1 region of the SDS/LDEA/Water phase diagram. The heavy line connects the maximum P^* values along the ridge. Note that this phase diagram is for the purified surfactant system.

were smaller than the hydrated sulfate ion. Instead a_0 would continue to decrease until one reached either a phase boundary or a purely nonionic system. Figure 8 shows the partial phase diagram and lines of constant relaxation time for aqueous solutions of SDS and Igepal 530. Igepal 530 is a commercial nonionic surfactant with a nonylphenol hydrophobe and a polyethylene oxide head group with a weight average of 6 ethylene oxide groups. The presence of the indentation in the phase boundary is clear, and the lines of constant τ_R are similar to those seen for the SDS/LDEA system. If Igepal 430, which has a smaller hydrophilic group of approximately 3 to 4 ethylene oxide groups, is used instead, no indentation is seen and τ_R and η_0 continue to increase until the system phase separates.

This trend is also seen in cationic/anionic systems, such as the TTAB/Na Salicylate and CPyCl/Na Salicylate systems studied by others. Originally, for these systems it was thought that the Na Salicylate acted simply as a strongly binding counter ion, thus reducing the electrostatic repulsion, which allows closer packing of the head groups (2). More recently, it has been shown that this is insufficient to produce viscoelastic (or birefringent) surfactant systems. Rather, the counterion must have a structure that allows it to orient itself with a hydrophobic moiety positioned into the micelle (5,30). That is, it must behave as a surface active solute. If the sodium salicylate is treated as a small anionic co-surfactant on the ternary phase diagram, the indentation in the phase boundary of the L_1 region becomes apparent. This is seen for the TTAB system on the partial phase diagram shown in Figure 9. Although the CPyCl/Na Salicylate system is not shown here, the phase boundary of its L_1 region exhibits a similar indentation around a 1:1 mole ratio.

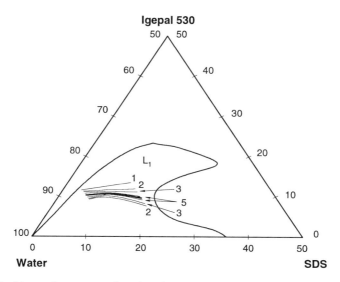

Figure 8. Lines of constant relaxation time, τ_R , mapped into the L_1 region of the SDS/Igepal 530/Water system at 25°C. The values for the lines are given in milliseconds. The heavy line connects the maximum τ_R values along the ridge.

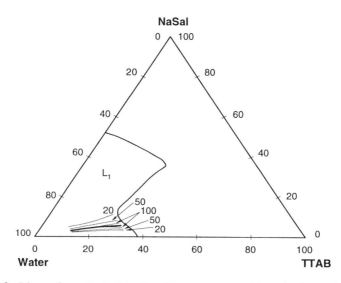

Figure 9. Lines of constant relaxation time, τ_R , mapped into the L_1 region of the TTAB/Na Salicylate/Water system at 25°C. The values for the lines are given in milliseconds. The heavy line connects the maximum τ_R values along the ridge.

Finally, Hoffmann (*31-33*) has correlated the interfacial tension of surfactant solutions with micellar structure and the observed rheology for a number of systems. This connection between interfacial tensions and the rheological properties is another strong indication that the packing of the surfactant head groups plays a major role in the relaxation process that appears to control these systems.

Conclusions

We have shown that the SDS/LDEA/Water system exhibits viscoelastic behavior typical of systems that are composed of elongated, rod-like micelles, which exhibit entangled polymer-like flow behavior. The relaxation time and zero-shear viscosity in the L_1 region were shown to go through a maximum as the surfactant concentration is increased, or as the ratio of LDEA to SDS is increased. When lines of constant η_0 and τ_R are plotted directly onto the phase diagram, there appears to be a relationship between the shape of the L_1 phase boundary and the location and values of these lines. Specifically, the lines of maximum τ_R and η_0 are found to terminate at or near the center of the indentation in the phase boundary. These rheological parameters also drop off rapidly as one moves away from the indentation in either direction. To determine the generality of this behavior, we examined two additional anionic/nonionic systems, and two cationic/anionic systems. Based on the results, we suggest that maxima in η_0 and τ_R can be predicted by the presence of this indentation in the L_1 phase boundary. We also showed that the lines of constant τ_R appear to follow the same paths as our estimated lines of constant critical packing parameter, suggesting the possibility of being able to predict the rheology of these surfactant solutions from more fundamental structure/property considerations for the micelles.

Acknowledgments

We would like to thank Professor Eric Amis and his colleagues at the University of Southern California for kindly running our samples on the MLR rheometer on very short notice. We would also like to thank Professor Charles Zukoski at the University of Illinois, Champaign-Urbana for the extended loan of his pulse shearometer. We also wish to acknowledge the helpful discussions about the critical packing parameter that we had with Professor Greg Warr of the University of Sydney that prompted this work.

Literature Cited

1. Hoffmann, H.; Rehage, H.; Schorr, W.; Thurn, H. In *Surfactants in Solution*; Mittal, K.L.; Lindman, B., Eds.; Plenum: New York, N. Y., 1984; Vol. 1, pp. 425-454.
2. Hoffmann, H.; Loebl, H.; Rehage, H.; Wunderlich, Ingrid. *Tenside Deterg.* **1985**, *22*(6), 290-298.
3. Hoffmann, H.; Rehage, H. In *Surfactant Solutions, New Methods of Investigation*; Zana, R., Ed.; Marcel Dekker: New York, 1987; Chapter 4, pp. 209-239.
4. Rehage, H.; Hoffmann, H. *J. Phys. Chem.* **1988**, *92*(16), 4712-4719.

5. Rehage, H.; Hoffmann, H. *Mol. Phys.* **1991**, *74*(5), 933-973.

6. de Gennes, P.G. *Scaling Concepts in Polymer Physics*; Cornell University Press: Ithaca, NY, 1979.

7. Doi, M.; Edwards, S.F. *The Theory of Polymer Dynamics*; Clarendon: Oxford, 1986.

8. Cates, M.E. *Macromolecules* **1987**, *20*, 2289-2296.

9. Cates, M.E. *J. Phys. France* **1988**, *49*, 1593-1600.

10. Cates, M.E.; Marques, C.M.; Bouchaud, J.P. *J. Chem. Phys.* **1991**, *94*, 8529-8536.

11. Turner, M.S.; Cates, M.E. *Langmuir* **1991**, *7*, 1590-1594.

12. Khatory, A.; Lequeux, F.; Kern, F.; Candau, S.J. *Langmuir* **1993**, *9*, 1456-1464.

13. Israelachvili, J.N. *Intermolecular and Surface Forces with Applications to Colloidal and Biological Systems*; Academic Press: London, 1985.

14. Aniansson, E.A.G.; Wall, S.N.; Almgren, M.; Hoffmann, H.; Zana, R. *J. Phys. Chem.* **1976**, *80*, 905.

15. Candau, S.J.; Merikhi, F.; Waton, G.; Lemaréchal, P. *J. Phys. France* **1990**, *51*, 977-989.

16. Lessner, E.; Teubner, M.; Kahlweit, M. *J. Phys. Chem.* **1981**, *85*, 3167.

17. Israelachvili, J.N. *Intermolecular and Surface Forces with Applications to Colloidal and Biological Systems*; Academic Press: London, 1985; p. 252.

18. Israelachvili, J.N.; Mitchell, D.J.; Ninham, B.W. *J. Chem. Soc., Faraday Trans. I* **1976**, *72*, 1525-1568.

19. Tanford, C. *The Hydrophobic Effect: Formation of Micelles and Biological Membranes*, 2nd ed.; John Wiley & Sons: New York, 1980; pp. 51-53.

20. Rosen, M.J.; Friedman, D.; Gross, M. *J. Phys. Chem.* **1964**, *68*, 3219-3225.

21. Hutchinson, E. *J. Colloid. Sci.* **1948**, *3*, 413.

22. Puvvada, S.; Blankschtein, D. *J. Phys. Chem.* **1992**, *96*, 5567-5579.

23. Puvvada, S.; Blankschtein, D. *J. Chem. Phys.* **1990**, *92*(6), 3710-3724.

24. Zuidema, H.H.; Waters, G.W. *Industrial and Engineering Chemistry* **1941**, *13*, 312-313.

25. Harkins, W.D.; Jordan, H.F. *J. Colloid Interface Sci.* **1930**, *52*, 1751-1772.

26. Buscall, R.; Goodwin, J.; Hawkins, M.; Ottewill, R. *J. Chem. Soc., Faraday Trans. I* **1982**, *78*, 2873-2887.

27. Hair, D.W.; Niertit, F.S.; Hodgson, D.F.; Amis, E.G. *Rev. Sci. Instr.* **1989**, *60*, 2780.

28. Hair, D.W.; Amis, E.G. *Macromolecules* **1989**, *22*, 4528.

29. Rathman, J.F.; Scamehorn, J.F. *Langmuir* **1986**, *2*, 354-361.

30. Smith, B.C.; Chou, L.; Lu, B.; Zakin, J.L. In this book, Chapter 26.

31. Oetter, G.; Hoffmann, H. *J. Dispersion Sci. Technol.* **1988**, *9*, 459-492.

32. Hoffmann, H. *Progr. Colloid Polym. Sci.* **1990**, *83*, 16-28.

33. Hoffmann, H. In *Organized Solutions: Surfactants in Science and Technology*; Friberg, S.E.; Lindman, B., Eds.; Marcel Dekker: New York, 1992; pp. 169-192.

RECEIVED August 22, 1994

Chapter 11

Effect of Alcohol Chain Length on Viscosity of Sodium Dodecyl Sulfate Micellar Solutions

Eric Y. Sheu, Michael B. Shields, and David A. Storm

Research and Development, Texaco, Inc., Box 509, Beacon, NY 12508

The effect of alcohols on sodium dodecyl sulfate (SDS) micelles in aqueous solutions was investigated using viscosity measurements. Alcohols of carbon chain length from C-3 (propanol) to C-8 (octanol) were systematically studied. For short chain alcohols, the alcohol solubility in SDS micelle was evaluated using a proposed simple model. The results obtained agreed with the NMR results. In the case of long chain alcohols, their effect on micellar growth was characterized using the model proposed by Kohler and Strnad (*J. Phys. Chem.*, **1990**, *94*, 7628-7634) *(1)*, from which the growth energy was calculated. The results were compared to both small angle neutron scattering and laser light scattering data.

Alcohols have been known to alter the micellar state by changing the solvent hydrophobicity or by being solubilized in the micelle *(2-7)*. Due to these alcohol-micelle or alcohol-solvent interactions, the micellar structure and the micellar solution properties may be modified upon addition of alcohols. Many studies have been dedicated to studying the alcohol/micelle systems *(2-10)*. One important issue on the effect of short chain alcohols is their partition coefficients in the micellar phase and in the bulk phase. Because short chain alcohols (carbon number less than or equal to four) are usually soluble both in water and in the micelles, the common approach for partition coefficient measurement is to measure the amount of alcohol solubilized in (or associated with) the micelle. Techniques such as NMR, ESR, dynamic light scattering, small angle X-ray and neutron scattering can be applied. If alcohols are dissolved in water, the solvent hydrophobicity will increase. This results in increase of the critical micelle concentration (CMC), because the micellization energy gained by the system decreases. This is the case when adding short chain alcohols. Following the alcohol addition, the micelles usually decrease the aggregation number and ultimately break, when the alcohol concentration exceeds a certain threshold value *(2)*.

0097–6156/94/0578–0167$08.00/0

On the other hand, if the alcohol molecules "insert" into the micelles as co-surfactant (this happens for long chain alcohols), the micellar surface curvature will decrease, provided that the number density of the micelles does not change. From simple packing consideration, this insertion will lead to micellar structural transitions *(11)*. For many surfactant systems (including the SDS micelles studied here), the minimum micelles (i.e, the micelles formed near the CMC) are known to be spherical, and gradually grow into an ellipsoidal or a spherocylindrical micelle, when concentration is increased or when electrolytes are added *(12-14)*. It is thus reasonable to anticipate that these micelles will grow into ellipsoids or spherocylinders when long chain alcohols are added.

Obviously, the main difference between the effect of short chain and long chain alcohols is the resulting micellar structure. Since the rheological properties largely depend on the structure of the micelles, viscosity measurement may be an appropriate technique to demonstrate this structural transition. However, viscosity measurement is a macroscopic technique, and it may not be easy to deduce the microscopic information from it unless a proper theory or model is established. To build a model for linking the measured viscosity to the microscopic structure often requires unknown parameters. The parameter values are then determined by fitting the model to the experimental data. In many cases, the model parameters do not have much quantitative physical meanings *(15-17)*. In order to avoid this problem, we established a model to evaluate the short chain alcohol (for carbon = 3 and 4) partition coefficients in such a way that all the parameters built in the model are measurable. In this model there are essentially no adjustable parameters. To achieve this, we carefully selected the parameters to be the viscosity or its rate of change along the surfactant or the alcohol concentration axis for water/alcohol, water/surfactant, and water/surfactant/alcohol systems. Counting all the above mentioned measurable quantities, there are six that can be used to build the model.

Experimentally, we measured the zero shear viscosity of alcohol/SDS micellar solutions as a function of both alcohol and SDS concentrations. The measurements were performed for alcohol carbon chain length from three (propanol) to eight (octanol). For propanol and butanol (short chain alcohols), we determined the amount of alcohol that associates with the SDS micelle using the model developed here. Since the equation formulated involves the viscosities of water/alcohol and water/SDS systems, we also measured the water/alcohol system as a function of alcohol volume fraction and the water/SDS system as a function of SDS concentration. As for pentanol to octanol (long chain alcohols), the subject of study was the micellar growth upon alcohol addition. In this case, the model proposed by Kohler and Strnad *(1)* was used, in which only the water/alcohol/SDS viscosity as a function of alcohol volume fraction is needed. Using this model, we determined the free energy responsible for micellar growth. The value was comparable to the results from the small angle neutron scattering and the dynamic light scattering.

Experimental

Samples. Sodium dodecyl sulfate was obtained from Sigma with > 99% purity and was used for sample preparation without further purification. All alcohols used were of HPLC grade, and the water was deionized with > 18 MΩ resistance. The sample preparation procedure for measuring viscosity as a function of alcohol at a given SDS concentration is as follows: (1) prepare a solution of known weight % SDS, (2) add a known amount of alcohol to the SDS solution, and (3) adds an appropriate amount of SDS according to the added alcohol volume to maintain the SDS concentration constant.   In this way, a series of samples of constant SDS concentration, but with different alcohol content can be prepared. All measurements and sample preparation were performed at T = 25 °C.

Viscosity Measurements. In order to accurately determine the zero shear viscosity, three viscometers were used to measure the viscosity. The first apparatus used was a Gilmen falling ball viscometer. The solution density was determined using a PAAR densitometer (model DMA45) with ±0.0001gm/cm^3 accuracy. Both glass and stainless steel balls were used in the measurement. The results obtained from both balls differed by less than 0.02%.   During measurement the viscometer was immersed into a temperature bath kept at 25±0.1 °C. The second viscometer used was a low shear rate Brookfield viscometer (Model LV8) with 0 to 1000 centipoise working range and a 0.1% accuracy of the selected scale. A cuvette geometry was used with 18 cm^3 fluid volume and 1 mm spindle-wall gap. A shear rate of 12 s^{-1} was chosen. The third viscometer used was a Rheometrics RFX viscometer with a working shear rate range from 10 to ~ 10,000 s^{-1}, depending on the SDS concentration. Since only the zero shear viscosity was interested, the data was taken from the Newtonian range (approximately from 100 to 800 s^{-1}, viscosities for shear rate lower than 100 s^{-1} exhibited poor reproducibility).

Theory

Solubility of Alcohol in Micelle (the Alcohol Partition Coefficient).   The evaluation of alcohol solubility in micelle (the partition coefficient) involves a three-component system containing water, micelles, and alcohol. In fact, there is another component. That is the surfactant monomers in the bulk. However, one can take water and the surfactant monomers as the bulk phase and treat it as the solvent component. This is a plausible assumption, since the monomer concentration in the bulk phase does not vary substantially for concentrations below or above the CMC. With this assumption, we derived the alcohol partition equation by first breaking the three-component system into to two two-component systems.   They are the water/alcohol and water/SDS systems. The relative viscosity of the water/alcohol systems can be expressed as

$$\eta_r^A = \frac{\eta_{AW}}{\eta_W} = 1 + f(\phi_A) \qquad (1)$$

where η_r^A is the relative viscosity with water being the solvent. η_{AW} is the viscosity of the water/alcohol and η_W is the water viscosity. f(ϕ) is a function representing the contribution of the added alcohol to the viscosity, and ϕ_A is the alcohol volume fraction. Similarly, we can write the relative viscosity for water/SDS system as,

$$\eta_r^S = \frac{\eta_{SW}}{\eta_W} = 1 + g(\phi_S) \qquad (2)$$

with η_{SW} being the viscosity of the SDS solution, and ϕ_S being the SDS volume fraction. With these two equations, we can then express the relative viscosity of the water/SDS/alcohol system with respect to the "water+alcohol solvent" as,

$$\eta_r = \frac{\eta_{SAW}(\phi_S, \phi_A)}{\eta_{AW}(\alpha_A \phi_A)} \qquad (3)$$

where η_{SAW} is the viscosity of the three-component solution and α_A is the fraction of alcohol in the bulk. Since the disperse phase in the water/SDS/alcohol system consists of both SDS and alcohol, equation 3 can be re-written as

$$\eta_r = \frac{\eta_{SAW}(\phi_S, \phi_A)}{\eta_{AW}(\alpha_A \phi_A)} = 1 + h(\phi_S K)\phi_S K \qquad (4)$$

where K is the solvation constant representing the scaled volume of the micelle after alcohols are solubilized into the micelles. Thus, K can be expressed in terms of ϕ_A, ϕ_S, and α_A as

$$\phi_A = \alpha_A \phi_A + (K-1)\phi_S \qquad (5)$$

Since η_r in equation 4 can not be experimentally determined, we formulated the measurable quantity $\eta^T = \eta_{SAW}/\eta_{SW}$ instead. This is done by substituting equation 1, 2 and 5 into 4,

$$\eta^T = [1 + f(\alpha_A \phi_A)]\frac{1 + h(K\phi_S)K\phi_S}{1 + g(\phi_S)\phi_S} \qquad (6)$$

Equation 6 is a rigorous equation for describing the relative viscosity of the three-component system with respect to the water/SDS solution viscosity. However, the functional forms for f, h, and g are not known. We therefore, need to make certain assumptions to simplify equation 6 in order to make it useful. We first assume g and h behave linearly at low SDS and low alcohol concentrations with corresponding slopes C_S and B_S. This is

$$g(\phi_S) \approx C_S\,\phi_S \tag{7}$$

and

$$h(K\phi_S) \approx B_S \tag{8}$$

The second assumption made was that the relative viscosity of the water/alcohol mixture, $\eta_r{}^A$, can be described by a quadratic equation, i.e.,
These assumptions were later found to be very reasonable for the system studied

$$\eta_r^A = 1 + f(\phi_A) = 1 + C_A\phi_A - B_A\phi_A^2 \tag{9}$$

here. With these assumptions, equation 6 can be re-written into a simple algebraic form,

$$\eta^T(\phi_A,\phi_S) = (1 + C_A\phi_A - B_A\phi_A^2)\frac{1 + B_S K\phi_S}{1 + C_S\phi_S} \tag{10}$$

Substituting equation 5 for K in equation 10 yields

$$\eta^T(\phi_A,\phi_S) = \frac{1 + B_S\phi_S}{1 + C_S\phi_S} + C_{AS}\phi_A + O(\phi_A^2) \tag{11}$$

where

$$C_{AS} = \frac{C_A\alpha_A(1 + B_S\phi_S)}{1 + C_S\phi_S} + \frac{B_S(1 - \alpha_A)}{1 + C_S\phi_S} \tag{12}$$

From equation 12 one gets

$$\alpha_A = \frac{C_{AS}(1 + C_S\phi_S) - B_S}{C_A(1 + B_S\phi_S) - B_S} \tag{13}$$

This is the equation to be used for evaluation of the alcohol dissolved in the bulk. $1 - \alpha_A$ is the fraction of the alcohol solubilized in the micelles. Equation 13 is a rather simple relation, and the parameters C_{AS}, C_A, C_S, B_S, can all be determined experimentally. Here we summarized their experimental meanings: (a) C_{AS}= initial slope of η_{SAW}/η_{SW} versus ϕ_A plot, (b) C_A= initial slope of η_{AW}/η_W versus ϕ_A plot, (c)

C_s= initial slope of η_{sw}/η_s versus ϕ_s plot, (d) B_s= initial slope of η_{saw}/η_{AW} versus ϕ_s plot. As long as one makes the above measurements, α_A can be determined rigorously.

Micellar Growth Upon Addition of Long Chain Alcohols. The analysis of the viscosity data for the long chain alcohol cases was based on the Kohler and Strnad theory *(1)*. Thus, only the relevant steps of this model will be summarized. The starting point of the theory is the "shape factor". In classical viscosity theory for disperse system, the shape factor is defined as the intrinsic viscosity, representing the viscosity of an individual particle. This can be determined by the slope of the relative viscosity at the dilute concentration limit. Mathematically, it is

$$[\eta] = \lim_{\phi \to 0} \frac{\eta_r - 1}{\phi} \tag{14}$$

where $[\eta]$ is the intrinsic viscosity characterizing the shape of an individual particle, η_r is the relative viscosity, and ϕ the volume fraction of the disperse phase. In principle, the particle shape information can be extracted by studying $[\eta]$. However, the micellar solutions differ from a typical disperse system. A definition of intrinsic viscosity like equation 14 does not represent the structure of an individual micelle. This is because the micellar size varies as a function of concentration. Below the CMC) surfactants are in monomer form and equation 14 will give the intrinsic viscosity of a monoer. Thus, the shape factor of a micelle can only be defined at a finite concentration above the CMC. Because the shape factor is concentration dependent, equation 14 for the intrinsic viscosity should be modified as,

$$[\eta] = [\eta(\phi)] = \frac{\eta_r(\phi) - 1}{\phi} \tag{15}$$

SDS micelles are known to grow from spheres to spherocylinders *(12-14)* with increasing concentration or with addition of salts. Likely, they will grow into spherocylinders when long chain alcohols are added. If this is the case, the shape factor derived by Jeffery *(18)* and Eisenschitz *(19)* can be used to determine the axial ratio p,

$$[\eta(\phi)] = \frac{4\beta^2}{15}\left\{\frac{14}{p^2(4p^2-10+3\alpha)} + \frac{3\beta}{p^2(p^2+1)[(2p^2-1)\alpha-2]} + \right.$$

$$\left. \frac{6}{(P^2+1)(2p^2+4-3p^2\alpha)} \quad \frac{4p^2+2-(4p^2-1)\alpha}{p^2(4p^2-10+3\alpha)[(2p^2+1)\alpha-6]}\right\} \tag{16}$$

where

$$\beta = p^2 - 1; \qquad \alpha = \frac{1}{p\sqrt{\beta}} \ln \frac{p+\sqrt{\beta}}{p-\sqrt{\beta}} \qquad p \geq 1 \qquad (17)$$

This is a rather complicated equation. A simpler, approximation form is (for polydisperse case),

$$[\eta(\phi)] = 0.25 \left\langle \frac{p^2}{\ln(0.63p)} \right\rangle_v \qquad p > 3 \qquad (18)$$

where v is the dried volume of a surfactant molecule. One can expand equation 18 at a reference p value, p_o, to include the polydispersity, which is known for the SDS micellar system *(13)*. Taking $p_o = <p^3>_v / <p^2>_v$ to account for the second and the third moment, equation 18 becomes

$$[\eta(\phi)] = 0.25 \left\langle \frac{p^2}{\ln[0.63<p^3>_v /<p^2>_v]} \right\rangle_v \qquad p > 3 \qquad (19)$$

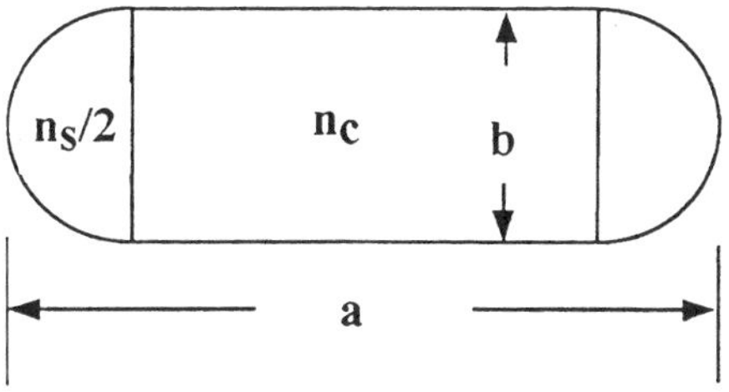

Figure 1. Schematic diagram of a spherocylindrical micelle.

The axial ratio p = a/b (see Figure 1) can be related to the minimum aggregation number n_s (this is the aggregation number when the micelle is spherical) and the aggregation number n,

$$p = \frac{2n}{3n_s} + \frac{1}{3} \qquad (20)$$

the next step is to relate n and n_s to the energy that is responsible for the micellar growth. This can be achieved by an association model, assuming a linear relation between n and the standard potential of a micelle, μ_M°

$$\mu_M^o(n) = n_s\mu_s^o + (n-n_s)\mu_c^o \qquad (21)$$

$$W_s^o = \mu_s^o - \mu_A^o - \gamma_s\mu_s^o \qquad (22)$$

$$W_c^o = \mu_c^o - \mu_A^o - \gamma_c\mu_s^o \qquad (23)$$

where μ_s^o and μ_c^o are the standard potentials of a surfactant molecule at the hemispherical caps and at the cylindrical section respectively (see Figure 1). γ_s and γ_c are the corresponding activity coefficients. W_s^o and W_c^o are the standard works for transferring a surfactant ion from the bulk to the hemispherical caps and to the cylindrical section of the micelle respectively, and μ_A^o is the standard potential of an alcohol molecule. Obviously, the energy involving the micellar growth is the difference between W_s^o and W_c^o, i.e.,

$$\Delta W = W_c^o - W_s^o \qquad (24)$$

ΔW is the energy to be determined via viscosity measurement. This energy is related to the shape factor by the following equations (1)

$$[\eta(\phi)] = \frac{1}{9} \frac{6\Gamma_A + 2\sqrt{\Gamma_A}}{\ln[1.68\sqrt{\Gamma_A} + 0.069]} \qquad [\eta(\phi)] > 4.5 \quad (25)$$

$$\Gamma_A = \frac{C_A}{\delta n_s^2 C_o} \qquad (26)$$

$$\Delta W = \frac{\ln\delta}{n_s} \qquad (27)$$

where C_A is the molar concentration of SDS in a micelle, $C_o = (N_A \ v)^{-1}$ (N_A = Avogadero's number and v the SDS molecular volume), and ΔW is in units of $k_B T$ (k_B = Boltzmann constant). The procedure to get ΔW is to first determine $[\eta(\phi)]$ by equation 15, then to use equation 25 to calculate Γ_A, and finally to use equation 26 and 27 to evaluate ΔW. In this analysis one needs to know v and n_s in order to compute ΔW. In the following section we shall discuss how to determine v and n_s.

IV. Results and Discussion

Short Chain Alcohols - The Alcohol Solubility.   Figure 2 shows the relative viscosity of the water/propanol mixture for propanol up to 0.35 volume fraction. Within this volume fraction, the relative viscosity was found to be linear with 0.998 correlation coefficient. The slope (C_A) obtained was 5.01 (it becomes quadratic at higher ϕ_A, as described by equation 9, but C_A remains the same). Figure 3 shows relative viscosity of the SDS/water system as a function of the SDS concentration. It is fairly linear for SDS up to ~ 0.06 volume fraction. The slope (C_S) obtained was 8.04. For the propanol case, B_S turned out to be very similar to C_S at low SDS and low alcohol concentration. We thus took $B_S=C_S$. From equation 13 one can see that the only quantity needed for calculating the partition coefficient is C_{AS}. Figure 4 and 5 shows the results for the 2 wt % and the 5 wt % SDS cases as a function of alcohol where the initial slopes are the C_{AS} values needed. With these C_{AS} values

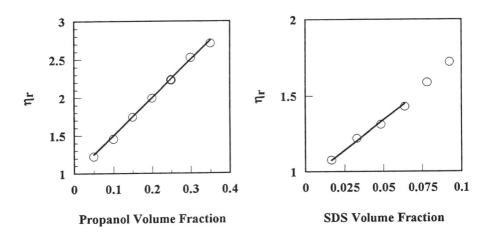

Figure 2. η_r of Propanol/water mixture as a function of propanol volume fraction.

Figure. 3 η_r as a function of SDS concentration.

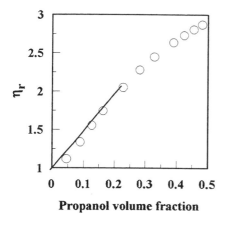

Figure 4. η_r of 2% SDS in water/ propanol solution as a function of propanol volume fraction.

Figure 5. η_r of 5% SDS in water/ propanol solution as a function of propanol voume fraction.

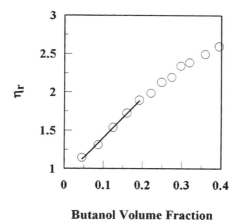

Figure 6. η_r of 5% SDS in water/ butanol solution as a function of butanol volume fraction.

and equation 13 we found that ~ 30% of the propanol solubilized in the micelles (i.e., $1-\alpha_A$) for the 2% SDS case and ~ 35% for the 5 wt% SDS solution. As for the butanol case, 43% alcohol was found to be solubilized in the micellar phase ($\alpha_A = 0.57$) (see Figure 6). In the butanol case $B_s \neq C_s$ and needs to be measured independently. We measured it by measuring η_r as a function of SDS concentration at 0.02, 0.05, 0.10, 0.15, and 0.20 butanol volume fractions. We then took the average value B_S thus obtained from 5 alcohol concentrations which differed by less than 2%. Table I gives the parameter values obtained from the propanol and butanol cases. The NMR results by Stilbs *(3)* are also tabulated.

To accurately determine α_A, one needs to carefully determine all four coefficients (i.e., C_A, B_S, C_S, and C_{AS}). This is because equation 13 is very sensitive to these four parameter values. We found that only when these four parameters are obtained within 3% error, can one get a reasonably accurate α_A.

Table I. Parameters values for SDS /alcohol systems

Systems	C_A	C_{AS}	B_S	C_S	α_A
2% SDS + Propanol	5.01	5.63	8.04	8.04	0.7
5% SDS+ Propanol	5.01	5.32	8.04	8.04	0.65
5% SDS+ Butanol	4.21	5.22	8.79	8.04	0.57
7% SDS+ Propanol (NMR)					0.68
7% SDS+ Butanol (NMR)					0.56

From the short chain alcohol results, it is clear that a substantial amount of the added alcohol molecules are solubilized or associate with the micelles. A qualitative argument for this phenomenon is that the short chain alcohols are soluble both in the aqueous phase, at the micellar-water interface, or even in the micellar core near the polar head, depending on the hydrophobicity of the alcohol. In fact, earlier ESR work *(20)* suggested that the short chain alcohols are mostly near the micelle-water interface due to the low HLB values. Since the viscosity measurement does not provide the location of the alcohol molecules, we can only conclude that there are nearly 50% of the short chain alcohol molecules hydrodynamically associate with the micelles. As for micellar breakage at high short chain alcohol concentration one can explain as follows. For any given SDS concentration, the total micellar surface is finite. Thus, when the alcohol concentration is increased to a critical amount that completely "dresses" the micelles, the hydrophilicity-hydrophobicity contrast is between the alcohol and the alcohol-containing water, which is certainly very small. In this situation, if the alcohol concentration is further increased, the micelles can no longer be sustained because of the small free energy gain on micellization. Thus, they are expected to break, which was indeed observed.

Long Chain Alcohols - The Micellar Growth. The main difference between the effect of long chain alcohols and short chain alcohols on the viscosity is the η_r behavior toward high alcohol concentration. As one can see from Figure 7, η_r turns upward (i.e, concave up) for the pentanol and the octanol cases, while it is concave down for propanol and butanol (See Figure 4 to 6). In fact, this is a good indication that the short chain alcohols tend to dress on the micelles while the long chain alcohol will insert into the micelles and serve as co-surfactants. Since the long chain alcohols insert into the micelle, they initiate micellar growth, while the short ones end up breaking the micelles.

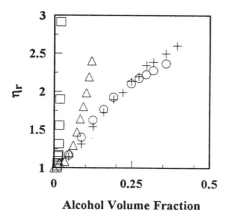

Figure 7. η_r for 5 % SDS as a function of propanol (circle); butanol (cross); pentanol (triangle); and octanbol (square).

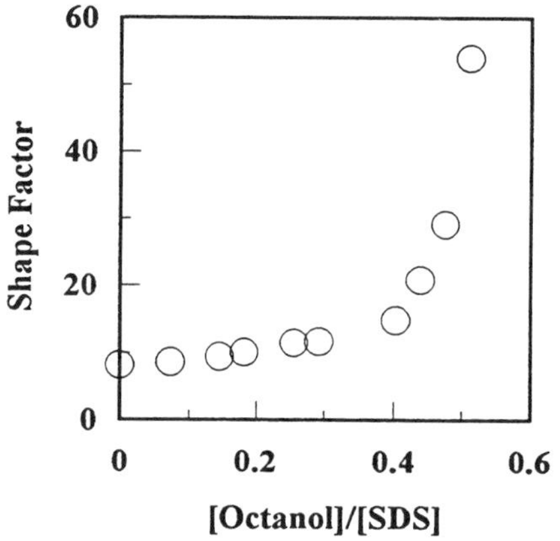

Figure 8. Shape factor for 5% SDS/octanol system using equation 15.

Figure 8 shows the shape factor for the octanol case in 5% SDS solution. [octanol]/[SDS] represents the molar ratio. To get ΔW we need v and n_s as described in equation 25 to 27. For SDS surfactant molecules, v has been reported (12) to be ~ 405 Å^3 and n_s can be determined by v and the hydrocarbon chain length of the SDS molecule. Taking 1.54 Å for C-C bond with 109.5 degree bond angle, and a 1.11 Å C-H bond length for the methyl group (21), one can easily get the carbon chain length for SDS to be 15.2 Å (this is different from 16.7 Å obtained by using Tanford's empirical formula (22)). This chain length corresponds to an aggregation number n_s = 42 based on spherical packing and v = 405 Å^3. Small angle neutron scattering confirms these estimates (12). Taking these two numbers and analyzing the viscosity data for the octanol case according to the procedure described in the previous section, the ΔW can be estimated.

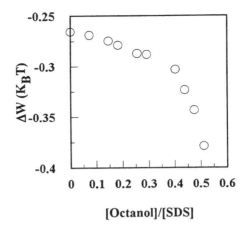

Figure 9. ΔW as a function of octanol to SDS molar ratio.

Figure 9 gives the ΔW obtained. Small angle neutron scattering for 8% SDS without alcohol *(12)* gives $\Delta W = -0.29$ k_BT compared to -0.26 k_BT obtained here for 5% SDS. We also compared this value with light scattering result for SDS + 0.8M NaCl *(13)*. The way we compared the two was to take the SDS concentration where both SDS/NaCl and SDS/Octanol have an aggregation number of ~ 2000 (0.058% for the SDS/NaCl system and p = 28 for the SDS/octanol system). The ΔW in these corresponding cases was -0.35 k_BT for SDS/Octanol and -0.37 k_BT for the SDS/NaCl system from the light scattering result. This comparison may not be accurate. However, it gives a qualitative comparison, and the results are on the same order. These results demonstrate that the model proposed by Kohler and Strnad *(1)* is in fact applicable for SDS/alcohol systems. Actually, this model may be applicable for all systems growing from spherical to spherocylindrical micelles. The advantage of using viscosity measurements is that it is a rather simple in-house technique, while NMR and small angle scattering are much more involved experimentally.

Some improvement may be made to the model. One can see from equations 16 and 18 that the determination of ΔW depends on the formulation of the polydispersity (i.e., the particle size distribution), which in turn depends on the formulation of the shape factor. Equation 18 is a convenient expression to analytically formulate the polydispersity and to link to Γ_A (equation 25). However, it is only applicable for p > 3 and $[\eta(\phi)] > 4.5$. This is a severe restriction for some critical cases, such as pentanol, where the micelle go from being broken (such as the butanol case) to rapid growth (the pentanol case). For this critical alcohol chain length we found that p was less than 3 at the low alcohol concentrations. Due to the restricted applicable p range of equation 25, we did not analyze the pentanol data.

Conclusion

We investigated the alcohol effect on SDS micellar solutions. For short chain alcohols we derived a simple formula to determine the alcohol partition coefficient between the micellar phase and the aqueous bulk phase by using viscosity measurements. The results obtained agree with the previous NMR data. For long chain alcohols, we characterized the micellar growth. It was achieved   by determination of the micellar growth energy using the theory proposed by Kohler and Strnad (1). The results are qualitatively consistent with some previous works using small angle neutron scattering and light scattering.

Literature Cited

1. Kohler, H.H., and Strnad, J.; *J. Phys. Chem.*, **1990**, *94*, 7628-7634.
2. Stilbs, P.; *J. Coll. Int. Sci.*, **1982**, *89*, 547-553.
3. Stilbs, P.; *J. Coll. Int. Sci.*, **1982**, *87*, 385-394.
4. Candau, S. and Hirsch, E.; *J. Coll. Int. Sci.*, **1982**, *88*, 428-436.
5. Treiner, C. and Mannebach, M-H.; *J. Coll. Int. Sci.*, **1987**, *118*, 243-251.
6. Hayase, K., Hayano, S., and Tsubota, H., *J. Coll. Int. Sci.*, **1984**, *101*, 336-343.
7. De Lisi, R., Genova, C., and Liveri, V.T., *J. Coll. Int. Sci.*, **1983**, *95*, 428-433.
8. Attwood, D., Mosquera, V., and Perez-Villar, V.; *J. Coll. Int. Sci.,* **1989**, *127*, 532-536.
9. Boström, Backlund, S., Blokhus, A.M., and Høiland, H.; *J. Coll. Int. Sci.*, **1989**, *128*, 169-175.
10. Elworthy, P.H., Florence, A.T., and McFarlane, C.B.; *"Solubilization by Surface Active Agents."* Chapman and Hall, London, **1968.**
11. Israelachvili, J.N., Mitchell, D.J., Ninham, B.W.; *J. Chem. Soc., Faraday Trans. II*, **1976**, *72*, 1525.
12. Sheu, E.Y., and Chen, S.H.; *J. Phys. Chem.*, **1988**, *92*, 4466-4474.
13. Missel, P.J., Mazer, N.A., Benedek, G. B., and Young, C.Y., *J. Phys. Chem.*, **1980**, *84*, 1044-1057.
14. Nagarajan, R.J., *J. Coll. Int. Sci.*, **1982**, *30*, 477.
15. H. Eiler, Koll. Z., 1941, **97**, 913.
16. M. Mooney, J. Coll. Int. Sci., 1951, **6**, 162.
17. I. M. Krieger and T. J. Dougherty, *Trans. Soc. Rheol.*, **1951**, *3*, 137-152.
18. Jeffery, G.B.; *Proc. R. Sco. London A*, **1923**, *102*, 163.
19. Eisenschitz, R.Z., *Phys. Chem. A*, **1933**, *163*, 133.
20. P. Baglioni, and L. Kevan, *J. Phys. Chem.*, **1987**, *91*, 1516.
21. March, J. *"Advanced Organic Chemistry - Reaction, Mechnisms, and Structure";* McGraw-Hill, New York, **1968**; pp22.
22. Tanford, C; *"The Hydrophobic Effect."* second edition, Wiley, New York **1980**.

RECEIVED August 4, 1994

Chapter 12

Control of Flow Properties in Surfactant Solutions via Photoreactions of Solubilizates

Thomas Wolff

Institut für Physikalische Chemie und Elektrochemie, Technische Universität Dresden, 01062 Dresden, Germany

Certain aromatic solubilizates have drastic influences on the viscosity of dilute aqueous micellar solutions, while other compounds do not show this effect. Whenever it is possible to photochemically convert solubilizates of the former type to compounds of the latter type, viscosities and flow properties may be varied in situ by exposing samples to appropriate light. These effects were found in cationic cetyltrimethylammonium bromide as well as in non-ionic Triton X-100 systems.

The rheological properties of dilute aqueous micellar solutions are determined by the volume fraction, size and shape of the micelles present. Early investigations have also shown that the structure and extent of the layer of hydration water (*1*) and the addition of salt and/or certain solubilizates can also influence viscosities and flow behavior (*2,3,4*). Several sorts of non-Newtonian flow including viscoelasticity were observed. These non-Newtonian features are generally ascribed to the existence of long wormlike micelles (*5,6*), while spherical micelles give rise to Newtonian flow.

Photochemical transformations of aromatic solubilizates can be performed in situ using light not absorbed by the surfactant. These transformations thus provide a means of varying and controlling viscosity, provided photoeduct and photoproduct induce different viscosities. Search for the origins of these "photorheological effects" is complicated by the specificity of the effects, i.e. any scaling concept will not help, since volume fractions are not affected.

Examples of photorheological effects, the strategy to elucidate microscopic origins, and some results thereof will be presented in this article.

Photorheological effects

Newtonian flow systems. In Table I viscosities of aqueous solutions of the cationic surfactant cetyltrimethylammonium bromide and of the non-ionic surfactant Triton X-

0097–6156/94/0578–0181$08.00/0

100 are listed in the absence and presence of solubilizates. Inspection reveals an increase of viscosity caused by the solubilizates (photoeducts) and either a decrease or a further rise of viscosity upon photochemical formation of photoproducts. This shows the specificity of the effects.

Table I. Influence of Solubilizates and Their Photochemical Conversions on Viscosities of Micellar Solutions

Surfactant	Solubilizate	Conversion (%)	Viscosity (mPas)
CTAB			
(0.25 mol/dm³)	-	-	1.76
	n-butylanthracene		
	(9.5 mmol/dm³)	0	3.42
		80	1.62
	(18.5 mmol/dm³)	0	10.19
		62	2.60
	N-methyldiphenyl=		
	amine (10 mmol/dm³)	0	2.50
		100	1.68
	(20 mmol/dm³)	0	11.50
		100	5.62
	3-stilbene carboxylic		
	acid (27 mmol/dm³)	0	32.03
		75	63.26
	phenylglyoxylic		
	acid (25 mmol/dm³)	0	18.00
		50	8.88
Triton X-100			
(0.25 mol/dm³)		-	5.61
	N-methyldiphenyl=		
	amine (20 mmol/dm³)	0	7.95
		100	9.44
	(50 mmol/dm³)	0	16.70
		100	27.68

values from references 7,8,9,10,11,12

The photochemical reactions involved are given by the schemes 1 through 4 showing that a common chemical structure responsible for the increase or decrease of viscosity cannot be recognized. It should be noted that the wavelengths of the light used for photochemical transformations are not absorbed by the surfactants. Irradiations of surfactant solutions without solubilizates, thus, do not affect viscosities and flow.

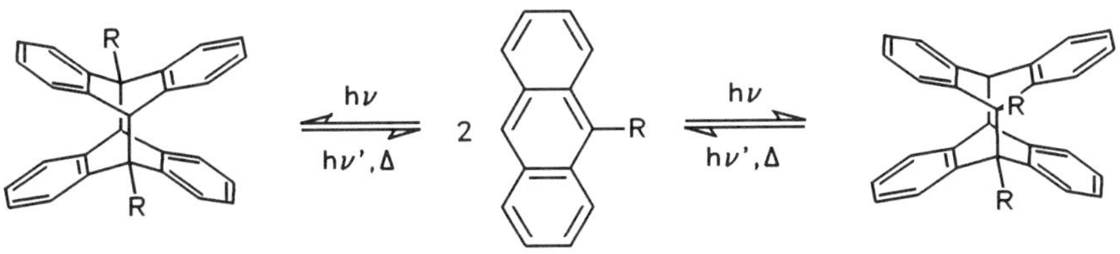

Scheme 1: Photodimerization of anthracene derivatives (*7,9,13,14*); *n*-butylanthrace-ne: R = C$_4$H$_9$; 9-anthracene carboxylic acid (9-AC): R = COOH; TFAE: R = CH(CF$_3$)OH

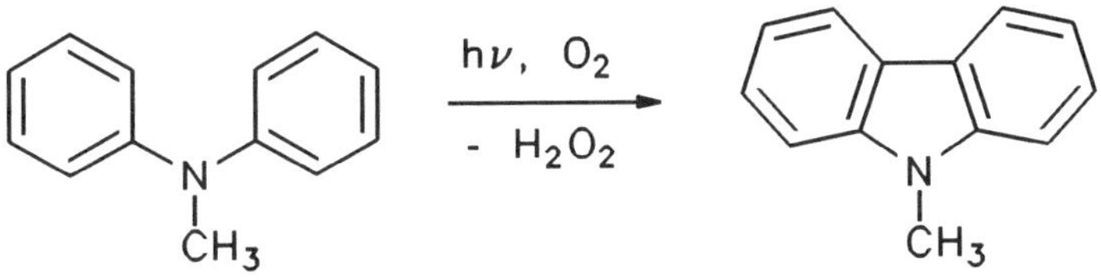

Scheme 2: Photocyclization of *N*-methyl-diphenylamine to *N*-methylcarbazole (*8*)

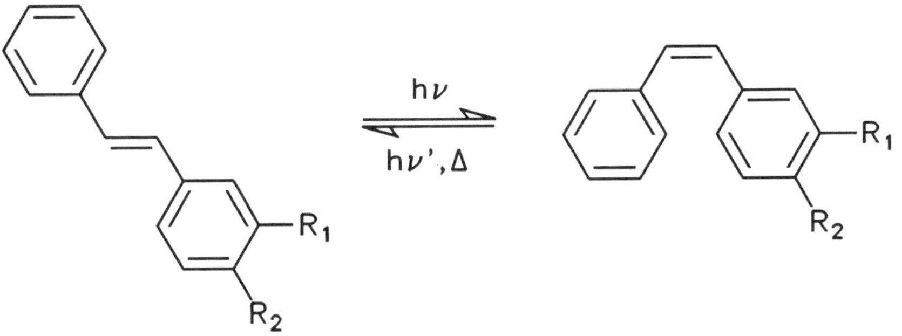

Scheme 3: Photoisomerization of stilbene derivatives; 3-stilbene carboxylic acid: R$_1$ = COOH, R$_2$ = H (*11*)

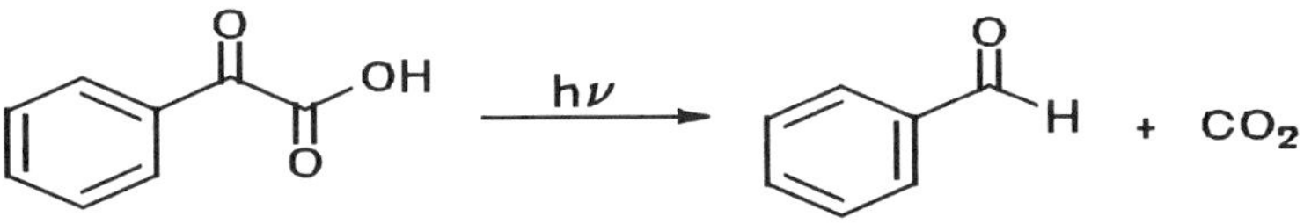

Scheme 4: Photodecarboxylation of phenylglyoxylic acid (*12*)

Non-Newtonian systems. Deviations from Newtonian flow in some cases are observed at high ratios of solubilizate to surfactant (exceeding those given in Table I). This was found e.g. in the CTAB system using the solubilizates 3-stilbene carboxylic acid (*11*) and phenylglyoxylic acid (*12*). These effects, however, can be ascribed to beginning phase separation since the solubilizate concentrations are close to solubility limits.

More interesting are systems which even at low solubilizate concentrations show all sorts of deviations from Newtonian flow (such as rheopexy, thixotropy and viscoelasticity), which are sensitive to photochemical reactions of the solubilizates. Compounds inducing such behavior in aqueous CTAB are 9-anthracene carboxylic acid (9-AC) and 2,2,2-trifluor-1-(9-anthryl)-ethanol (TFAE):

9-AC   TFAE

0.15 Molar CTAB solutions containing 9-anthracene carboxylic acid (9-AC) exhibit Newtonian flow at 22.5 mmol/dm³ 9-AC (and below), rheopectic flow at 37.5 mmol/dm³, and thixotropic flow above 37.5 mmol/dm³ (*9*). Rheopectic as well as thixotropic flow is accompanied by viscoelasticity, which can be visually observed in swirled solutions. The complicated dependence of viscosity on concentration is depicted in Figure 1 giving the viscosity as a function of shear time at $\dot{\gamma} = 65$ s^{-1}. The three lowest concentrations give rise to typical rheopectic flow: the curves exhibit a zero time viscosity (see insert of Figure 1), increase and level off after 90 s. The curves for 40 and 60 mmol/dm³ do not show a zero time viscosity but the viscosity is proportional to the shear time - a behavior indicating elastic deformation of the solution rather than flow. Finally, at 75 mmol/dm³, pure thixotropy is observed, i.e. the viscosity decreases from a high zero time viscosity and levels off after some time of shear.

The curves shown in Figure 2 are typical examples for the influence of partial photochemical conversion of 9-AC to dimers according to Scheme 1. Flow curves after irradiation (photoconversion) always exhibit fewer non-Newtonian features than those for solutions containing 9-AC monomers only.

Similar behavior is observed when TFAE is the solubilizate instead of 9-AC (*13,14*). In both cases there is a qualitative resemblance of results obtained using compounds bearing phenolic or aromatic acid groups, such as naphthol or salicylic acid and their salts (*2,5*). TFAE belongs to a rare class of compounds which induce viscoelasticity in aqueous CTAB without being acidic and without addition of salt.

Strategies and Results

To understand the microscopic origins of the observed effects we need to know the size and shape of micellar aggregates. Besides detailed rheological studies this

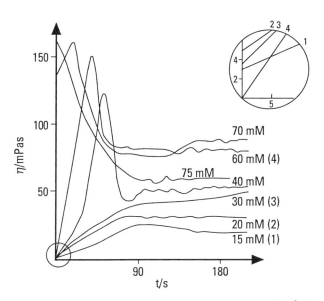

Figure 1.   Viscosity η vs. shear time at a shear rate $\dot{\gamma} = 65$ s^{-1} for aqueous solutions of cetyltrimethylammonium bromide (CTAB) at various concentrations in the presence of 9-anthracene carboxylic acid (9-AC); molar ratio CTAB:9-AC = 2.5; temperature 25 °C. (Adapted from ref. 7.)

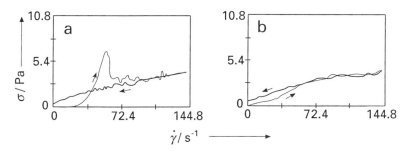

Figure 2.   Flow curves (shear stress σ vs shear rate $\dot{\gamma}$) at 25 °C for aqueous solutions of 22.5 mmol/dm^3 cetyltrimethylammonium bromide (CTAB) and 9.9 mmol/dm^3 9-anthracene carboxylic acid (9-AC); (a) before, (b) after 5 % photoconversion. (Reproduced with permission from ref. 7. Copyright 1989 American Chemical Society.)

requires light scattering experiments. However, since the positions of scattering particles have to be independent of each other (*15*), these measurements must be performed at surfactant concentrations significantly below those used for producing photorheological effects. We cannot suppress electrostatic interactions in ionic micellar systems by the addition of salt since this would alter all the photorheological effects. The concentration dependence of micelle sizes and shapes thus presents a problem. Therefore and in order to have independent evidence for any strange light scattering result, further techniques such as electron microscopy and ESR spin probe investigations are needed occasionally.

Table II. Results of rheological and light scattering experiments: micelle weights M, micelle lengths L, relaxation time constants τ (*7-9, 11-15*); in parentheses: values after photoconversion

$\dfrac{c(\text{so})}{[c(\text{su})+c(\text{so})]}$	$M_{LS}/10^4$g/mol	L_{LS}/nm	L_{DLS}/nm	L_{VIS}/nm	τ_1/s	τ_2/s
		CTAB				
-	3.0	-	-	-	-	-
		Triton X-100				
-	6.1	-	-	-	-	-
		n-butylanthracene in CTAB, globular micelles				
0.037	120(120)	-	-	-	-	-
		N-methyldiphenylamine in CTAB, globular micelles				
0.14	97(99)	-	-	-	-	-
		N-methyldiphenylamine in Triton X-100, globular micelles				
0.14	5.9(5.8)	-	-	-	-	-
		9-anthracene carboxylic acid in CTAB, thread-like micelles				
0.2	2.4(1.7)	5.2(3.7)	n.m.	32(24)	n.m.	n.m.
		TFAE in CTAB, thread-like micelles				
0.1	71(64)	105(98)	-	-	-	-
0.2	-	-	32(13)	-	-	-
0.25	-	-	-	102(92)	34(34)	0.16(0.05)

$c(\text{so}),c(\text{su})$: concentrations of solubilizate and surfactant, respectively; indices pertain to values determined from static (LS) and dynamic light scattering (DLS) and viscometry (VIS), respectively. LS- and DLS values at $c(\text{su}) = 1\text{-}4$ mmol/dm^3, VIS-values for 9-AC at $c(\text{su}) = 150$ mmol/dm^3, VIS- and τ-values for TFAE at $c(\text{su}) = 250$ mmol/dm^3.

Selected results from light scattering and rheological investigations are collected in Table II. On inspection we note that in the Newtonian systems (in which micelle shapes cannot deviate much from sphericity) the changes in viscosity (Table I) are not accompanied by significant variations in micelle weights or sizes (or the changes are so minor that they cannot account for the viscosity effects). We further note that micelle weights are extremely high as compared to the systems without solubilizates, i.e. these induce increases of aggregation numbers by several orders of magnitude (*7,8,9*). Since globular micelles of that size are hard to imagine, electron micrographs of rapidly frozen CTAB solutions containing *N*-methyldiphenylamine were taken (*16*). In these, very large micelles are seen which appear as clusters of smaller, disk-like aggregates forming an overall globular shape. The smallest entities on the micrographs agree in size with the light scattering results (Table II). Most aggregates are much larger. This suggests that the aggregates increase in size via cluster formation upon raising the concentration from that used for the light scattering samples (1-4 mmol/dm³) to that applied for the rheological as well as the electron micrograph samples (250 mmol/dm³). The electron micrographs, thus, represent the situation in the samples used for the photorheological experiments rather than the light scattering results.

The question remains what causes the photochemically induced viscosity changes in the Newtonian systems if not a variation in micelle size. As there is nothing but water and micelles (containing solubilizates) and as the latter are not affected, we have to consider the water. The existence of more or less extended layers of structured hydration water surrounding colloids is discussed in the literature (*17, 18, 19*). If there is such a thing it may well contribute to viscosity effects: solubilizates in the more viscous sample induce a more rigid structure of the micellar surface, which gives rise to a more extended (or more viscous) shell of hydration water; solubilizates in the less viscous samples reside a little deeper in the micelles, so that the ordering effect on the micelle surface is reduced. Transformations of the micelle hydration shell will not be seen in static light scattering experiments because of too small a difference in refractive index. Support for this hypothesis was gained from the addition of urea, known to have a"structure breaking" influence on water (*17*), which did reduce viscosities but not micelle sizes (*8*). Independent evidence arose from ESR spin probe investigations: ESR signals of stable nitroxide radicals are sensitive to the viscosity of their immediate environment (*20*), i.e. the relative intensities of the observed triplet signals vary. Spin probes located in the layer of hydration water indicate a higher local viscosity in the systems exhibiting the higher bulk viscosity (*21*), proving the above hypothesis. This technique also provides evidence for the hypothesis that solubilizates in less viscous samples are preferredly located a little deeper in the micelles: the coupling constants in the ESR signals reflect the polarity of the environment of the probes; as compared to solubilizate free micelles, these coupling constants (for probes located in layer of hydration water) are affected only in the more viscous samples in which the solubilizates are supposed near the surface (*21*).

Easier to understand are the microscopic solution structures responsible for the photorheological effects in the non-Newtonian systems. Here - in contrast to the Newtonian systems - changes of sizes and shapes of micelles upon phototransformations of solubilizates are observed (Table II). In the presence of monomers of 9-AC or TFAE one has to assume long thread- or worm-like micelles which - entangled or crosslinked - give rise to the non-Newtonian features. Accordingly, electron micrographs of viscoelastic CTAB solutions containing 9-AC show thread-like aggregates and indicate some crosslinks or entanglements (*16*). The data in Table II reveal a reduction in the length of micellar threads upon photoconversion of either 9-AC or TFAE, independent of the method. This is in agreement with the observation that the non-Newtonian features diminish upon irradiation (photodimerization) indicating that the photorheological effects are caused by removing the monomers.

For the solubilizate 9-anthracene carboxylic acid more details of the interaction with the micelles are known. A substantial fraction of the monomers reacts with cetyltrimethylammonium bromide, i.e. the anions of the aromatic acid form ion pairs with the cationic surfactant head groups according to Scheme 5. Thereby hydrobromic acid is set free deceasing the pH of the solutions to values around 2 (9). The photodimers are not capable of forming analogous ion pairs since the pH increases upon irradiation. Thus, the photorheological effect is due in part to the inherent generation and consumption of electrolyte (HBr), which affects micelle size and

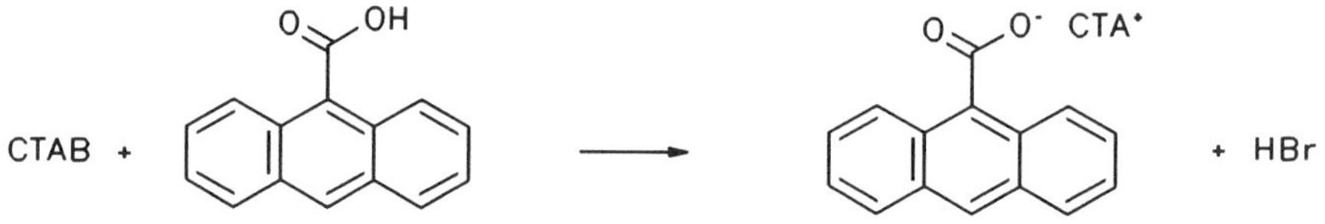

Scheme 5: Ion pair formation by 9-AC and CTAB

shapes. Other aromatic acids such as 3- and 4-stilbene carboxylic acid, however, do not act in this way. pH-Measurements in these cases as well as in systems containing phenylglyoxylic acid clearly show that ion pairs are not formed and that these compounds are solubilized in their non-dissociated forms (11).

In TFAE, however, different, not yet identified interactions with CTAB must occur, since the pH of the solutions is the same as in TFAE free aqueous CTAB. Some details of the viscoelastic solution structure follow from oscillating shear experiments (14) allowing the calculation of relaxation spectra, which give the distribution of relaxation time constants (22). Examples are shown in Figure 3. The spectra exhibit two relaxation maxima corresponding to two distinct relaxation mechanisms. In an example at 0.05 mol/dm^3 TFAE on Figure 3a, it can be seen that the longer time constant (34 s) is not affected by the photodimerization of TFAE while the shorter time constant is reduced from 0.16 to 0.05 s after photoconversion of only a few % of the TFAE monomers (see also Table II). Following Cate's theory (6) we can attribute the slower relaxation process to breaking of thread-like micelles and the faster one to translational diffusion of micelles along their contour length. It is thus sound that the latter process becomes faster upon photodimerization of TFAE since the micelles become shorter (Table II).

Discussion and Outlook

All the experiments described were performed at surfactant concentrations at which small spherical micelles are present in the absence of solubilizates, i.e. below the second critical micelle concentration. The observed effects, thus, are due to the action of the solubilizates only, which were present at low concentrations, below 1 % by weight. The rheological and light scattering experiments have enabled us to elucidate the microscopic changes of micelle size and shape accompanied with the solubiliza-

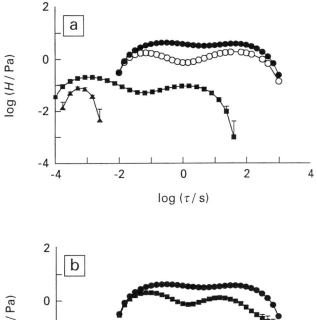

Figure 3. (a) Relaxation spectra $H(\tau)$ for aqueous 0.25 molar CTAB containing 0.05 mol/dm³ TFAE (circles), 0.04 mol/dm³ TFAE (squares), and 0.03 mol/dm³ TFAE (triangles); open circles: irradiated solution at 3 % photoconversion. (b) Relaxation spectra for aqueous solutions containing CTAB and TFAE at a molar ratio CTAB:TAFE = 5; circles: 0.25 mol/dm³ CTAB; squares: 0.2 mol/dm³ CTAB; triangles: 0.15 mol/dm³ CTAB. (Reproduced with permission from ref. 14. Copyright 1993 Academic Press.)

tion of photoeducts and with their photochemical conversion. Information on all of the interactions of solubilizates (or their photoproducts) and surfactant has been rare up to now. A combination of several spectroscopic methods may add to clarification, such as ^{19}F-NMR on samples containing TFAE or Raman spectroscopic investigations on the state of water in the hydration layer, cf. (*23*) and (*24*).

The observed photorheological effects open a new method for investigating micelle formation kinetics: a rearrangement of micelles can be started by flash irradiation in a running (rotating or oscillating) viscometer, which then registers the change in viscosity as a function of time.

In future investigations it will be interesting to determine whether or not similar effects can be observed in surfactant aggregates differing from micelles - such as vesicles and biocolloids.

Finally, it should be mentioned that many of the solubilizates showing photorheological effects also influence transition temperatures in lyotropic liquid crystalline phases appearing at higher surfactant concentrations. In these systems phase transitions can be switched isothermally simply by exposing the samples to appropriate light (*25*).

Conclusions

Rheological properties of surfactant solutions can be varied in situ when solubilized compounds are photochemically converted. Microscopically the effects can originate from changes of the size of worm-like micellar aggregates. There are, however, systems (exhibiting Newtonian flow throughout), in which photochemically induced viscosity variations take place without a change in micelle size. In these cases transformations of the micellar hydration shell are likely to be the cause of the photorheological effects.

Acknowledgments

Financial support by the Deutsche Forschungsgemeinschaft and the Fonds der Chemischen Industrie is gratefully acknowledged.

Literature Cited

1.   Ekwall, P.; Mandell, L.; Solyom, P. *J. Colloid Interface Sci.* **1971**, *35*, 519.
2.   Nash, T. *J. Colloid Interface Sci.* **1958**, *13*, 134.
3.   Wan, L. S. C. *J. Pharm. Sci.* **1966**, *55*, 1395.
4.   Gravsholt, S. *J. Colloid Interface Sci.* **1976**, *57*, 575.
5.   Rehage, H.; Hoffmann, H. *Mol. Phys.* **1991**, *74*, 933.
6.   Cates, M. E. *J. Phys. Chem.* **1990**, *94*, 371.
7.   Wolff, T.;Emming, C.-S.; Suck, T. A.; von Bünau, G. *J. Phys. Chem.* **1989**, *93*, 4894.
8.   Wolff, T.; Emming, C.-S.; von Bünau, G. *J. Phys. Chem.* **1991**, *95*, 3731.
9.   Wolff, T.; Suck, T. A.; Emming, C.-S.; von Bünau, G. *Prog. Colloid Polym. Sci.* **1987**, *73*, 18.
10. Wolff, T.; Schmidt, F.; von Bünau, G. *J. Photochem. Photobiol., A: Chem.* **1989**, *48*, 435.

11. Kerperin, K. J.; Wolff, T. *Ber. Bunsenges. Phys. Chem.* **1993**, *97*, 36.
12. Kerperin, K. J.; Wolff, T. *J. Surf. Sci. Techn.* **1992**, *8*, 349.
13. Wolff, T.; Bott, R.; Kerperin, K. J. *Colloid Polym. Sci.* **1992**, *270*, 1222.
14. Wolff, T.; Kerperin, K. J. *J. Colloid Interface Sci.* **1993**, *157*, 185.
15. Kerker, M. *The Scattering of Light*; Plenum: New York, 1985.
16. Wolff, T.; Emming, C.-S.; von Bünau, G.; Zierold, K. *Colloid Polym. Sci.* **1992**, *270*, 822.
17. Drost-Hansen, W. In *Chemistry of the Cell Interface, Part B*; Brown, H. D., Ed.; Academic: New York, 1971.
18. Luck, W. A.; Schiöberg, D.; Siemann, K. *J. Chem. Soc., Faraday Trans. 2*, **1980**, *76*, 136.
19. Halle, B.; Piculell, L. *J. Chem. Soc., Faraday Trans. 1*, **1986**, *82*, 415.
20. Ottaviani, M. F.; Baglioni, P.; Martini, G. *J. Phys. Chem.* **1983**, *87*, 3146.
21. Baglioni, P.; Wolff, T., to be published.
22. Ferry, J. D. *Viscoelastic Properties of Polymers, 2nd ed.;* Wiley: New York, NY, 1970
23. Ben Hassine, B.; Gorsane, M.; Pecher, J.; Martin, R. H. *Bull. Chem. Soc. Belg.* **1985**, *94*, 425.
24. Guillaume, B. C. R.; Yogev, D.; Fendler, J. H. *J. Chem. Soc., Faraday Trans.* **1992**, *88*, 1281.
25. Wolff, T.; Seim, D.; Klaussner, B. *Liq. Cryst.* **1991**, *9*, 839.

RECEIVED July 8, 1994

Chapter 13

Microstructure and Rheology of Nonionic Trisiloxane Surfactant Solutions

M. He[1,3], R. M. Hill[2], H. A. Doumaux[1], F. S. Bates[1], H. T. Davis[1], D. F. Evans[1], and L. E. Scriven[1]

[1]Department of Chemical Engineering and Materials Science, University of Minnesota, 421 Washington Avenue Southeast, Minneapolis, MN 55455
[2]Central Research, Dow Corning Corporation, 2200 West Salzburg, Auburn, MI 48611

The nonionic trisiloxane surfactant, $M(D'E_{12})M$ (= $((Me)_3SiO)_2Si(Me)\text{-}(CH_2)_3(OCH_2CH_2)_{12}OH)$, forms clear isotropic solutions in water at all concentrations between 10 and 43 °C. Both water-rich and surfactant-rich solutions are low viscosity Newtonian liquids. However, at intermediate concentrations, the solutions are non-Newtonian and viscoelastic. This rheological behavior along with supporting data from small-angle X-ray and neutron scattering, and pulse-gradient NMR measurements points to a progressive change in surfactant microstructure across the concentration range. This change is parallel to the progression of liquid crystal phase behavior at lower temperatures. The results are interpreted in terms of a model in which small spherical micelles formed at low concentrations transform progressively to entangled worm-like micelles, branched interconnected worm-like micelles, and then a random bilayer structure which persists to 100 % surfactant.

In a previous study (*1*), we showed that a nonionic trisiloxane surfactant, $((Me)_3SiO)_2Si(Me)(CH_2)_3(OCH_2CH_2)_{12}OH$, which is denoted as $M(D'E_{12})M$, shows unusual phase behavior in forming clear isotropic solutions across the entire concentration range and over a wide temperatures range: see Figure 1. In the temperature range between 10 and 43 °C, this surfactant is completely miscible with water. But at temperatures below 10 °C, a hexagonal liquid crystal phase, H_1, was found at about 50 wt% surfactant, and a lamellar liquid crystal phase, L_α, at about 70 wt%. At temperatures above 43 °C the liquid-liquid immiscibility called the cloud point or the lower consolute temperature (LCT) is found. The LCT varies with concentration; the minimum in the LCT curve is about 5% and 43 °C. A few other trisiloxane and alkyl polyoxyethylene surfactants such as $M(D'E_{18})M$ (*1*), $C_{12}E_8$ and $C_{12}E_6$ (*2,3*) also form clear isotropic solutions in water at all concentrations, but only in a much narrower temperature range well above room temperature.

[3]Current address: Unilever Research United States–Edgewater Laboratory, 45 River Road, Edgewater, NJ 07020

0097–6156/94/0578–0192$08.90/0

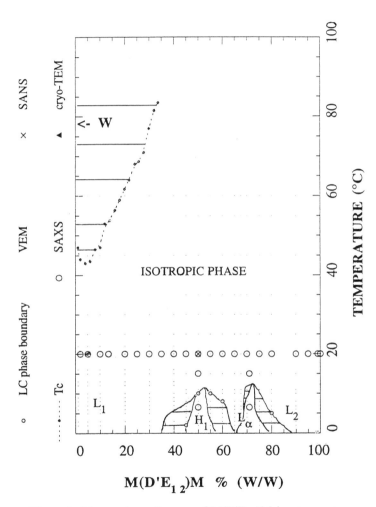

Figure 1. Binary phase diagram of M(D'E$_{12}$)M / water system.

We undertook this study to determine how the surfactant microstructures present in such isotropic solutions evolve as the concentration changes across such a wide range. Possibly at low concentrations small globular micelles are formed (*4*). But the microstructures cannot remain small globular micelles all the way to 100 % surfactant. Most nonionic surfactants form clear isotropic solutions in two or more disconnected regions of concentration and temperature (variously labelled as the L_1, L_2 and L_3 regions (*2,3*)). This study should also bear on the evolution of microstructure within and between such regions. The microstructures of the sponge phase, L_3 (*5, 50*), and of bicontinuous microemulsions (*6, 7*) have been intensely studied. But parallel studies of other isotropic surfactant solutions at midrange and surfactant-rich compositions have not been made. It is possible that bicontinuous microstructures are formed in these regions also. We felt that the study reported here could help to broaden our understanding of the general rules that govern the evolution of microstructures within such isotropic solution regions. We chose the M(D'E12)M system especially because it offers the opportunity to study this evolution at an experimentally easy-to-access temperature around 25 °C.

The rheological properties of dilute micellar isotropic solutions is related closely to the shape, size and concentration of micelles (*9-12*), and therefore is a sensitive probe of solution microstructure. However, the relationship between rheology and microstructure at higher surfactant concentrations for isotropic solutions are not well understood. In what follows, we report the rheological behavior of the M(D'E$_{12}$)M solutions at compositions spanning the entire concentration range. Shear-viscosity and dynamic modulus were measured as a function of steady shear-rate and oscillatory angular frequency. We also report supporting data from small-angle X-ray and neutron scattering and pulse field gradient spin-echo NMR.

Materials and Methods

The surfactant ((Me)$_3$SiO)$_2$Si(Me)(CH$_2$)$_3$(OCH$_2$CH$_2$)$_{12}$OH is denoted here as M(D'E$_{12}$)M, in which M = (CH$_3$)$_3$SiO-, D'=-(CH$_3$)Si, and the polyoxyethylene group E$_{12}$ = -CH$_2$CH$_2$CH$_2$(OCH$_2$CH$_2$)$_{12}$OH. The preparation of this surfactant was reported elsewhere (*1,13*). M(D'E$_{12}$)M has a polydisperse oxyethylene chain-length distribution with M_n/M_w around 1.2, and so E$_{12}$ represents the average degree of ethoxylation. Aqueous solutions of M(D'E$_{12}$)M were prepared by weight using distilled, deionized water. Samples of known composition were contained in 25 ml 2.5-cm-ID capped vials and sealed with paraffin wax (parafilm®, American Can Company). After quickly hand-shaking the vials for about one minute to ensure adequate mixing, the vials were allowed to stand for more than three hours to equilibrate and defoam. The resultant liquid mixtures were clear and colorless as observed through transmitted white light, and isotropic as observed between crossed polars. Viscosity changes of the samples was first estimated by the drainage time upon inverting the vials, and viscoelasticity was detected by observing the movement of small air bubbles trapped in the samples (*14*). By these visual assessments, the viscosity and viscoelasticity of the samples at mid-range of concentrations were found to be higher than those of water-rich or surfactant-rich samples.

Steady-state and dynamic rheological measurements were performed on a

Rheometrics RFS II fluids spectrometer (Rheometrics, Inc., Piscataway, NJ) supported by the RECAP III and RHIOS software (Rheometrics, Inc.). Transducers with torques of 10 g-cm and 100 g-cm were used. The couette test fixture was used in the steady-shear viscosity and dynamic modulus measurements. The cone-and-plate test fixture was specifically used for the viscosity measurement at high shear rate over 1700 to 4500 seconds^{-1}. For the couette test fixture, the radius of the cup was 34 mm and that of the bob was 32 mm. The length of the bob was 33 mm. The gap between the cup and bob was around 6 mm, and the cup contained roughly 17 ml of solution for each measurement. For the cone-and-plate test fixture, the cone angle was 0.021 rad. and plate diameter was 50 mm. The space between the cone and plate at center was set at 0.05 mm for each measurement.

Steady-shear viscosity was measured at shear-rates from 0.1 to 4500 seconds^{-1} at fixed temperature of 25 ± 0.5 °C. The viscosity measured at each shear-rate was averaged for clock- and counter-clockwise directions from 7 measurements in 25 seconds. An interval time of 15 to 30 seconds between different shear-rate was found to be adequate for the $M(D'E_{12})M$ solutions to relax between each measurement. Step-shear rate tests revealed that the typical relaxation time for the solutions from 100 rad/sec to zero shear was less than 10 seconds.

Dynamic elastic and loss moduli, G' and G", were measured at 25 ± 0.5 °C with the rheometer operating in the oscillatory mode ($0.01 \leq \omega \leq 100$ seconds^{-1}). The strain amplitude for each measurement was set below 10%. We determined that this was in the linear viscoelastic region by scanning G' and G" over strain amplitudes of 1 to 100 percent.

Small angle X-ray scattering (SAXS) and small angle neutron scattering (SANS), were performed under the similar conditions as detailed in a previous report (*1*). For SAXS, the X-ray generator was a rotating anode with Cu-K$_\alpha$ wavelength of 1.54 Å selected by means of Nichol filters. The sample-to-detector distance was 68.2 cm. The detectable wave vector q range was from 0.02 Å^{-1} to 0.3 Å^{-1}, where q = $(4\pi / \lambda) \sin(\theta / 2)$ and θ is the scattering angle. The scattering data accumulated over 120 minutes to 360 minutes were corrected for background scattering by subtracting the scattering intensity of water and the empty capillary. For the small angle neutron scattering (SANS) experiments, the wave length of the neutron beam was selected to be 6.0 Å. The sample-to-detector distance was set at 15.34 or 14.34 m for low q measurements (0.003 Å^{-1} < q < 0.03 Å^{-1}), and 4.25 or 3.57 m for high q measurements (0.01 Å^{-1} < q < 0.12 Å^{-1}). The scattering data accumulated over 20 minutes to 40 minutes were corrected for background scattering by subtracting the scattering intensity of beam block and an empty quartz cell. Then the scattering data were calibrated and converted to absolute intensity by using a silica gel standard (SIL-B2, NIST) as reference.

Pulsed field gradient spin-echo NMR (PFGSE NMR) was performed on a Nicolet 1180 NMR spectrometer operating at 300 MHz with a 90-δ-180-δ pulse sequence (*15,16*) to measure the self-diffusion coefficients of both $M(D'E_{12})M$ and water in solutions. By varying the field gradient pulse duration while maintaining a constant field gradient interval, the self-diffusion coefficient of a proton was determined by fitting the echo amplitude decay, A, using the equation A = A$_0$ exp[-

$D\gamma^2G^2\delta^2(\Delta - (\delta / 3))$]. Here G was the field gradient strength, γ the proton gryomagnetic ratio, δ the variable duration and Δ the interval between field gradient pulses. The field gradient was calibrated by neat water whose self-diffusion coefficient is known to be 2.299×10^{-9} m^2/sec at 25 °C (*17*). The field gradient strength was set between 0.2 and 0.5 T/m. $M(D'E_{12})M$ solutions were prepared as introduced above, and then flame-sealed in 5-mm NMR tubes (Wilmad Glass Company, Inc., Buena, NJ). In the measurement process, the tubes were put in the NMR sample chamber with temperature controlled at 25 ± 0.2 °C.

Results and Analysis

The isotropic solutions of $M(D'E_{12})M$ in water evolve from water-rich to surfactant-rich without any detectable phase transitions across the entire concentration range at 25 °C. We here present four threads of evidence supporting the hypothesis that the microstructures evolve from spherical micelles to entangled worm-like micelles to random bilayers as $M(D'E_{12})M$ concentration rises from 0 to 100 wt%. The evidence consists of (i) rheology, (ii) small-angle X-ray and neutron scattering, (iii) pulse field gradient spin-echo NMR, and (iv) analogy with the liquid crystal phase behavior of the system below 10 °C.

Rheology. The viscosity of $M(D'E_{12})M$ solutions between 0.6 and 100 wt% surfactant was measured at steady shear-rates from 0.1 to 4500 seconds^{-1}. At shear-rates below 500 seconds^{-1}, the viscosity of the solutions at all the compositions was found to be shear-rate independent (Newtonian). In Figure 2, the effective zero-shear viscosity measured at a shear-rate of 1.0 seconds^{-1} is plotted as function of $M(D'E_{12})M$ concentration. The zero-shear viscosity begins to rise sharply at about 20 wt% surfactant, and reaches a maximum at 50 wt%, then lowers gradually as the surfactant concentration rises further. The zero-shear viscosity at the maximum is about three orders of magnitude higher than that below 20 wt%, and about 10 times higher than that of the neat liquid surfactant. In Figure 3, the steady shear viscosity is plotted as a function of shear rate. At shear-rates above 500 seconds^{-1}, the solutions at compositions between 40 and 80 wt% $M(D'E_{12})M$ are shear-thinning. In contrast, solutions outside this composition range are Newtonian at shear-rates up to 4500 seconds^{-1}.

Shear storage modulus, G', and shear loss modulus, G", of the solutions were measured at angular frequencies, ω, from over 0.01 to 100 seconds^{-1}. Figure 4a and 4b show plots of log(G') *vs.* log($a_c\omega$) for some of the solutions studied. Here a_c is a concentration-dependent scaling factor, and is 1 in Figure 4a and 4b. Also drawn in the figures are straight lines of slope = 0.5, 1 and 2. Comparing the lines with the results at frequencies below 50 seconds^{-1} shows that the data exhibits three discrete slopes at low frequency. At midrange concentration between 45 and 50 wt% $M(D'E_{12})M$, which corresponds to the maximum zero-shear viscosity, G' is approximately proportional to $\omega^{0.5}$ (slope = 0.5). On either side of the midrange concentration, i.e., at 40 wt% or 60 to 70 wt% surfactant, G' is proportional to ω (slope = 1). Further away from the midrange concentration, at surfactant concentration below 35 wt% and above 90 wt%, G' is proportional to ω^2 (slope = 2).

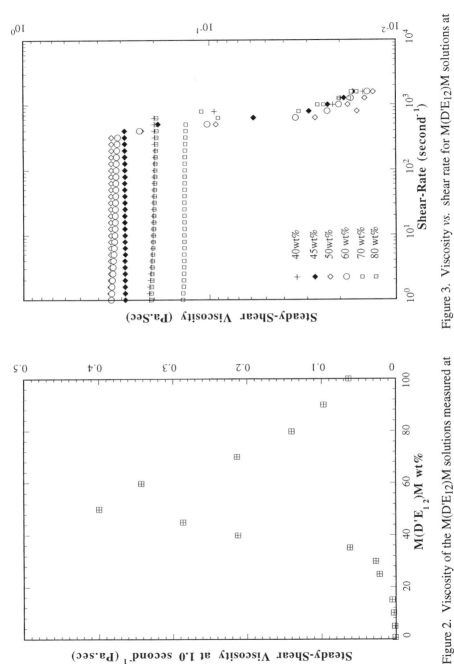

Figure 3. Viscosity vs. shear rate for M(D'E$_{12}$)M solutions at midrange of concentrations.

Figure 2. Viscosity of the M(D'E$_{12}$)M solutions measured at shear rate of 1 second^{-1} vs. the surfactant concentration.

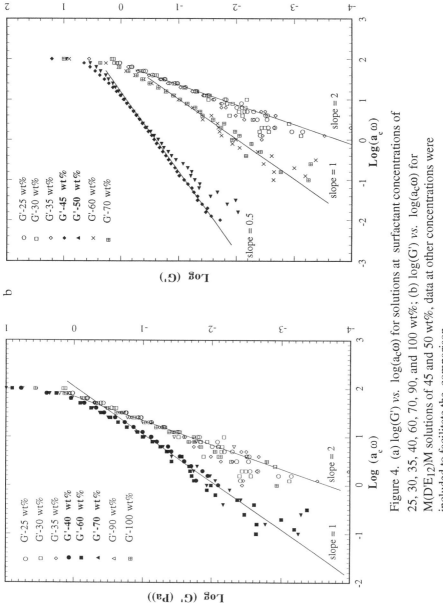

Figure 4. (a) log(G') vs. log($a_c\omega$) for solutions at surfactant concentrations of 25, 30, 35, 40, 60, 70, 90, and 100 wt%; (b) log(G') vs. log($a_c\omega$) for M(D'E$_{12}$)M solutions of 45 and 50 wt%, data at other concentrations were included to facilitate the comparison.

The frequency dependence of the dynamic loss modulus, G", is shown in Figure 5a and 5b as $\log(G")$ vs. $\log(a_c\omega)$. In this case, to facilitate the analysis, we shifted the curves horizontally by a concentration-dependent scaling factor a_c, so that all the curves overlapped at frequencies above 50 seconds^{-1} but were differentiated at lower frequencies (the values of a_c are shown in the figures). Figure 5a and 5b show that G" is proportional to ω (slope =1) for all the compositions except 45 and 50 wt% $M(D'E_{12})M$. At these two concentrations, G" is proportional approximately to $\omega^{0.5}$ at ω less than 50 seconds^{-1}. Also the values of a_c at midrange of concentrations are much higher than those at surfactant-rich and surfactant-lean concentrations. Effectively this means that the G" at midrange of concentrations is much higher than those at low or high concentrations if compared at same frequency.

Extracted from Figure 4 and Figure 5, the slopes of $\log(G')$ or $\log(G")$ vs. $\log(a_c\omega)$ at frequencies below 50 seconds^{-1} are plotted in Figure 6 against $M(D'E_{12})M$ concentration. In both low and high surfactant concentration regions, the slopes for G' and G" are 2 and 1 respectively, which can be interpreted by the Maxwell model ([18]). At midrange of concentrations, the slopes for G' and G" are lowered significantly from 2 and 1. This non-Maxwell behavior suggests that the solutions in the midrange of concentrations are more elastic.

Both this non-Maxwell behavior and the maximum zero-shear viscosity showed by the solutions in the midrange of concentrations may result from microstructures different from that in either surfactant-lean or surfactant-rich solutions. This is supported by the fact that the solutions at midrange of concentrations above 40 wt% but below 80 wt% showed shear-thinning at shear-rates above 500 seconds^{-1}. In contrast, the surfactant-lean and the surfactant-rich solutions were Newtonian up to a shear-rate of 4500 seconds^{-1}. The shear-thinning could be due either to the breaking-down of a network-type of microstructure or to the alignment of large anisotropic micelles in the shear direction ([8]).

For a 50 wt% solution, shear-induced birefringence was not observed by hand-shaking a vial containing the solution. However, the dynamic shear experiment described above may change part of the 50 wt% isotropic solution to hexagonal liquid crystal domains ([19]). Although the exact condition of shear which may induce this phase transition is not clear, the appearance of a liquid crystal phase after the dynamic shear indicates that large aggregates may exist in the 50 wt% isotropic solution before the shear.

Results of some recent works also support our speculation that worm-like micelle and bilayer network types of microstructures could be responsible to the rheological properties displayed by the solutions of the midrange of concentrations ([10,11, 20-24]).

In summary the solutions can be categorized into three concentration regions based on their steady and dynamic rheological properties:

(1) surfactant-lean solutions (<40 wt% $M(D'E_{12})M$);
(2) surfactant-rich solutions (>80 wt%);
(3) midrange solutions (between 40 and 80 wt%).

The solutions of the first two categories are Newtonian and their viscoelasticity can be

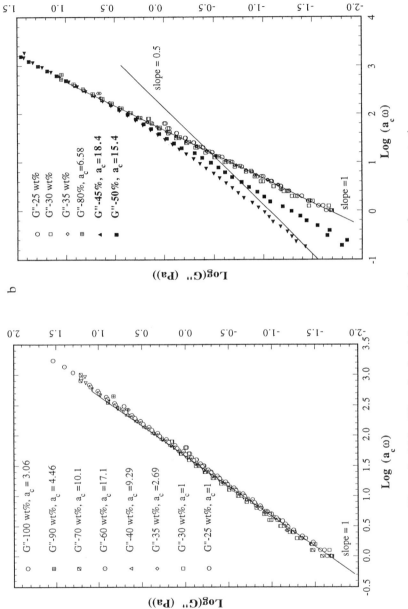

Figure 5. (a) $\log G''$ vs. $\log(a_c\omega)$ for the solutions at surfactant concentrations of 25, 30, 35, 40, 60, 70, 90, 100 wt%; (b) $\log G''$ vs. $\log(a_c\omega)$ for the solutions at surfactant concentrations of 45, 50 wt%.

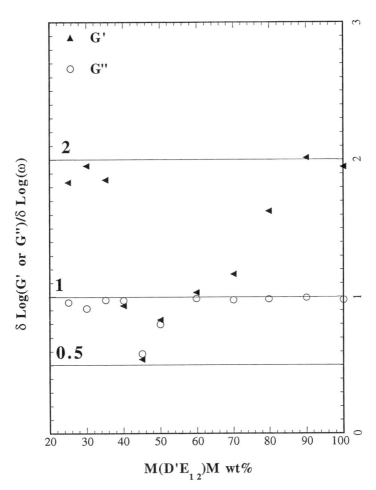

Figure 6. Slopes of Log(G') and log(G") *vs.* log(ω) at ω < 50 second^{-1} for the solutions of different compositions.

described by the Maxwell model. Those of the third category are non-Newtonian and are more elastic. Thus the rheology data presented above indicates that the microstructures in the midrange of concentrations are significantly different from those in the surfactant-rich and surfactant-lean concentration ranges.

Small-Angle X-ray (SAXS) and Neutron Scattering (SANS). SANS spectra of 1.92 and 10 % $M(D'E_{12})M$ solutions are shown in Figure 7a and b. These SANS spectra can be simulated by the Fourier transformation of the radial distribution function of hard-spheres (*25* and references therein), with the size of the hard spheres being the only adjustable parameter. The contrast difference between the sphere and D_2O solvent is calculated as that between neat $M(D'E_{12})M$ liquid and D_2O, namely $(\rho_m - \rho_s)^2 = 0.0038 \times 10^{24}$ cm^2/Å^6. So the effect of D_2O penetration into the hydrophilic headgroup is not included here. Figure 7 shows that the calculated scattering function of hard-spheres fits the experimental SANS data well. The asymptotic oscillatory behavior shown by the calculated curves at q > 0.1 Å^{-1} can be damped by incorporating a small amount of polydispersity in size (R), i.e, R = 35 ± 2 Å (*26*). Therefore it is reasonable to conclude that up to at least 10 wt% $M(D'E_{12})M$ the solutions are composed of spherical micelles with the size of the micelles almost unvarying. In a previous study (*1*), small spherical micelles were also found in a 5 wt% solution by cryo-TEM.

The SAXS spectra of the solutions at midrange concentrations and above show a broad correlation maxima. The Bragg d-spacing (d_{max}) of the correlation maxima derived from SAXS spectra is shown as a function of surfactant concentration in Figure 8. As a general trend, d_{max} diminishes as $M(D'E_{12})M$ concentration rises. However, in the concentration range between 35 and 45 wt%, and between 70 and 80 wt% $M(D'E_{12})M$, d_{max} varies little and forms two composition-independent plateaus at about 68 Å and 55 Å respectively.

The neat $M(D'E_{12})M$ liquid was vacuum-dried at 25 millitorr at 80 °C for an hour in order to get rid of the residual amount of water absorbed from air, and then flame-sealed in the SAXS capillary within a few minutes. The SAXS spectrum of neat liquid $M(D'E_{12})M$ at 25 °C shown in Figure 9 exhibits a broad correlation maximum at d_{max} of 49.7 Å, suggesting that the liquid is locally microstructured even with water absent. Wide angle X-ray diffraction of the neat liquid reveals a broad intensity maximum at d-spacing around 4.6 Å, which likely results from the correlation among the neighboring molecules.

As the surfactant concentration rises from 10 to 35 wt%, either the size or the number density of micelles must increase. The rheology data presented above and the NMR data discussed below both point toward an increase in size of the microstructure. Here we analyze the implication of the scattering data. If small micelles are present and only the number density of micelles rises when the surfactant concentration increases, then the average inter-micellar distance, which is proportional to d_{max} (*27*), should decrease monotonically. However this conflicts with the d_{max} plateaus shown in Figure 8. Also, the spherical close-packing limit is about 70 volume % surfactant – if small micelles are present at concentrated concentrations around 70 %, the solution viscosity would be quite high, similar to that of cubic liquid crystal phase I_1 (*28*)

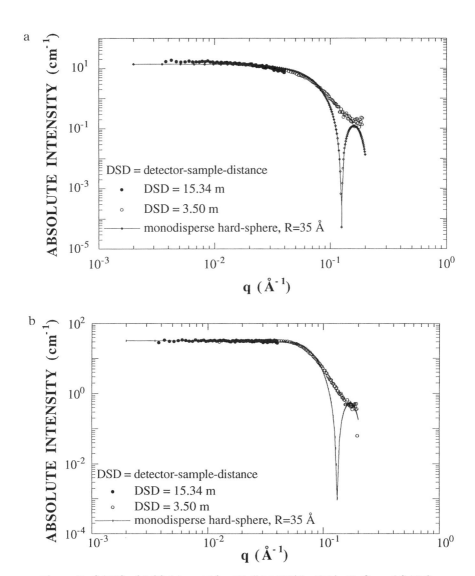

Figure 7. SANS of 1.92 (a) and 10 wt% (b) M(D'E$_{12}$)M in D$_2$O, and SANS modelled by the hard sphere model at the same compositions.

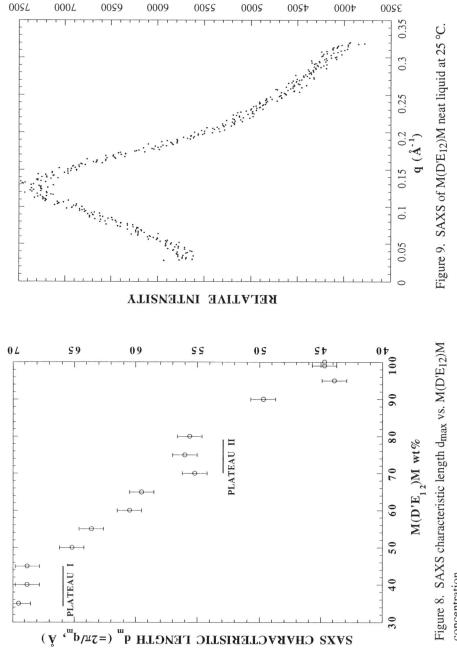

Figure 9. SAXS of M(D'E$_{12}$)M neat liquid at 25 °C.

Figure 8. SAXS characteristic length d$_{max}$ vs. M(D'E$_{12}$)M concentration.

which is composed of densely packed micelles. Although the viscosity of the solutions at midrange of concentrations is higher than those of the surfactant-rich and surfactant-lean solutions, it is much lower than has been reported for cubic liquid crystal phases (*29*). Therefore, we believe that larger aggregates, rather than small micelles, are present above 35 wt%.

In the space-filling model of midrange microemulsions (*30,31,38*), a characteristic length ξ which is proportional to d_{max} (*32*) can be expressed as

$$\xi = \frac{C \, \phi_o \, \phi_w}{C_s \, A} \tag{1}$$

C equals 5.82 in the Talmon-Prager model and 6 in the de Gennes-Taupin model, and therefore depends on the detailed model microstructure. ϕ_o and ϕ_w are the volume fractions of oil and water respectively, C_s the surfactant number-concentration, and A the surfactant head-group area.

If the hydrophobic domains of $M(D'E_{12})M$ solutions are liquid-like, then the microstructure of $M(D'E_{12})M$ at midrange concentrations is analogous to that of midrange microemulsion, and may have a characteristic length, ξ, proportional to (ϕ_s ϕ_w) / (C_s A) with ϕ_s as the volume fraction of the surfactant hydrophobic moiety. Because C_s is proportional to ϕ_s, the characteristic length can be further expressed as

$$\xi = C' \frac{\phi_w}{A} \tag{2}$$

where C' is also a constant that depends on detailed model microstructure.

It is evident from equation 2 that for ξ, or effectively d_{max}, to form a plateau region within a composition range, the surfactant head-group area, A, must be proportional to the volume fraction of water ϕ_w. This implies that the surfactant packing density per unit area (or the capacity, α, defined by Talmon *et al.* (*30,38*)) of surfactant monolayer or bilayer has to rise as surfactant concentration rises. The head-group area A and surfactant packing density, of course, relate directly to the curvature C of the surfactant monolayer and hence the microstructure (*33*). So we think that in the concentration range corresponding to the d_{max} plateaus, the curvature, and therefore the microstructure of the surfactant aggregates change significantly. In the composition range of the first plateau between 35 and 45 wt% $M(D'E_{12})M$, we suppose that spherical micelles transform to worm-like micelles (with comparable cross sectional diameter); while in that of the second plateau between 70 and 80 wt% $M(D'E_{12})M$, worm-like micelles transform to random bilayers. Although this deduction is consistent with the rheology results presented above and NMR results given below, it can not be definitively proven from these results alone.

For the low viscosity neat liquid $M(D'E_{12})M$, the surfactant volume fraction is too high to allow even densely packed spherical micelles. The SAXS correlation length d_{max} (49.7 Å) of the neat liquid is close to the 'extended' length of this surfactant which is 43.7 Å. We do not describe this length as 'all-trans' because, unlike the linear alkane case, the bond orientation that gives the most stretched-out or extended length for siloxane moieties is not all-trans. In a parallel study of the neat liquids of monodispersed alkyl polyoxyethylene surfactants (C_mE_n) (*34*), SAXS also yielded broad correlation maxima at d-spacings close to these surfactant's 'all-trans' length. Early studies (*35-37*) found similar results for some neat liquids of alcohols and fatty acids. Warren (*37*) argued that the alcohols and fatty acids in liquid state remain roughly in straight configuration and roughly parallel to each other locally.

The correlation peaks at the d-spacing close to a molecular stretched-out chain length must result from the segregation of the hydrophilic/hydrophobic portions of the molecules. We interpret this to mean that the neat liquids of $M(D'E_{12})M$, C_mE_n, alcohols and acids discussed above associate locally in such a way as to form alternating hydrophobic and hydrophilic domains with characteristic size of a molecule's extended chain length. This microstructure could be bicontinuous, or continuous in either hydrophobic or hydrophilic domains. The low viscosity of the neat liquids may suggest that the microstructure continuously fuses and breaks down – or effectively, there is significant structure fluctuation.

This model microstructure for neat liquid $M(D'E_{12})M$ is not fundamentally different from the distorted, warped bilayers proposed by Vonk *et al.* (*39,51*) for midrange bicontinuous microemulsions, except that water and oil are absent in either the hydrophobic or the hydrophilic domains.

Pulse Field Gradient Spin-Echo NMR (PFGSE NMR). NMR self-diffusion coefficients of $M(D'E_{12})M$ and H_2O, measured at 25 °C for the isotropic solutions at different concentrations, are shown in Figure 10. The surfactant self-diffusion coefficient, D_s, is the smallest at midrange of concentrations. The water self-diffusion coefficient, D_w, decays monotonically as the surfactant concentration rises. It lowers by roughly a factor of four as the surfactant concentration rises from 5 to around 50 wt%, and then gradually approaches a plateau as the concentration rises further.

An increase in either the number or the size of *small* surfactant aggregates such as micelles could explain the monotonic slowing-down of $M(D'E_{12})M$ self-diffusion as the surfactant concentration rises from 5 to 40 wt% (*40-44*). If this explanation holds correct also for the surfactant self-diffusion at higher concentration, D_s should be lowered further. However this is not consistent with the minimum D_s found at 40 wt% $M(D'E_{12})M$ and the subsequent rising of D_s at higher concentrations.

Therefore another self-diffusion mechanism other than micellar and monomer diffusion in continuous water media is required to explain the D_s at midrange and high surfactant concentrations. Suggested by the rheology and the SAXS results above, $M(D'E_{12})M$ above 40 wt% in water possibly forms *large* aggregates such as entangled, long worm-like micelles (*9*), or certain types of bicontinuous microstructures (*45*). The self-diffusion of such large surfactant aggregates is too slow to be detected by this technique. The diffusion represented by D_s would then have to be the *self-diffusion of surfactant molecules* in the surfactant-rich domains. This type of self-diffusion is strongly dependent on the shape and size of the surfactant microstructure (*44,46*), and provides us a valuable clue to the detailed nature of that microstructure. Such an analysis has been carried out in terms of a surfactant obstruction factor A_s, defined as the ratio of the effective (measured) surfactant self-diffusion coefficient D_S and an imagined D_S^0 which is free of obstruction (*46, 42*). The A_S of a few typical microstructures are listed below:

(1) For large bilayer aggregates such as large oblate micelles, bicontinuous sponge-like microstructure, and long-range stacked bilayers, $A_S \approx 2/3$;

(2) For aggregates comprised of cylinders such as entangled worm-like micelles or parallel cylindrical micelles, $A_S \approx 1/3$.

The above approach does not count the details of a microstructure, and the relation between obstruction factor and solution composition (*46*). But it may still serve to roughly establish bench marks to distinguish the surfactant microstructures that are locally comprised of cylinders from those of bilayers.

For isotropic solutions of different concentrations at 25 °C, the obstruction factor A_S is calculated simply by D_S/D_S^0 (*42*), where D_S^0 is the self-diffusion coefficient of the $M(D'E_{12})M$ neat liquid at 25 °C. Figure 11 shows a plot of the obstruction factor A_S *vs.* $M(D'E_{12})M$ concentration. This figure shows that the A_S rises from roughly 1/3 to 2/3 as $M(D'E_{12})M$ concentration rises from 25 to 85 wt%. A_S at concentration below 25 wt%, i.e. 5 and 10 wt% $M(D'E_{12})M$ means little because the surfactant forms small spherical micelles in this composition region (see SAXS results above). The progression pattern of A_S between 25 and 85 wt% is consistent with the microstructure evolving from being locally comprised of cylinders to being comprised of bilayers.

The lowering of D_W as surfactant concentration rises can be explained by the excluded volume effect (*40,41*). Figure 10 shows that D_W at surfactant concentrations above 70 wt% does not drop much in comparison with lower concentrations. Nevertheless, a drastic drop of D_W by roughly an order of magnitude is expected if *small* inverse micelle is formed with its core comprised of water. This is because in the core of an inverse micelle with characteristic size at around 43 Å, and strongly associated with the polyoxyethylene head-groups, water should diffuse at a rate roughly comparable with that of the surfactant molecules ($D_S \sim 10^{-11}$ m²/sec, see Figure 10). In fact, the D_W of the 90 wt% $M(D'E_{12})M$ solution is 3.2×10^{-10} m²/sec, which is about 30 times higher than the D_S. In other studies, it was found that for a 90 wt% $C_{12}E_8$ solution at 66 °C D_W is around 8×10^{-10} m²/s, which is more than eight times higher than the D_S (*40-43*). This coincidence indicates that for these surfactant-rich isotropic solutions, water diffuses much *faster* than is consistent with small inverse micelles. Thus the fact of Ds « Dw argues that in the $M(D'E_{12})M$-rich solutions small inverse micelles are not favored. In contrast the random bilayer network model proposed from the X-ray results above is consistent with these D_W values at surfactant-rich region, because water diffuses in this model structure in a network of hydrophilic domains.

In another point of view, some suggested that the water self-diffusion is greatly influenced by the local movement of polyoxyethylene head-group. Because the energy barrier to the internal rotation of polyoxyethylene chain is low, and hydrogen bonding among water molecules may be weaker in such a molten polyoxyethylene media, D_W is expected to be greater than D_S in the surfactant-rich isotropic liquids described above. Then small, discrete inverse micelles in this case seem to be possible. However, to explain the low viscosity of these surfactant-rich solutions, one has to argue that the

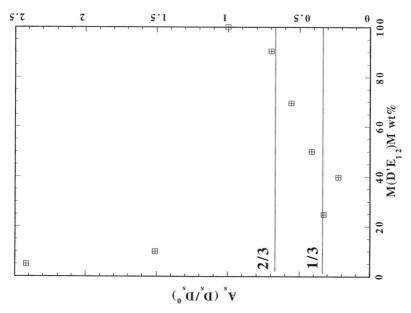

Figure 11. M(DE$_{12}$)M obstruction factor vs. M(D'E$_{12}$)M concentration.

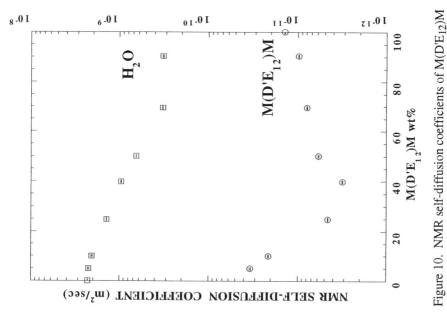

Figure 10. NMR self-diffusion coefficients of M(D'E$_{12}$)M and water in the solutions of different compositions vs. surfactant concentration.

inverse micelles continually fuse and break down. Under such high inverse micelle density when water is absent or only a few percent, the fusing and breaking-down of small inverse micelles actually suggest the tendency of forming *large* network-type microstructures. It may seem odd to describe the neat liquid surfactant or surfactant-rich solution as containing microstructures similar to bicontinuous sponge phase. Nevertheless, a bicontinuous sponge phase, L_3, has been found in extremely water-rich region of many surfactant-water binary systems (*1-3, 47*). Thus, by imagining an inversion of the hydrophilic and hydrophopic portions of the L_3 phase, we believe it is possible that the surfactant-rich solution (or L_2) has a bicontinuous microstructure similar to the L_3 except with much higher hydrophobic volume fraction. In summary, the NMR results provides two clues to the microstructures at midrange and surfactant-rich concentrations:

(1) In the midrange of concentrations between 40 and 80 wt% $M(D'E_{12})M$, large aggregates are formed, and a transformation is indicated from cylindrical aggregates to bilayer aggregates;

(2) at low water content and neat surfactant liquid, small, discrete inverse micelles are not favored, rather a randomly fluctuating bilayer network is more consistent with the NMR D_w data.

Discussion

Concentration Driven Microstructure Progression

Small Micelles at Low Surfactant Concentrations.   At 25 °C in solutions below 10 wt% $M(D'E_{12})M$ forms small spherical micelles with a radius close to 35 Å. The shear-viscosity of solutions in this composition range is independent of shear-rate up to 4500 rad/sec, and the limiting behavior of the dynamic modulus G' or G" fits the Maxwell model (*18*).

Other studies shows that some alkyl polyoxyethylene surfactants C_mE_n form micelles in surfactant-lean isotropic solutions. The size and shape of the micelles are influenced by temperature, concentration and surfactant molecular structure. $C_{12}E_8$, whose packing constraint parameter S is close to that of $M(D'E_{12})M$ (*1*), forms spherical micelles at surfactant concentrations below 20 wt% at 20 °C (*40-43*). This agrees with what is found here for $M(D'E_{12})M$ micelles.

Large Aggregates at Midrange Surfactant Concentrations and Above.   At surfactant concentrations between 35 wt% and 45 wt%, the SAXS and NMR results suggest that $M(D'E_{12})M$ forms large aggregates locally comprised of cylindrical micelles. The rheology results show that solutions of 40 to 50 wt% $M(D'E_{12})M$ have the highest zero-shear viscosity and viscoelasticity, which is most consistent with the presence of entangled long worm-like micelles.

The reason for the lowering of viscosity and lessening of the viscoelasticity at solution concentrations above 50 wt% requires an explanation. Recent works (*10,11,23*) indicate that increasing connection (or branch) points among the worm-like micelles effectively lowers the topological constraint due to entanglement, and causes viscosity and viscoelasticity reduction. Thus we suspect that the entangled long wormlike micelles found in the 50 wt% M(D'E$_{12}$)M solution may fuse with each other and begin to form a *multi-connected tubular network* as surfactant concentration increases. Meanwhile, the shape of the cross section of these long, polymer-like micelles could transfer from near-circle to near-elliptical with increasing surfactant concentration.

NMR and SAXS results presented above are consistent with a microstructural transition from cylinders to bilayers in the surfactant concentration between 70 and 80 wt%. The concept of a multi-connected tubular network also sheds light on this structural transition. The multi-connected tubular network, as demonstrated in Figure 12, is composed of surfactant monolayers forming tubules. This tubular network can transform to a random bilayer network by progressively changing monolayer to bilayer in the vicinity of tubule junctions. As the junction portion swells and the tubule portion shrinks, the whole microstructure can convert to a random bilayer network: see also Figure 12. Therefore, the multi-connected tubular network may serve as an intermediate state for the transition between cylindrical monolayers and sheet-like bilayers (*11,24*).

Based on the combination of the rheology, SAXS / SANS, and NMR results presented above, we can now reexamine the phase diagram of this surfactant in Figure 1. The various techniques all indicate structural changes taking place at compositions corresponding to the hexagonal and lamellar phase regions. So the specific transformations appear to be a sphere-to-cylinder change in the temperature range directly above the H$_1$ region, and a cylinder-to-bilayer change in the temperature range directly above the L$_\alpha$ region. Thus the local ordering of these liquid crystal phases may be preserved at higher temperature in isotropic liquid (or 'melts'), although long-range order is lost.

As the surfactant concentration rises above 80 wt%, we proposed that the microstructure is continuous and resembles the random bilayers present at 80 wt% solution. However, the microstructure in the solutions above 80 wt% M(D'E12)M must become progressively more disordered, and continue to fuse and break down, because the solutions become Newtonian and show the limiting behavior of G' and G" at low ω limit.

The microstructure progression as surfactant concentration raises from 0 wt% to 100 wt% at 25 °C is summarized below in Table I.

Table I. Microstructure Evolution as Function of Surfactant Concentration

Microstructure	Concentration (M(D'E$_{12}$)M wt%)	Evidence
spherical micelle	below 10 wt%	SANS, Cryo-TEM
elongated micelle	between 10 to 35 wt%	(speculation)
entangled long worm-like micelle	between 35 to 45 wt%	RM, NMR, SAXS
multi-connected tubular network	between 50 and 70 wt%	RM, SANS
bilayer network	between 70 and 80 wt%	SAXS, NMR, RM
disordered and fluctuating bilayers	above 80 wt%	SAXS
SANS = small angle neutron scattering;	NMR = pulse-gradient spin-echo NMR	
SAXS = small angle X-ray scattering;	RM = rheology measurement	

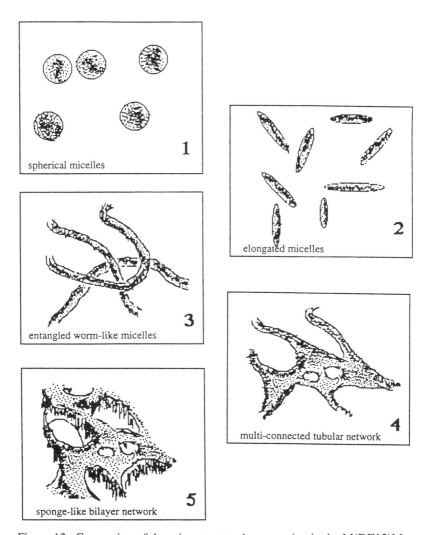

Figure 12. Conception of the microstructural progression in the M(DE12)M isotropic solutions as concentration ranges from surfactant-lean to surfactant-rich.

Microstructure Progression Patterns in Surfactant / Water Binary Systems of Isotropic Solutions

It should be noted that the pattern of microstructure progression in the isotropic phase across the composition range from surfactant-lean to surfactant-rich would be expected to be strongly dependent on temperature. An example is polyoxyethylene surfactant in water which often forms the sponge-like bicontinuous phase, L_3, from midrange concentration toward surfactant-lean region in a narrow island in the temperature-composition binary phase diagram (*1-3, 47,48*). At midrange of concentrations, the L_3 phase has always been found close to and sometimes even contiguous with the surfactant-rich isotropic phase L_2 in the surfactant/water phase diagram. How does the microstructure evolve from L_3 to L_2 in the path shown in Figure 13 ? Is the microstructure of L_2 related to that of L_3 ? We cannot find in the literature satisfactory answers to these questions.

As a general trend, mean curvature of nonionic surfactant bilayers or monolayers lowers when surfactant concentration or temperature rises. If this trend holds here, sponge-like bilayers of L_3 at low surfactant concentration should evolve to the fluctuating and disordered bilayers of L_2 at high surfactant concentration. Spherical and worm-like micelles should be absent in this evolution, and small, discrete inverse micelles are not favored as discussed above. In fact, we do not know what is the fundamental difference in the microstructures of L_3 and L_2.

A related problem is the microstructure progression in isotropic solutions at temperatures just below the clouding temperature T_c as the surfactant concentration rises. The NMR surfactant self-diffusion coefficients of $C_{12}E_8$ water solutions at 69.2 °C do not have the minimum at midrange of concentrations (*40-43*). As discussed above, without spherical and worm-like micelles involved in the microstructural progression we would not have found the minimum D_s presented in Figure 10. Tiddy (*49*) suggested that disc-like micelles were formed in some dilute isotropic C_mE_n solutions at temperatures just below T_c. As surfactant concentration rises, the size of the disc-like micelles may grow and the shape of them fluctuate. Thus in this microstructural evolution, spherical and worm-like micelles may also be absent.

Summary

At 25 °C, M(D'E$_{12}$)M in water forms a single isotropic phase at all concentrations. Both concentrated and dilute solutions are Newtonian, while the solutions at midrange of concentrations are viscoelastic and non-Newtonian. We interpret this behavior in the light of small-angle scattering and NMR self-diffusion results to indicate an evolution of microstructures with increasing surfactant concentration. As the surfactant concentration rises from 0 to 100 wt%, microstructures evolve from spherical micelles to locally-ordered random bilayers. Between these two extremes entangled or multi-connected worm-like micelles, and random bilayer networks are formed at midrange of concentrations.

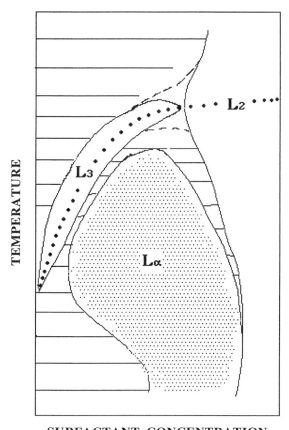

SURFACTANT CONCENTRATION

Figure 13. Typical binary phase diagram of a polyoxyethylene surfactant in water, which is composed of a sponge-like bicontinuous phase L_3; the microstructure evolution in the route marked by dark dots is discussed in the text.

Acknowledgments

This research was sponsored by the Center for Interfacial Engineering and the Department of Chemical Engineering and Materials Science, University of Minnesota, Minneapolis, MN. Dow Corning Corporation at Midland, MI., and National Institute of Standards and Technology at Gaithersburg, MD. also provided valuable technical and financial support. M. He thanks Dr. E. W. Kaler for his help in SANS modelling and numerical simulation.

Literature Cited

1.   He, M.; Hill, R. M.; Lin, Z.; Scriven, L. E.; Davis, H. T. *J. Phys. Chem.* **1993**, *97*, 8820.
2.   Mitchell, D. J.; Tiddy, G. J. T.; Waring, L.; Bostock, T.; MacDonald, M. P. *J. Chem. Soc. Faraday Trans. I.* **1983**, *79*, 975.
3.   Conroy, J. P.; Hall, C.; Leng; C. A.; Rendall, K.; Tiddy, G. J. T.; Walsh, J.; Lindblom, G. *Progr Colloid Polym Sci.* **1990**, *82*, 253.
4.   Degiorgio, V., "Nonionic Micelles" in *Physics of Amphiphiles: Micelles, Vesicles and Microemulsions;* Degiorgio, V.; Corti, M., Eds.; North-Holland: New York, NY, 1985.
5.   Strey, R.; Jahn, W.; Porte, G.; Bassereau, P. *Langmuir* **1990**, *6*, 1635.
6.   Xie, M.; Zhu, X.; Miller, W.; Bohlen, D. S.; Vinson, P. K.; Davis, H. T.; Scriven, L. E. "Generic Patterns in the Microstructure of Midrange Microemulsions" in*Organized Solutions* ; Friberg, S. E.; Lindman, B., Eds.; Surfactant Science Series 44; Marcel Dekker, Inc.: New York, NY, 1992.
7.   Chang, N. J.; Billman, J. F.; Licklider, R. A.; Kaler, E. W. "On the structure of Five-Component Microemulsions" in *Micellar Solutions and Microemulsions* ; Chen, S.-H.; Rajagopalan, R., Eds.; Springer-Verlag: New York, NY., 1990.
8.   Hoffmann, H.; Rehage, H. "Rheology of Surfactant Solutions" in*Surfactant Solutions* ; Zana, R., Ed.; Surfactant Science Series 22; Marcel Dekker Inc.: New York, NY., 1987.
9.   Clausen, T. M.; Vinson, P. K.; Minter, J. R.; Davis, H. T.; Talmon, Y.; Miller, W. G. *J. Phys. Chem.* **1992**, *96*, 474.
10. Khatory, A.; Lequeux, F.; Kern, F.; Candau, S. J. *Langmuir* **1993**, *9*, 1456.
11. Khatory, A.; Kern, F.; Lequeux, F.; Appell, J.; Porte, G.; Morie, N.; Ott, A.; Urbach, W. *Langmuir* **1993,** *9*, 933.
12. Lin, Z.; He, M.; Scriven, L. E.; Davis, H. T.; Snow, S. A. *J. Phys. Chem.* **1993**, *97*, 3571.
13. Hill, R. M. "Interactions between Siloxane Surfactants and Hydrocarbon Surfactants" in *Mixed Surfactant Systems* ; Holland, P. M.; Rubingh, D. N., Eds.; ACS symposium Series 501; American Chemical Society: Washington DC., 1992.
14. Saul, D.; Tiddy, G. J. T.; Wheeler, B. A.; Wheeler, P. A.; Willis, E. *J. Chem. Soc. Faraday Trans. I.* **1974**, *70*, 163.
15. Bodet, J.-F.; Bellare, J. R.; Davis, H. T.; Scriven, L. E.; Miller, W. G. *J. Phys. Chem.* **1988**, *92*, 1988.

16. Lindman, B.; Stilbs, P.; Mosley, M. E. *J. Colloid Interface Sci.* **1981**, *83*, 569.
17. Mulb, R. *J. Phys. Chem.* **1973**, *77*, 685.
18. Ferry, J. D. *Viscoelastic Properties of Polymers* ; 3rd edition, John Wiley and Sons, Inc.: New York, NY, 1980.
19. He, M. *Organized Solution of Siloxane Surfactants* ; Ph. D thesis, University of Minnesota, 1993.
20. Makhloufi, R.; Cressely, R. *Colloid Polym. Sci.* **1992**, *270*, 1035.
21. Snabre, P.; Porte, G. *Europhys. Lett.* **1990**, *13*, 641.
22. Drye, T. J.; Cates, M. E.; *J. Chem. Phys.* **1992**, *96*, 1367.
23. Appell, J.; Porte, G.; Khatory, A.; Kern, F.; Candau, S. J. *J. Phys. II* **1992**, *2*, 1045.
24. Vinches, C.; Coulon, C.; Roux, D. *J. Phys. II France* **1992**, *2*, 453.
25. Kaler, E. W. *J. Appl. Cryst.* **1988**, *21*, 729.
26. Vrij, A. J., *J. Chem. Phys.* **1979**, *71*, 3267.
27. Chen, S. -H.; Sheu, E. Y. "Interparticle Correlations in Concentrated Charged Colloidal Solutions – Theory and Experiment" in *Micellar Solutions and Microemulsions* ; Chen, S. -H.; Rajagopalan, R., Eds.; Springer-Verlag: New York, NY., 1990.
28. Bleasdale, T. A.; Tiddy, G. J. T. "surfactant Liquid Crystals" in *The structure, Dynamics and Equilibrium Properties of Colloidal Systems* ; Bloor, D. M.; Wyn-Jones, E., Eds.; Kluwer Academic Publishers: London, 1990.
29. Strom, P.; Anderson, D. M. *Langmuir* **1992**, *8*, 691.
30. Talmon, Y.; and Prager, S.; *J. Chem. Phys.* **1978**, *69*, 2984.
31. De Gennes, P. G.; Taupin, C. *J. Phys. Chem.* **1982**, *86*, 2294.
32. Zemb, T. "Scattering of Connected Networks" in *Neutron, X-ray and Light Scattering* ; Lindner, P.; Zemb, Th., Eds.; North-Holland: Amsterdam, 1991.
33. Israelachvili, J. N.; Mitchell, D. J.; Ninham, B. W. *J. Chem. Soc. Trans. II.* **1976**, *72*, 1525.
34. He, M.; Scriven, L. E.; Davis, H. T. "Thermotropic Phase Behavior of Polyoxyethylene Surfactant (C_mE_n)"; manuscript in preparation.
35. Stewart, G. W.; Morrow, R. M. *Phys. Rev.* **1927**, *30*, 232.
36. Morrow, R. M. *Phys. Rev.* **1928**, *31*, 10.
37. Chapman, D. *The Structure of Lipids* ; Mtthuen: 1965; and references therein.
38 Talmon, Y.; and Prager, S. *J. Chem. Phys.* **1982**, *76*, 1535.
39. Vonk, C. G.; Billman, J. F.; Kaler, E. W. *J. Chem. Phys.* **1988**, *88*, 3970.
40. Nilsson, P.-G.; Lindman, B. *J. Phys. Chem.* **1983**, *87*, 4756.
41. Nilsson, P.-G.; Wennerstrom, H.; Lindman, B. *J. Phys. Chem.* **1983**, *87*, 1377.
42. Nilsson, P.-G.; Lindman, B. *J. Phys. Chem.* **1984**, *88*, 4764.
43. Nilsson, P.-G.; Wennerstrom, H.; Lindman, B. *Chemica Scripta* **1985**, *25*, 67.
44. Lindblom , G.; Wennerstrom, H. *Biophysical Chemistry* **1977**, *6*, 167.
45. Scriven, L. E. "Equilibrium Bicontinuous Structures" in *Micellization, Solubilization, and Microemulsions* ; Mittal, K. L., ed.; Plenum Press: New York, NY., 1978; Vol 2.
46. Anderson, D. M.; Wennerstrom, H. *J. Phys. Chem.* **1990**, *94*, 8683.

47.   Hill, R. M.; He, M.; Davis, H. T.; Scriven, L. E. "Comparison of the Liquid Crystal Phase Behavior of Four Super-wetters" *Langmuir*, in press.
48.   Strey, R.; Schomacker, R.; Roux, D.; Nallet , F.; Olsson, U. *J. Chem. Soc. Faraday Trans.* **1990**, *86*, 2253.
49.   Tiddy, G. J. T. *Phys. Rep.* **1980**, *57*, 1.
50.   Ott, A.; Urbach, W.; Langevin, D.; Hoffmann, H. *Langmuir* **1992**, *8*, 345.
51.   Billman, J. F.; Kaler, E. W. *Langmuir* **1990**, *6*, 611.

RECEIVED August 8, 1994

Chapter 14

Rheology of Sucrose Ester Aqueous Systems

C. Gallegos, J. Muñoz, A. Guerrero, M. Berjano

Departamento de Ingeniería Química, Universidad de Sevilla, c/ Prof. García González, s/n 41012 Sevilla, Spain

The overall objective of this research was to study the influence that sucrose ester concentration and temperature exert on the viscous and viscoelastic behavior of micellar and lamellar liquid crystal phases occurring in aqueous systems containing a sucrose ester. Four sucrose esters prepared from different fatty acids were used. The relationship between Newtonian specific viscosity and temperature for sucrose laurate systems was adequately described by a modified Goodwin model. Entangled micellar systems and liquid crystalline phases showed non-Newtonian viscous behavior, although a limiting viscosity at low shear rates was always found. The flow curves at different temperatures were superposed by calculating a shift-factor from the Ellis model. The development of lamellar liquid crystals coincided with the occurrence of measurable normal stresses.

Sucrose esters are nonionic surfactants manufactured from sugar and vegetable oil. They are known to be useful for various applications (*1*). Because of their low toxicity, sucrose esters are widely used as food emulsifiers (*2-3*) and in the cosmetic (*4*) and pharmaceutical industries (*5*). The properties of sucrose esters depend dramatically on the degree of esterification of the sucrose (*6*). This influences their applications.

Sucrose esters may form different micellar and liquid crystalline phases in the presence of aqueous and nonaqueous solvents (*7*). The existence and extent of these phases depend on surfactant concentration, temperature and the hydrophilic-lipophilic balance (HLB) of the sucrose esters.

The different association structures that develop in the above mentioned binary systems significantly affect their rheological properties. This paper is concerned with the viscous and viscoelastic behavior of aqueous systems containing a high HLB sucrose ester (sucrose laurate, palmitate, oleate or stearate). Specific subjects that will be discussed are: a) Influence of sucrose ester concentration and temperature on the viscous behavior of globular micelles, entangled micelles and liquid crystalline phases; b) Linear viscoelastic behavior of an entangled micellar phase; and c) Non-linear viscoelasticity (normal stresses) of a lamellar liquid-crystalline phase.

Experimental

Sucrose esters (SE) were supplied by Mitsubishi Kasei Corporation (Japan) and used as received. The four sucrose esters investigated were those of lauric (L-1695), palmitic (P-1570), stearic (S-1570) and oleic (O-1570) acids. The first two digits correspond to the HLB value of the emulsifier. The second two digits refer to the minimum purity (%wt) of the fatty acid used. Aqueous systems of each surfactant up to 50%wt SE were prepared.

Rheological measurements were carried out in a Haake Rotovisco RV-20 rheometer (Germany), using a CV-20N measuring head. Steady flow measurements were conducted using different Mooney-Ewart sensor systems. Their radii ratios were always less than 1.1. Shear rate was varied between 0.1 and 300 s^{-1}. All the samples were presheared at 300 s^{-1} for 10 minutes.

Oscillatory flow measurements were carried out, at 50°C, only on samples containing sucrose stearate, up to a maximum concentration of 15%wt SE.

Normal stress measurements were conducted, at 50°C, using a cone and plate sensor system (4°, 45 mm) on samples with a sucrose stearate concentration ranging between 10 and 45%wt surfactant. All the samples were presheared at 300 s^{-1} for 10 minutes.

Results and Discussion

Non-entangled Micellar Systems. The experimental results demonstrate that aqueous binary systems, with sucrose laurate contents up to 45%wt, show Newtonian viscous behavior over the whole temperature range studied. This behavior has been related to the presence of globular micelles.

Although the absolute viscosity decreases as temperature increases, the specific viscosity, η_{sp} , is related to temperature according to equation 1 for surfactant concentrations up to 45%wt.

$$\ln\left[\frac{\eta_{sp}}{\eta_{sp}(T_{ref})}\right] = 10^3 k_1\left[\frac{1}{T} - \frac{1}{T_{ref}}\right] - 10^6 k_2\left[\frac{1}{T^2} - \frac{1}{T_{ref}^2}\right] \tag{1}$$

where $T_{ref} = 298K$.

This equation may be derived from the activated diffusive relaxation model, described by Goodwin (8), and has been successfully used to describe the temperature dependence of anionic surfactant solutions (9). This model establishes that, in disperse systems at a very low shear gradient, the relative viscosity is defined by a linear superposition of stresses, some due to hydrodynamic interactions and others to the structural relaxation of the uniform distribution of micelles in the quiescent micellar dispersion. This equation is:

$$\eta_{rel} = q_{rel} + J\exp(E/kT) \tag{2}$$

The contribution due to the hydrodynamic interactions is expressed as follows

$$q_{rel} = (1 - k_p \phi)^{-[\eta]k_p} \tag{3}$$

where k_p is the crowding factor, $[\eta]$ the intrinsic viscosity, and ϕ the volume fraction of the micelles. The parameter J is defined as

$$J = \hbar / 16\pi\eta_w a^3 \tag{4}$$

where $\hbar$ is Planck's constant divided by 2π, η_w is the dynamic viscosity of the continuous phase and a is the mean micellar radius. As previously reported for anionic surfactant solutions (9), if the activation energy for long-range diffusive motion is expressed by:

$$E = E_o - \frac{A}{T} \tag{5}$$

where the parameters E_o and A do not change with temperature within the range studied, then the Goodwin equation becomes:

$$\ln\left[\frac{\eta_{rel} - q_{rel}}{\eta_{rel}(T_{ref}) - q_{rel}}\right] = \left(\frac{E_o}{k}\right)\left[\frac{1}{T} - \frac{1}{T_{ref}}\right] - \left(\frac{A}{k}\right)\left[\frac{1}{T^2} - \frac{1}{T_{ref}^2}\right] \tag{6}$$

Since experimental viscosity values fit equation 1 fairly well, the value of q_{rel} may be expected to be unity. Then equation 6 yields equation 1, where $k_1 = (E_o / k) \cdot 10^{-3}$ and $k_2 = (A / k) \cdot 10^{-6}$. The values of these parameters are shown in Table I.

A dramatic increase in the values of these parameters is observed for concentrations higher than 25%wt sucrose laurate. This increase coincides with a significant change in the slope of the reduced viscosity as a function of the sucrose ester concentration (10). This change in slope has been related by other authors (11) to micelle overlapping.

The fit of specific viscosity values for different surfactant concentrations to equation 1 is shown in Figure 1. For sucrose ester concentrations of 20%wt or greater, a concentration dependent maximum is observed in the viscosity versus temperature curve.

This relationship between specific viscosity and temperature can be related to an

TABLE I. Values of Goodwin's Parameters for Sucrose Laurate Newtonian Aqueous Systems

Concentration (%wt)	$k_1(K)$	$k_2(K^2)$
10	11.7	1.91
15	11.3	1.82
20	13.5	2.12
25	12.2	1.92
30	16.8	2.60
35	19.4	2.98
40	27.1	4.09
45	28.4	4.22

increase in micellar size as temperature rises, as previously reported for other nonionic surfactants, such as polyoxyethylene alkyl ethers (*12*). This micellar growth is associated with the progressive breakdown of hydrogen bridge-bonds between water molecules and surfactant hydroxyl groups with temperature. The presence of a maximum specific viscosity, which appears at lower temperatures as surfactant concentration increases, results from a balance between two different processes: an increase in specific viscosity with temperature due to a larger micelle size, and a decrease in specific viscosity due to the fact that the increased thermal motion tends to disorient the particles and, consequently, weakens the strength of the micelle-micelle attractive interaction. An increase in surfactant concentration implies a progressive decrease in micellar hydration, which means that the preferential contribution of micellar growth to the specific viscosity is gradually displaced towards lower temperatures.

Entangled Micellar Systems and Lamellar Liquid-crystalline Phases. Subsequent increases in concentration produce a dramatic change in the viscous behavior of the sucrose laurate system. Similar complex viscous behavior may also be found using other SE (sucrose oleate, sucrose palmitate, sucrose stearate) but at much lower surfactant concentrations. For instance, systems containing 1 and 2%wt sucrose stearate show a power-law decrease in viscosity with shear rate, although a tendency to a constant viscosity region (η_∞) occurs at high shear rate. At higher sucrose stearate concentrations the flow curve shows constant viscosity values at low shear rates, up to a critical value that depends on the surfactant concentration. Above this critical shear rate a shear-thinning behavior is observed. These results fit the Carreau model A (*13*) fairly well, as is observed in Figure 2.

$$\frac{\eta}{\eta_o} = 1/[1 + (t_1\dot{\gamma})^2]^S \qquad (7)$$

where $\dot{\gamma}$ is the shear rate, η_o is the limiting viscosity at low shear rates and t_1 is a characteristic time of the system.

In Table II limiting viscosities η_o, critical shear rates $\dot{\gamma}_c$, and the values of the parameter S (related to the slope of the shear-thinning region) are shown as a function of sucrose stearate concentration. It can be observed that the limiting viscosity remains constant up to 10%wt SE. Between 10 and 35%wt SE, a nearly linear increase in the limiting viscosity occurs. Moreover, the critical shear rate, reciprocal of parameter t_1 in equation 5, rises sharply with surfactant concentration up to 10%wt. Then the rate of increase slows so that a smooth maximum is reached around 30-35%wt SE. Finally, no significant differences are observed in the parameter S with increasing surfactant concentration below 35%wt sucrose stearate. However, around that and at higher concentrations a decrease in S is noticed.

Oscillatory shear experiments were also carried out on these samples. First, strain sweep tests were carried out to search for the linear viscoelastic domain. There is a significant linear viscoelastic range for systems having less than 15%wt SE. Moreover, the sample containing 1%wt SE exhibited dramatically lower values of G' and G".

The results obtained in frequency sweep tests show that, at low frequency, G' and G" become proportional to the frequency squared and to the frequency,

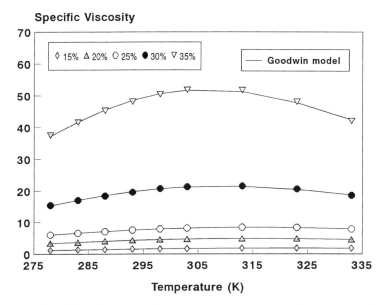

Figure 1. Specific viscosity values versus temperature for sucrose laurate aqueous solutions at different concentrations.

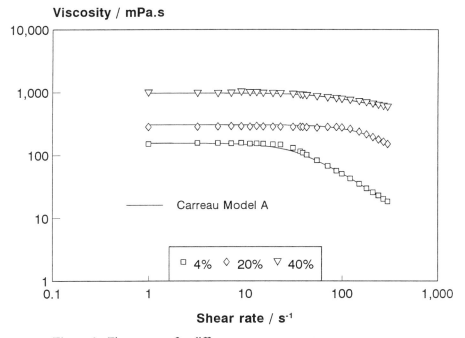

Figure 2. Flow curves for different sucrose stearate aqueous systems.

TABLE II. Values of the Parameters of Carreau's Model A as a Function of Sucrose Stearate Concentration

Concentration (%wt SE)	η_o (mPa·s)	$\dot{\gamma}_c$ (s^{-1})	S
3	154	20	0.49
4	154	45	0.46
5	156	76	0.51
10	138	172	0.48
15	261	179	0.50
20	289	185	0.52
25	425	192	0.55
30	598	211	0.57
35	788	190	0.43
40	946	27	0.18
45	1005	25	0.20

respectively. Both straight lines crossover at a frequency that decreases as SE concentration increases. From these results, it can be inferred that in the low frequency range, the dynamic data fit the Maxwell model. Nevertheless, the results obtained at higher frequencies with this rheometer do not allow us to extend the validity of the Maxwell model outside the above mentioned frequency range.

As is observed in Figure 3, the limiting viscosity values obtained from dynamic tests are slightly higher than those derived from steady-state flow curves. This may be explained by taking into account the different shear history of the samples. Moreover, although the limiting viscosity for the 2%wt SE system could not be reached at steady-state shear, it was possible to obtain the limiting viscosity from dynamic tests.

Moreover, if the steady shear and complex viscosities are compared in order to ascertain if the Cox-Merz rule is being followed (14), it is clear enough that the steady-state viscosity is progressively lower than the complex viscosity as shear rate and frequency increase. However, both viscosities tend to coincide at low shear rate and frequency.

Finally, normal stress measurements were carried out on these samples. As regards the influence of sucrose stearate concentration, Figure 4 shows values for the first normal stress difference (N_1) against surfactant concentration at several shear rates. Below 15 no data are shown due to the low values obtained. Regardless of the shear rate considered, N_1 exhibits similar values from 15 to 25%wt SE, reaches a peak around 30-35%wt and drops at higher SE concentrations.

The influence of the surfactant concentration on the rheological behavior of these systems may be discussed by taking into account the aggregation structure sequence of the sucrose stearate in water. On that basis, the whole concentration range can be divided in three regions:

a) The first one extends up to about 10%wt, although the system with the lowest concentration studied (1%wt) exhibits specific properties: thus, a limiting viscosity is not reached or estimated at the lowest shear rates available. Moreover, the steady-state viscosity and the dynamic viscoelastic functions are significantly lower than those at

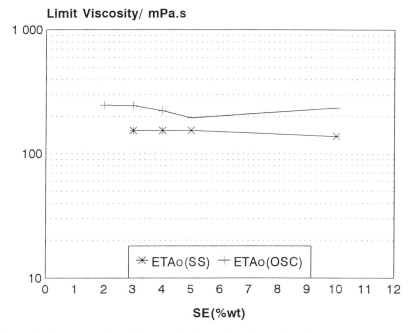

Figure 3. Steady and oscillatory limiting viscosity values for sucrose stearate micellar solutions.

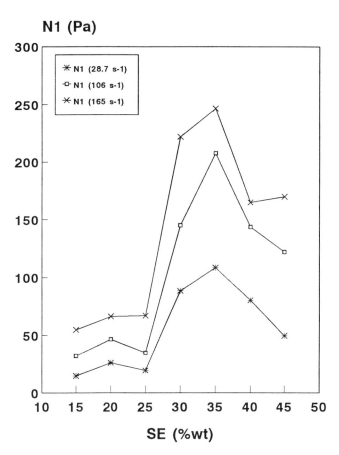

Figure 4. First normal stress difference values versus sucrose stearate concentration.

higher sucrose stearate concentrations. This behavior may be attributed to the presence of non-globular micelles.

The systems containing sucrose stearate concentrations between 2 and 10%wt SE exhibit the following characteristics: the limiting viscosity, G' and G" values are not significantly influenced by the SE concentration; and the critical shear rate for the onset of shear-thinning increases with SE concentration.

The dramatic increase in the dynamic moduli and in the steady-state viscosity produced when the surfactant concentration rises from 1 to 2 may be attributed to an increase in the number and size of micelles, which in turn gives rise to the formation of entanglements between non-globular micelles. The shear-resistance of these entanglements in the micellar region increases with SE concentration.

b) The beginning of the second composition range (10-35%wt sucrose stearate) is indicated by a dramatic decrease in the slope of the critical shear rate versus surfactant concentration. Moreover, the limiting viscosity increases linearly with surfactant concentration, although the parameter S does not change when compared to the previous range. This suggests that the structural breakdown mechanisms have not changed.

In relation to the dynamic viscoelastic response, the linear viscoelastic domain is dramatically reduced for concentrations higher than 15%wt. This may be attributed to a significant presence of lamellar liquid-crystalline portions in the dispersion (as demonstrated with polarizing microscopy), which influences the bulk structure.

c) The transition from dispersion to the lamellar liquid crystal region is readily monitored by the appearance of a maximum of N_1 at 35%wt. In fact, upon the formation of liquid crystals, the N_1 values at steady state decrease sharply. This maximum is due to the appearance of a dramatic time-dependent decrease in the normal stress values in the transition from the $L_1 + L_\alpha$ to the L_α region in the phase diagram. The fact that the maximum N_1 is far sharper as shear rate increases may be due to an increasing structural destruction in the liquid-crystal phase. The beginning of this range is also confirmed by the dramatic drop in both the critical shear rate and the parameter S of the Carreau model A. The latter reveals that the structural destruction mechanisms have changed when a fully developed monophasic lamellar liquid crystal is formed.

Influence of Temperature. Figure 5 shows apparent viscosity values versus shear rate for a system containing 15%wt sucrose palmitate at different temperatures. As may be observed, limiting viscosity values decrease with an increase in temperature, although crossovers among flow curves at different temperatures can be observed, because the critical shear rate for the appearance of shear-thinning viscous behavior increases with temperature for a specific concentration.

The increase in the critical shear rate with temperature may indicate that the shear-resistance of these networks is also dependent on temperature. This resistance has been related to micellar flexibility (*11*). Thus, entangled micelles are aligned or destroyed more easily by the application of shear as their flexibility increases. Furthermore, micelles are less flexible as temperature rises. This explains why $\dot{\gamma}_c$ increases with temperature.

Furthermore, from Figure 5 we may predict that the flow curves, obtained at different temperatures, can be superposed by calculating a shift-factor, a_T, from the experimental viscosities. The shift-factor values obtained from Bueche's theory (*15, 16*) are not useful to superpose the experimental flow curves. Consequently, the shift-factor for the superposition of the flow curves of these systems has been empirically calculated from the Ellis model,

$$a_T = \frac{\tau_{1/2A}\, \eta_{oB}}{\tau_{1/2B}\, \eta_{oA}} \qquad (8)$$

taking into account that "n" Ellis model parameter values are independent of temperature.

Figure 6 shows the master flow curve, using the shift-factor obtained from this equation, including the experimental flow curves at different temperatures and for sucrose palmitate concentrations ranging from 5 to 35%wt. This superposition was carried out by defining a dimensionless viscosity estimated from the ratio between the viscosity at a particular shear rate and the limiting viscosity for the same sucrose palmitate concentration and temperature. This dimensionless viscosity was plotted versus the product of shear rate and the shift-factor. Figure 6 shows that the superposition is excellent for all the concentrations included.

The temperature-dependence of the shift-factor is associated with phase transitions as temperature rises. Thus, for sucrose oleate/water systems two different regions may be observed when log a_T is plotted versus the reciprocal of the temperature, except at 35%wt surfactant. The relationship between both variables may be expressed by an Arrhenius-like equation for each of those regions:

$$a_T = Ae^{-E_a/RT} \qquad (9)$$

where E_a is the activation energy of the viscous flow. The activation energy values for both regions, at temperatures lower or higher than 25°C, at different sucrose oleate concentrations are shown in Table III.

The temperature-dependent activation energy values, obtained for the systems below 35%wt sucrose oleate, reflect the change in the aggregation structure at a temperature around 30°C. At this temperature weakly birefringent lamellar liquid crystal textures were observed by optical microscopy.

TABLE III. Activation Energy Values for Equation 9

Concentration (%wt)	E_A (kJ/mol)	
	T ≤ 25°C	T > 25°C
10	64.7	94.1
15	68.2	83.3
20	89.7	113
25	73.8	127
30	93.5	121
35	105	105

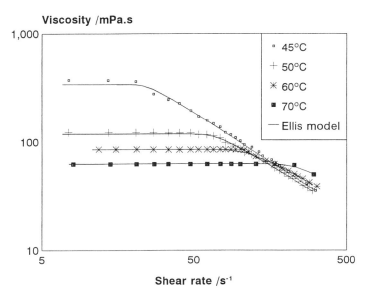

Figure 5. Flow curves of sucrose palmitate aqueous systems: Influence of temperature.

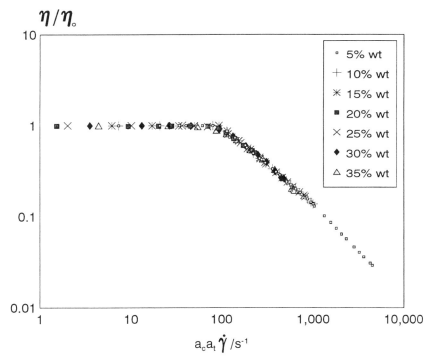

Figure 6. Master flow curve for sucrose palmitate aqueous systems.

Conclusions

From the experimental results we may conclude that, for Newtonian systems, the specific viscosity may be related to temperature using a modified Goodwin model.

An increase in surfactant concentration always results in the formation of entanglements between non-globular micelles. Further increases in SE concentration yield both higher critical shear rates for the onset of shear-thinning flow behavior and significant increases in the linear viscoelastic functions. Moreover, the shear-resistance of these micellar entanglements increase with temperature.

The presence of a lamellar liquid crystal phase reduces both the linear viscoelastic range and the critical shear rate for the onset of shear-thinning behavior and gives rise to measurable normal stresses.

The flow curves obtained at different temperatures can be superposed by calculating a shift-factor from the Ellis model. The variation of this shift-factor with temperature is related to the occurrence of temperature-dependent phase transitions.

Acknowledgment

This work is part of a research project sponsored by the CICYT, Spain (research project ALI90-0503). The authors gratefully acknowledge its financial support.

Literature Cited

1. Gallegos, C.; Berjano, M. *Alimentación, Equipos y Tecnología.* **1991**, *10*, 153-159.
2. Yuki, A.; Matsuda, K.; Nishimura, A. *J. Jpn. Oil Chem. Soc.* **1990**, *39*, 236-244.
3. Clark, D.C.; Wilde, P.J.; Wilson, D.R.; Wustrieck, R. *Food Hydrocolloids.* **1992**, *6*, 173-186.
4. Desay, N.B. *Cosmetics and Toiletries.* **1990**, *105*, 99-107.
5. Schenk, P.; Ausborn, M.; Bendas, F.; Nuhn, P.; Arndt, D.; Meyer, H.W. *J. Microencapsulation.* **1989**, *6*, 95-103.
6. Harrigan, K.A.; Breene, W.M. *Cereal Foods World.* **1989**, *34*, 261-267.
7. Herrington, T.M.; Sahi, S.S. *J. Am. Oil Chem. Soc.* **1988**, *65*, 1677-1681.
8. Goodwin, J.W. In *Surfactants*; Tadros, Th.F., Ed.; Academic Press: Orlando, 1984; pp. 133-151.
9. Guerrero, A.; Gallegos, C.; Flores, V.; Gómez, C. *J. Am. Oil Chem. Soc.* **1988**, *65*, 1964-1970.
10. Berjano, M.; Guerrero A.; Muñoz, J.; Gallegos, C. *Colloid Polym. Sci.* **1993**, *271*, 600-606.
11. Imae, T.; Sasaki, M.; Ikeda, S. *J. Colloid Interface Sci.* **1989**, *127*, 511-521.
12. Lindman, B.; Wennerström, H. *J. Phys. Chem.* **1991**, *95*, 6053-6054.
13. Carreau, P.J.; De Kee, D.; Daroux, M. *Can. J. Chem. Eng.* **1979**, *57*, 135-140.
14. Cox, W.P.; Merz, E.H. *J. Polym. Sci.* **1958**, *28*, 619-622.
15. Bueche, F. *J. Phys. Chem.* **1954**, *22*, 603-609.
16. Bueche, F. *J. Phys. Chem.* **1954**, *22*, 1570-1576.

RECEIVED July 8, 1994

Chapter 15

Rheological Study of Polycrystalline Lyotropic Mesophases in the Cesium n-Tetradecanoate–Water System

Peter K. Kilpatrick[1], Saad A. Khan[1], Akash Tayal[1], and John C. Blackburn[2]

[1]Department of Chemical Engineering, North Carolina State University, Raleigh, NC 27695
[2]Charleston Research Center, Westvaco, 5600 Virginia Avenue, North Charleston, SC 29406

The dynamic elastic (G') and viscous (G") moduli of polycrystalline samples of lyotropic mesophases comprised of cesium n-tetradecanoate (CsTD) and water were measured by rheometry using the oscillatory shear technique. The phases and microstructural morphologies studied were (1) a smectic phase comprised of a hexagonal array of surfactant rodlike aggregates, (2) a ribbon phase comprised of a monoclinic array of deformed rodlike aggregates, and (3) a lamellar phase comprised of alternating layers of water and surfactant bilayer. Frequency-dependent measurements of the moduli show hexagonal and biaxial ribbon phase samples to behave as elastic, gel-like materials whereas the lamellar phase samples behave as viscous fluids. Apparent yield stress values were constructed for all samples by plotting $G'\gamma_0$ versus % strain (γ_0) at a frequency of 1 rad/sec. Values of apparent yield stress varying from 100-4500 Pa were obtained. The variation in apparent yield stress with CsTD concentration was consistent with the variation in mesophase equilibria and with the corresponding dimensionality of order in the phase.

Lyotropic surfactant mesophases *(1-3)*, comprised of ionic single-chain surface-active agents, are important in a variety of physical phenomena and commercial industries, varying from the classical soap boilers pot to emulsification to phase transformations in biological membranes and its impact on phagocytosis and cellular locomotion. These mesophases are characterized by both supermolecular ordering due to

0097–6156/94/0578–0229$08.00/0

aggregation of the surfactant molecules into topologically ordered aggregates and by super-aggregate ordering into one- and two-dimensional crystals (so-called liquid crystals). The type and structure of mesophase formed depends strongly on the curvature adopted by the interface between the aggregate and the surrounding solvent (often water). This curvature is in turn strongly controlled by the relative inventory of surfactant and solvent; i.e. at low concentration of surfactant and water, the interface tends to curve strongly towards the interior of the surfactant aggregates and spherical aggregates or micelles are usually formed. At high concentration of surfactant, the interface tends to either be planar (as with lamellar mesophases) or curves towards the solvent phase. At intermediate concentrations, a variety of different phase structures and types can arise (4,5). These transitional phases and structures give rise to dramatically different rheological properties. In fact, one of the real challenges in the soap-making industry is the marked increase in viscosity and stiffness of hexagonal and viscous isotropic mesophases relative to the lamellar phase, even though the lamellar phase is higher in solute concentration.

In this paper, we report on rheological measurements performed on binary mixtures of cesium n-tetradecanoate (CsTD) and water. The cesium soap was selected in contrast to the sodium soap because it has a low Krafft temperature (ca. 285 K) and hence its liquid crystalline phases can be studied at ambient temperatures. Samples varying in composition from 54 -88 wt% CsTD were selected for study, corresponding to lyotropic mesophase structures varying from hexagonal smectic, to ribbon phase, to lamellar phase. The hexagonal phase is essentially a two-dimen-sional crystal comprised of uniaxial rodlike aggregates of surfactant oriented on a hexagonal close-packed lattice. The ribbon phase is also a two-dimensional crystal but the aggregates are biaxial and the unit cell is monoclinic. Finally, the lamellar phase is a simple one-dimensional crystal comprised of alternating layers of water and surfactant bilayers. All of these phases are polycrystalline, i.e. comprised of small, micron-sized domains with intervening grain boundaries. The goal of this study was to determine if the rheological properties as measured in the oscillatory shear experiment gave an indication of mesophase type and structure and trended with CsTD concentration in ways consistent with mesophase behavior.

Experimental

Materials. Cesium n-tetradecanoate was synthesized by reacting equimolar amounts of cesium hydroxide (99.98%, Carus Chemical Co., LaSalle, IL) and n-tetradecanoic acid (99.5%, gold label, Aldrich Fine Chemicals, Milwaukee, WI). Each reagent was dissolved in HPLC grade methanol (Fisher Scientific, Raleigh, NC) which was removed from the final reaction product by rotary evaporation. The purified surfactant was dried at 5 torr and 40°C and stored in a vacuum dessicator over calcium sulfate. The purity of the surfactant was determined to be > 98% by elemental analysis. The melting temperature of the dry surfactant was determined by DSC to be 290°C, in agreeement with the literature. The water used was distilled in glass and passed through a four-stage Millipore cartridge filtration system. Liquid crystalline samples were prepared by weighing out desired amounts of surfactant and water into glass vials, mixing thoroughly with an aluminum spatula, and sealing with a screw cap and teflon insert. After several days equilibration, these polycrystalline liquid crystalline samples were ready for rheological measurements.

Rheological Measurements. The rheological properties of these CsTD-water liquid crystalline samples were measured using dynamic oscillatory shear experiments. A Rheometrics Mechanical Spectrometer (RMS 800) with a parallel plate geometry was used to perform the experiments. In this measurement, a sinusoidal deformation $\gamma = \gamma_0 \sin \omega t$ was imposed on the sample at a fixed frequency ω and maximum strain amplitude γ_0. The resulting stress τ_{yx} has components given by (6,7)

$$\tau_{yx} = G' \gamma_0 \sin \omega t + G'' \gamma_0 \cos \omega t$$

In this equation, the stress component in-phase with the deformation defines the storage or elastic modulus G', whereas the stress component out-of-phase with the strain defines the loss or viscous modulus G''. The dependence of the loss and storage moduli on frequency of oscillation and strain amplitude are known to be sensitive indicators of complex microstructure (8-13).

Results and Discussion

Phase Behavior and Liquid Crystalline Phase Identification. The sequence of liquid crystalline phases in the CsTD-water system and the morphology and phase symmetry of the individual phases has been previously reported (5). The phases were identified and the phase boundaries delineated by a combination of optical polarizing microscopy and quadrupole nmr spectroscopy. Figure 1 shows a partial temperature-composition phase diagram in the region of interest (50-90 wt % CsTD). At ambient temperature (ca. 300 K) the primary phases identified are H_α, uniaxial rodlike aggregates of surfactant molecules packed on a smectic hexagonal lattice, R, biaxial surfactant aggregates of ellipsoidal cross section packed on a smectic monoclinic lattice, and L_α, lamellar bilayer sandwiches of surfactant molecules alternating with water. The lamellar phase is a one-dimensional smectic crystal, the hexagonal phase is a two-dimensional smectic crystal, and the ribbon phase is a two-dimensional smectic

crystal comprised of biaxial aggregates. The composition of samples on which rheological measurements were performed is shown in Figure 1 and provided in Table I.

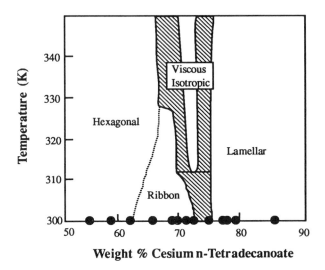

Figure 1: Phase diagram (temperature-weight %) of the CsTD-H$_2$O binary
system showing regimes of mesophase behavior. The filled circles represent the
locations at which rheological measurements were performed.

Rheological Measurements. Two types of rheological experiments were performed:
one in which the loss (G") and storage (G') moduli were measured as a function of the
frequency of oscillation ω and one in which these moduli were measured as a function
of strain amplitude γ_0. Prior to any measurements and after loading the samples in the
rheometer, the samples were subjected to an oscillatory shear at a frequency of 1 rad/s
and a strain amplitude of 1.5% for 40-60 minutes to generate a reproducible shear
history. Typical profiles of G' and G" as generated by the frequency sweep experiment
are shown in Figure 2 for representative hexagonal H$_\alpha$, biaxial ribbon R, and lamellar
L$_\alpha$ phase samples. All of these experiments were performed in the linear viscoelastic
regime at strain amplitudes γ_0 < 0.8%. The H$_\alpha$ and R samples exhibit gel-like
behavior, i.e. an elastic modulus G' which shows a weak dependence on oscillation
frequency and that is larger in magnitude than the viscous modulus G". The lamellar
L$_\alpha$ phase sample, however, exhibits rheological behavior intermediate between a
viscous fluid and a gel, i.e. the elastic and viscous moduli are very nearly equal in
magnitude and increase monotonically with increasing oscillation frequency. Typical
strain sweep experiments are shown in Figure 3 for the same samples at an oscillation

Table I: Compositions, Phase Morphology, and Apparent Yield Stresses of CsTD-Water Liquid Crystalline Samples

Composition (wt% CsTD)	Phase Morphology	Yield Stress (Pa)
54.3	Hexagonal (H_α)	1000
59.7	Hexagonal (H_α)	1180
63.3	Hexagonal (H_α)	1340
66.3	Biaxial Ribbon (R)	1290
69.4	Biaxial Ribbon (R)	2200
69.5	Biaxial Ribbon (R)	1470
71.1	Biaxial Ribbon (R)	4450
72.2	Biaxial Ribbon (R)	4220
74.7	Biphasic (R-L_α)	810
77.3	Lamellar (L_α)	825
77.8	Lamellar (L_α)	645
78.9	Lamellar (L_α)	268
84.9	Lamellar (L_α)	122

frequency of 1 rad/s. These experiments provide several pieces of information. First, the elastic moduli exhibit plateau values at very low strains (<1% strain), the linear viscoelastic regime. Second, G' is larger than G" by a factor of 5-7 in this regime with the H_α and R samples, indicating the presence of an elastic network. With the lamellar samples, the elastic and viscous moduli are of comparable magnitude at low strain, indicating the more viscous nature of these samples. This is consistent with the frequency-dependent measurements. Third, with the H_α and R samples, both moduli decrease with increasing strain and exhibit a crossover of the moduli at a strain of approxi-mately 30-50%. Such behavior suggests that the mesophase microstructure breaks down with increasing strain (*10,11*). However, the lamellar L_α phase exhibits crossover of the loss and elastic moduli at very low strain amplitude (< 1 %), if at all, suggestive of a very weak internal structure.

To explore whether or not these materials exhibit apparent yield stresses, the product of the elastic modulus and the strain were plotted versus the strain amplitude. A maximum in G'γ_0 is suggestive of a material yielding at that stress level for that particular frequency ω. With experiments conducted at low frequencies, this maximum corresponds to a 'true' yield stress (*14*). Figure 4 shows a plot of the apparent yield stress which is the maximum value of G'γ_0 as a function of increasing

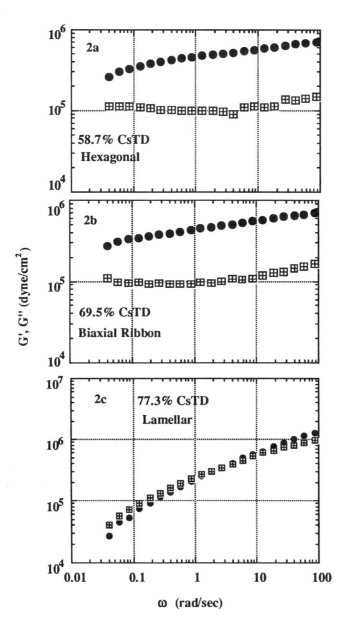

Figure 2: Elastic Modulus (G', filled circles) and Viscous
Modulus (G", squares with crosses) of (a) $H\alpha$ phase, (b) R
phase, and (c) $L\alpha$ phase as a function of oscillatory
frequency. The strain amplitude γ_0 was 1%.

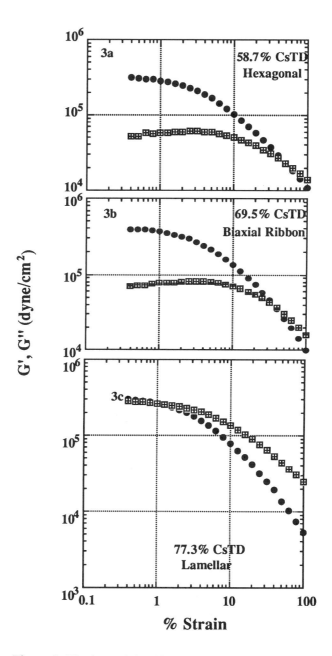

Figure 3: Elastic Modulus (G', filled circles) and Viscous Modulus (G", squares with crosses) of (a) Hα phase, (b) R phase, and (c) Lα phase as a function of strain amplitude. The oscillatory frequency ω was 1 rad/sec.

CsTD concentration of representative samples in the hexagonal, biaxial ribbon, and lamellar phases. All three types of materials exhibit an apparent yield stress, but with different magnitudes. The yield stresses of these liquid crystalline materials are dependent on material microstructure or phase morphology. As the CsTD concentration is increased from 55-66%, the morphology of the liquid crystalline phase changes from a smectic hexagonal phase to a biaxial ribbon phase (5), as shown in Figure 1. Although deuterium NMR is sufficiently sensitive to detect the onset of biaxial character in the aggregates, there is no indication from the rheology that a phase boundary has been crossed at 63-64 % CsTD. However, as concentration is increased further to 70% CsTD, a regime in which the axial ratio of the elliptical cross section of the surfactant aggregates is now appreciable, the apparent yield stress now begins to increase sharply, from 1400 to 4500 Pa. Clearly, the biaxial ribbon phase has a denser and/or stiffer microstructure than the hexagonal phase at concentrations of 70-73 % CsTD. This is consistent with an ellipsoidal deformation of the hexagonal phase spherocylinders to form the ribbon-shaped surfactant aggregates in the biaxial ribbon phase. The ribbon-shaped aggregates have a closer distance of approach than the cylinders in the hexagonal phase.

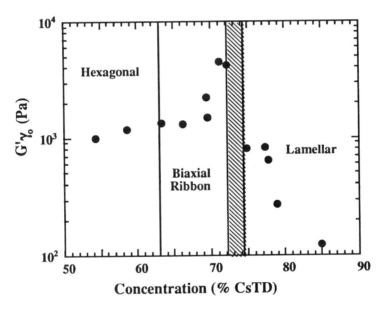

Figure 4: Stress (Pa) computed as the product of elastic modulus (G') and strain amplitude (γ_0) as a function of CsTD concentration. All experiments were performed at an oscillatory frequency of 1 rad/s.

As the CsTD concentration is increased further to 74.9%, corresponding to a biaxial sample of ribbon and lamellar phases coexisting, the yield stress falls from a maximum of 4450 Pa in the ribbon phase to 830 Pa (See Table I). Thus apparently in two-phase samples, the rheology is dominated by the lamellar phase. As surfactant concentration is further increased in the lamellar phase region, the apparent yield stress falls from 800 to 100 Pa, a factor of 50 lower than in the biaxial ribbon phase. This is dramatic and consistent with the change in microstructure from a two-dimensional biaxial liquid crystal to a one-dimensional uniaxial liquid crystal. The yield stresses and surfactant compositions of all of the samples studied are collected in Table I. It should be mentioned that a similar trend is observed if we plot the plateau (low strain) elastic modulus (obtained from a strain sweep experiment) as an increasing function of CsTD concentration.

Conclusions

The elastic and viscous moduli of lyotropic liquid crystalline phases comprised of cesium n-tetradecnoate and water have been measured by the oscillatory shear method as a function of oscillatory frequency and strain amplitude. The liquid crystalline phases studied are (1) a smectic phase of surfactant cylindrical aggregates organized on a two-dimensional hexagonal lattice (designated H_α), (2) a smectic phase of ribbon-shaped aggregates with biaxial symmetry organized on a two-dimensional monoclinic lattice (designated R), and (3) a smectic phase comprised of surfactant bilayers alternating with water in a lamellar geometry (designated L_α). The frequency dependence of the elastic (G') and viscous (G") moduli of H_α and R phases are indicative of gel-like or network-like microstructure, while the rheology of the L_α phase is intermediate between that of a viscous fluid and a gel. All three phases exhibited apparent yield stresses as gauged by maxima in plots of the stress ($G'\gamma_0$) as a function of strain amplitude (γ_0). These yield stresses varied from order 100-800 Pa in the L_α phase to 1000+ Pa in the H_α phase to 2000-4000 Pa in the R phase. This is consistent with an increase in both dimensionality and density of structure in these three mesophases.

Acknowledgments

This research was supported in part by grants from the National Science Foundation (CPE-8404599) and by discretionary grants from Colgate-Palmolive and the 3M Company.

Literature Cited

1. Ekwall, P. In *Advances in Liquid Crystals*, G. H. Brown, Ed., Academic, New York, NY, 1975; pp. 1-142.
2. Tiddy, G. J. T. *Physics Reports* **1980**, *57*, 1-46.
3. Tiddy, G. J. T.; Walsh, M. F. In *Aggregation Processes in Solution*, Wyn-Jones, E.; Gormally, J., Eds., Elsevier, Oxford; 1983; Chapter 7.
4. Kekicheff, P.; Cabane, B. *Journal Physique* **1987**, *48*, 1571-1583.
5. Blackburn, J. C.; Kilpatrick, P. K. *J. Colloid Interface Sci.* **1992**, *149*, 450-471.
6. Bird, R. B.; Armstrong, R. C.; Hassager, O. *Dynamics of Polymeric Liquids, Volume 1: Fluid Mechanics*, 2nd ed., Chapter 3, 10, Wiley, New York; 1987
7. Prud'homme, R. K. In *Electronic Materials Handbook, Volume 1: Packaging*, ASM International; 1990; pp. 838-853.
8. Winter, H. H. *Colloid Polymer Sci.* **1987**, *75*, 104-110.
9. Scanlan, J. C.; Winter, H. H. *Macromolecules* **1991**, *24*, 47-54.
10. Knoll, S.; Prud'homme, R. K. *Soc. Petroleum Eng. Preprints* **1987**, SPE 16283, 1-5.
11. Aranguren, M. I.; Mora, E.; DeGroot, J. V.; Macosko, C. W. *J. Rheology*, **1992**, *36*, 1165-1182.
12. Khan, S. A.; Plitz, I. M.; Frantz, R. A. *Rheologica Acta* **1992**, *31*, 151-160.
13. Khan, S. A.; Zoeller, N. J. *J. Rheology* **1993**, *37*, 1225-1235.
14. Yang, M. C.; Scriven, L. E.; Macosko, C. W. *J. Rheology* **1986**, *30*, 1015-1029.

RECEIVED July 8, 1994

Chapter 16

Optical Probe Studies of Surfactant Solutions

G. D. J. Phillies[1], J. Stott[1,2], K. Streletzky[1], S. Z. Ren[1,3], N. Sushkin[1], and C. Richardson[1]

[1]Department of Physics, Worchester Polytechnic Institute, Worcester, MA 01609
[2]Department of Physics, Case Western Reserve University, Cleveland, OH 44106
[3]Department of Physics, University of South Dakota, Vermillion, SD 57069

We discuss what optical probe diffusion, a specialized modification of quasi-elastic light scattering, can reveal about surfactant systems. In Triton X-100 solutions, optical probe diffusion has been used to examine the micelle radius, aggregation number, degree of hydration, and effect of temperature on micelle size. Preliminary results on probe diffusion in solutions of ionic surfactants (SDS, CTAB) are noted. Data on viscoelastic solutions shows that optical probe methods give information on structural relaxations in complex fluids.

Our objective is to discuss applications of optical probe methods to the study of surfactant systems. Mesoscopic probes have previously been used successfully to study polymer solutions (*1*), cell cytoplasm (*2*), and other complex fluids (*3*). Results on optical probes in polymer solutions were instrumental in rejecting (*1*) reptation/scaling models (*4,5*) of semidilute chain dynamics, reptation hitherto having been the canonical treatment of the problem. Optical probes in cell cytoplasm established (*2*) that the solvent viscosity in cytoplasm is close to the viscosity of pure water.

The next section examines probe diffusion and theories for the concentration dependence of the diffusion coefficients. A following section applies general results to Triton X-100 solutions; properties of near-minimal micelles are obtained. Results on probe diffusion in solutions of anionic and cationic surfactants are given. Finally, we treat mesoscopic probes in concentrated, viscoelastic solutions; here optical probes give information about long-time stress relaxations. A summary closes the paper.

Fundamental Principles Underlying the Technique.

Quasi-elastic light scattering is a non-invasive, non-destructive physical means for determining transport coefficients in liquids (*6*). In conventional experiments

0097–6156/94/0578–0239$08.00/0

an equilibrium sample of the fluid under study is illuminated with a laser; the intensity of the scattered light is measured. The scattering intensity $I(t)$ fluctuates around its average value; the time scale of the fluctuations is determined by the time required for scattering centers to diffuse a light-wavelength distance.

Instrumentally, $I(t)$ is used to compute the intensity-intensity temporal correlation function

$$C(\tau) \equiv \langle I(t)I(t+\tau) \rangle = \int_0^S dt I(t)I(t+\tau), \qquad (1)$$

S being the duration of the experiment. For a dilute suspension of scattering particles, $C(\tau)$ is composed of a static baseline B and a a time-dependent term $\exp(-2Dq^2\tau)$, where q is the scattering vector and D is the translational diffusion coefficient of the scatterers (7).

Here we are concerned with non-dilute solutions, in which direct forces (e.g., excluded volume, electrostatic, van der Waals) and hydrodynamic interactions between the diffusing scatterers substantially perturb the motion of the scatterers. In concentrated solutions, three physically distinct, unequal translational diffusion coefficients can be identified. The mutual diffusion coefficient D_m describes the relaxation of concentration gradients. The self diffusion coefficient D_s describes diffusion of a single scattering particle through a uniform background concentration of scatterers. The probe diffusion coefficient D_p describes motions of probe particles, typically of mesoscopic size, through solutions of a background (matrix) species.

Light scattering spectra of binary macromolecule solutions measure D_m of the scattering macromolecules. We are here concerned with optical probe experiments, which are performed on ternary systems containing a solvent, a matrix species (here, the matrix is the micelles and the free surfactant), and an intensely-scattering probe species (here, polystyrene latex spheres). The system is arranged so that light scattering spectroscopy can be used to monitor the diffusion of the probe species. In general, light scattering spectra of ternary systems contain two relaxations, which cannot be ascribed one-to-one to the two macrocomponents (8). However, *if the probe species is dilute*, and *if the probe species dominates the scattering spectrum*, light scattering spectra of probe:matrix:solvent systems are predicted to contain only one relaxation, determined by D_p, the self-diffusion coefficient of the probes through the unseen matrix (9). While the matrix has its own diffusion coefficient D_m, under probe conditions D_m does not contribute to $C(\tau)$.

The effect of direct and hydrodynamic interactions on D_m and D_p has an extremely extensive literature which we only summarize here. Direct interactions (e.g., hard-sphere, Debye-Huckel) are treated by standard references in statistical mechanics (10). Hydrodynamic interactions reflect solvent-mediated velocity-velocity couplings; the velocities of two nearby Brownian particles are partly correlated. Equivalently, the velocity $\mathbf{V}_a$ of a Brownian particle a is in part determined by the direct forces $\mathbf{F}_b$ placed on all the other Brownian particles, so that

$$\mathbf{V}_a = \sum_{b=1}^{N} \mu_{ab} \cdot \mathbf{F}_b, \qquad (2)$$

where μ_{ab} is the mobility tensor linking a and b ($a = b$ is allowed). The lowest-order term in μ_{aa} is the drag tensor $\mathbf{I}f^{-1}$, $\mathbf{I}$ being the identity tensor and f being the drag coefficient. The lowest-order term with $a \neq b$ is the Oseen tensor

$$\mu_{ab} = f^{-1}\frac{3}{4}\frac{a_m}{R}(\mathbf{I} + \hat{\mathbf{R}}\hat{\mathbf{R}}), \tag{3}$$

where a_m is the micelle radius, and R and $\hat{\mathbf{R}}$ are distance and unit vector from a to b. Higher-order, three-particle, and multi-size hydrodynamic interactions have been obtained by several authors (*11-14*).

The effect of direct and hydrodynamic interactions on diffusion coefficients was treated by Carter and Phillies (*15*) and Phillies (*16*), among others. For the mutual diffusion coefficient of a solution of hard spheres, Carter and Phillies (*15*) demonstrate

$$D_m = D_o(1 - 0.90\phi). \tag{4}$$

D_o is the infinite-dilution single-particle diffusion coefficient; $\phi = 4\pi a_m^3 N/(3V)$ is the solute volume fraction for N solute particles in a total volume V. For D_p of a probe of radius a_p in a matrix solution of spherical micelles, Phillies et al. (*17*) find

$$D_p = D_{p0}\left[1 - \left(\frac{15}{4}(\frac{a_p}{a_p + a_m}) - (\frac{5}{2} - \frac{11}{8}(\frac{a_m}{a_p})^2)(\frac{a_p}{a_p + a_m})^3\right)\phi\right], \tag{5}$$

D_{p0} being the diffusion coefficient of an isolated probe of radius a_p. Eq. 5 does not depend on the probe concentration because the probes are extremely dilute. For spherical probes, the zero-concentration diffusion coefficients D_{m0} and D_{p0} both follow the Stokes-Einstein equation

$$D_0 = \frac{k_BT}{6\pi\eta a}, \tag{6}$$

where k_B is Boltzmann's constant, T is the absolute temperature, η is the solvent viscosity, and a is the sphere (probe or micelle) radius.

Suppose one measures D_m and D_p experimentally at a series of low surfactant concentrations. Phenomenologically, low concentration data will follow

$$D_m = D_{m0}(1 + k_mc) \tag{7}$$

and

$$D_p = D_{p0}(1 + k_pc), \tag{8}$$

for a solution having surfactant mass concentration c. These equations involve four phenomenological quantities, namely two zero-concentration limits, D_{m0} and D_{p0}, and two limiting slopes, k_m and k_p. In contrast, the fundamental equations 4 and 5 are entirely determined by three fundamental parameters, namely a_m, a_p, and an aggregation number M which (in appropriate units) converts c to a micelle number density N/V. All of the other quantities in eqs. 4 and 5, e.g., D_{m0}, D_{p0}, and ϕ, can be expressed in terms of the three fundamental physical parameters. One thus has the highly desirable circumstance that the experimentally accessible phenomenological quantities overdetermine the significant physical parameters.

How does one proceed from data to physical parameters describing a micelle? If the surfactant in a micelle had its neat density, the volume of surfactant in a micelle could be computed directly from the aggregation number M. However, water of hydration can increase the micelle radius above the radius expected from the amount of surfactant in the micelle, so one cannot reliably calculate M from a_m. In our fits, a_m and a_p are determined primarily by the zero-concentration intercepts of the two diffusion coefficients, while the slopes k_m and k_p determine the volume fraction of micelles ϕ as a function of mass concentration c of surfactant. From ϕ and a_m the number concentration N/V of micelles can be determined; from N/V and c the average number of surfactant molecules in a micelle may be calculated. The difference between the volume of a nicelle and the volume of surfactant in the same micelle is the volume of solvent in the micelle; from the volume of solvent and the solvent density, the micelle hydration can be computed. The above steps are conceptual, and do not describe our fitting algorithm.

Our procedure below was to measure $D_m(c)$ and $D_p(c)$ and use a non-linear least-squares fit to determine a_m, a_p, and M simultaneously. The program did not determine D_{p0}, D_{m0}, k_p, or k_m directly. One could back calculate, e.g., D_{p0} from a_p, but we did not do so.

Nonionic Surfactant Solutions: Triton X-100

Triton X-100 (TX100) (p-(1,1,3,3-tetramethylbutyl)-phenoxy-poly(oxyethy-lene) glycol) is a commercially available nonionic surfactant. Its(18) critical micellar concentration is low, ≈ 0.6g/L. Physical studies of TX100 micelles have used NMR (19,20), scattering (20-22), and viscosimetric (21) methods.

A major objective of earlier works (19-22) was to determine the size and aggregation number of minimal TX100 micelles. Studies cited by Robson and Dennis (23) for our conditions reported aggregation numbers M in the range 100-155 monomers. Paradies et al. (22) interpreted light and x-ray scattering data to obtain a micelle radius $a_m \approx 49$Å. Kushner and Hubbard (21) found an intrinsic viscosity $[\eta] = 0.55 dL/g$ and inferred that their micelles were hydrated to the level of 1.2 g of water per g of TX100. Brown et al. (20) found that light scattering intensities followed hard sphere behavior for $c \leq 50$ g/L but deviated from this behavior at larger c.

We (24,25) measured mutual and probe diffusion coefficients in TX100 solutions. We covered a temperature range 15-55 C for TX100 concentrations up to 100 g/L. The probes were 670 Å (nominal diameter) polystyrene latex spheres. Probe studies in TX100:water were complicated by the not-quite- negligible scattering of the TX100 micelles. To separate micelle and probe scattering, spectra were fit both to a cumulant series and to a sum of two two-cumulant-exponentials, the two approaches agreeing at low c with the latter approach being superior at larger c.

Our results at 25 C are in the literature (24) and thus only sketched here. In summary, for $c < 40$ g/L D_m is linear in c; at the same temperature, D_p is linear for $c \leq 100$ g/L. We find $a_p = 360$Å, $a_m = 45$Å, and $M \approx 100$ monomers, corresponding to $65 \pm 5 \cdot 10^3$ Da of surfactant per micelle. If one assumes that the surfactant and the water in a micelle each have their neat-liquid densities, one calculates a hydration of 2.6 g of water per g of TX100.

This hydration level is quite high, higher than previous literature results (*21*). It should be emphasized that our measurements are purely hydrodynamic, so that our hydration value includes not only water bound thermodynamically to the micelle but also any water entrained by the rugose micelle surface. For example, a structure containing a ca. 30 $\mathring{A}$ core composed largely of surfactant and a ca. 15 $\mathring{A}$ outer layer containing a modest number of strands of surfactant and a great deal of water would be compatible with our findings.

To study the possibility of micelle growth, we added an extra parameter to the model, writing $a_m = a_{m0}(1+k'c)$. With an extra parameter, we obtain $a_p = 365\mathring{A}$, $a_m = 43.5\mathring{A}$, and 63 500 Da of surfactant per micelle, values quite close to those obtained above. The slope k' was -0.002 L/g, corresponding to a 10% shrinkage of the micelles over the observed concentration range. We believe that k' is zero to within experimental error.

The above pair of fits shows the accuracy of the analysis. Errors in single measurements of D_m and D_p are typically several percent; the scatter in independent estimates of probe and micelle radii have roughly the same error. M is inferred from the slopes $\partial D/\partial c$; determining a slope is more difficult than determining a diffusion coefficient, so the error in M is larger than the error in the probe and micelle radii.

Our results are now being extended to examine temperature behavior of TX100 micelles. As seen in Figure 1, D_{po} scales with temperature slightly more slowly than T/η, implying that the surfactant coating on each probe becomes thicker at large T. Figure 2 plots the slope k_p of eq. 8 as a function of solution temperature; k_p becomes substantially more negative with increasing T. Combining Fig. 2 with measurements of of $D_m(c)$ at different temperatures yields a complete quantitative interpretation of our results in terms of the temperature dependences of the micelle radius and degree of hydration. The micelle aggregation number increases between 15 and 50° C. These results will be reported elsewhere.

Ionic Surfactant Solutions

Our studies show that optical probe methods may be applied in solutions of ionic surfactants. Extensive data were obtained on diffusion by mesoscopic probes through solutions of cetyltrimethylammonium bromide (CTAB):NaBr and sodium dodecyl sulfate (SDS): NaCl. Relative to nonionic surfactants, solutions of ionic surfactants have an additional important variable, namely the ionic strength I of the solution, which is determined by the concentration of background electrolyte and monomeric surfactant. I enters both by altering probe-micelle direct interactions and by altering solution free energies, thereby changing micelle size and/or shape. At very large electrolyte concentrations (1.0 M), added salt may also obscure the interpretation of D_p by promoting aggregation of the probes, which are usually extremely stable in surfactant solutions.

Figure 3 shows D_p of 2400 $\mathring{A}$ polystyrene spheres in solutions of CTAB:Na Br:H_2O. We have not done a systematic study using different radius probes, but there appear to be fundamental advantages to using larger probes, notably that

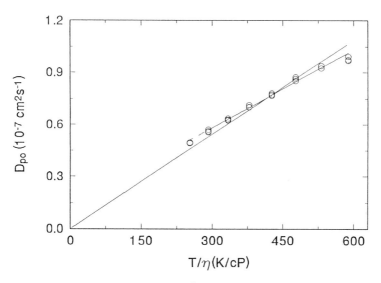

Figure 1. Diffusion coefficient of 670 $\mathring{A}$ diameter spheres in TX-100:water, in the limit of zero TX-100 concentration, as a function of T/η.

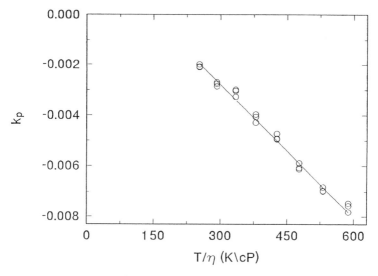

Figure 2. Linear slope k_p (eq. 8) of the concentration dependence of D_p for 670 $\mathring{A}$ diameter polystyrene latex probes in TX-100:water, plotted against solution temperature. k_p is independent of solvent viscosity; the temperature dependence seen here arises from changes in micelle radius or hydration.

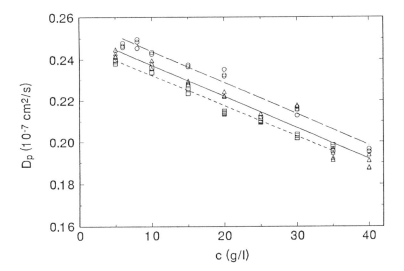

Figure 3. D_p of 2400 Å diameter polystyrene spheres in solutions of CTAB:NaBr:H$_2$O against CTAB concentration at NaBr concentrations of (squares) 0.01 M, (triangles) 0.05 M, and (circles) 0.20 M.

theoretical terms in a_m/a_p and $a_p/(a_p + a_m)$ become very nearly 0 or 1, respectively. As expected from theory, D_p falls with increasing surfactant concentration. D_p increases with increasing NaBr concentration, presumably because increasing the NaBr concentration increases the electrostatic screening, thereby reducing probe-micelle interactions. A quantitative interpretation of data such as that shown in Fig. 3 requires improvements in the above theoretical treatments, including the correct incorporation of electrostatic forces. In particular, standard Debye-Huckel theory (which gives the potential experienced near a macroion by a point charge in the solvent) has to be extended to describe interactions between two charged dielectric spheres of unequal radius. In order to get quantitative agreement with our experimental data, it may also be necessary to allow for effects arising from coupling of small ion dynamics to macroparticle motion (*26*).

Probes in Viscoelastic Liquids

Aqueous solutions of ionic surfactants often show intense viscoelastic properties, reasonably interpreted as stemming from the formation of entangled thread-like micellar arrays. Rehage and Hoffman (*27*) have shown that cetylpyridinium chloride:sodium salicylate solutions, allowed several days to equilibrate, show pronounced viscoelastic behavior. Brown et al. (*28*) report that CTAB:sodium naphthalenesulfonate solutions form threadlike micelles, whose spectra have the features characteristic of dilute and semidilute random-coil polymers; the solutions are also viscoelastic. Imae (*29*) studied static and dynamic light scattering, electrophoretic mobility, and spinnability (thread forming) characteristics of hexadecyltrimethylammonium salicylate(HTAS): sodium salicylate solutions. HTAS by itself forms short, rodlike micelles; addition of sodium salicylate provokes large changes in light scattering and viscoelastic properties.

We obtained preliminary spectra of polystyrene spheres in concentrated cetyl-pyridinium chloride:sodium salicylate solutions. Probe diffusion in a highly viscoelastic medium results in strongly non-exponential spectra. Repeated measurements spread over several days indicate that we have stable probes in fully equilibrated solutions.

Is probe diffusion sensitive to viscoelastic effects? As an example of the sorts of spectra that can be obtained from simpler viscoelastic systems, we present data on probes in dilute and concentrated long-chain polymer solutions. Figure 4a shows the time correlation function for 670 Å diameter probes in 1 g/L of 1.15 MDa hydroxypropylcellulose(HPC): water at 25 C, with 0.2 vol% TX-100 added to stabilize the probes. The solution is somewhat more viscous ($\eta = 2.72$mPaS) than water; $C(\tau)$ is a Williams-Watts function

$$C(\tau) = A_o \exp(-\Theta\tau^\beta) \qquad (9)$$

with $\beta = 0.9$, $\Theta = 790$s$^{-\beta}$. Figure 4b shows the same probes, now dispersed in 35 g/L of 300 kDa HPC: water. This solution is highly viscous ($\eta \approx 3,000$cP). At short times, $C(\tau)$ is visibly a Williams-Watts function. At $\tau \approx 2$ s, $C(\tau)$ changes to a power law, seen as straight lines on the log-log plots. Power-law decays are symptomatic of mode-mode coupling effects, and reflect the locally viscoelastic nature of the solution's response to a diffusing probe.

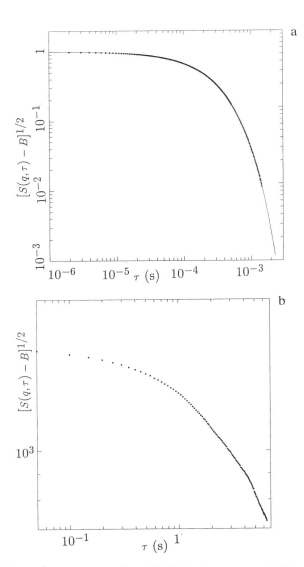

Figure 4. (a) 670 Å probes in 1g/L 1.15 MDa hydroxypropylcellulose:water, showing Williams-Watts relaxation of the normalized field correlation function. (b) 670 Å probes in 35 g/L 300 kDa hydroxypropylcellulose:water, showing long-time power-law decays (straight-line regions) of the field correlation function.

Conclusions

In the above, we discussed optical probe diffusion, a variant on conventional light scattering methods in which one observes the diffusion of mesoscopic probe particles through a complex fluid. The properties of the fluid are inferred from the effects of the fluid on the motion of the probe. In relatively dilute solutions of near-spherical micelles, modern hydrodynamic theory (*15,16*) predicts D_p in terms of probe and micelle radii and micelle concentration and hydration.

Detailed results were given for probes in near-dilute aqueous Triton X-100 solutions. From the linear concentration dependences of D_p and the micelle diffusion coefficient D_m, we were able to infer micelle radii, aggregation numbers, and levels of hydration. Our values for a_m and N are in good agreement with the literature. Preliminary data were given for probes in solutions of ionic surfactants. Here experiment is slightly ahead of theory; we are now extending the hard-sphere models used to analyse TX100 solution data to include effects important in aqueous solutions.

Under some conditions, surfactant mixtures form threadlike entangling micelles whose mechanical properties resemble the properties of polymer solutions. Our probes are physically stable under these conditions. Experiments on probes in concentrated polymer solutions reveal that optical probe spectra contain non-exponential phenomena and memory-function effects, e.g. power-law decays, related to the rheological properties of these complex systems.

In solutions, surfactants self-assemble into a wide variety of structures and phases, including spherical micelles, extended micelles, vesicles, thready viscoelastic systems, liquid crystals, microemulsions, and lamellar phases. The preliminary studies described above indicate that optical probe diffusion provides a novel hydrodynamic technique for studying these complex fluids.

Acknowledgments

The partial support of this work by the National Science Foundation under Grant DMR91-15637 is gratefully acknowledged.

Literature Cited

1. Phillies, G. D. J., *J. Phys. Chem.* **1989**, *93*, 5029.
2. Luby-Phelps, K.; Mujumdar, K.; Mujumdar, R.B.; Ernst, L.A.; Galbraith, W.; Waggoner, A.S. *Biophysical Journal* **1993**, *65*, 236.
3. (a) Phillies, G. D. J. *J. Phys. Chem.* **1981**, *85*, 2838; (b) Kiyachenko, Y. F.; Litvinov, Y. I. *Pis'ma Zh. Eksp. Teor. Fiz.* **1985**, *42*, 215; (c) Wiltzius, P.; van Saarloos, W. *J. Chem. Phys.* **1991**, *94*, 5061; (d) Phillies, G. D. J.; Clomenil, D. *J. Phys. Chem.* **1992**, *96*, 4196.
4. deGennes, P.G. *Scaling Concepts in Polymer Physics*, Cornell U. P., Ithaca, N. Y. (1979).
5. Doi, M.; Edwards, S. F. *The Theory of Polymer Dynamics*, Clarendon Press, Cambridge (1986).
6. Phillies, G. D. J. *Analytical Chemistry* **1990**, *62*, 1049A.
7. Pecora, R., Ed. *Dynamic Light Scattering*, Plenum, New York, NY (1985).
8. Phillies, G. D. J. *J. Chem. Phys.* **1974**, *60*, 983.

9.   Ullmann, K.; Ullmann, G. S.; Phillies, G. D. J. *J. Colloid Interface Sci.*, **1985**, *105*, 315.

10.   McQuarrie, D.A. *Statistical Mechanics*, Harper&Rowe, New York, NY (1976).

11.   Kynch, G.J. *J. Fluid Mech.* **(1959)**, *5*, 193.

12.   Batchelor, G. K. *J. Fluid Mech.* **(1976)**, *74*, 1.

13.   Phillies, G.D.J. *J. Chem. Phys.* **(1982)**, *77*, 2623.

14.   Mazur, P.; van Saarloos, W. *Physica A* **(1982)**, *115*, 21.

15.   Carter, J.M.; Phillies, G.D.J. *J. Phys. Chem.* **1985**, *89*, 5118.

16.   Phillies, G.D.J. *J. Colloid Interface Sci.* **1987**, *119*, 518.

17.   Phillies, G. D. J.; Stott, J., Ren, S. Z., *Journal of Physical Chemistry*, **1993**, *97*, 11563.

18.   Gonick, E.; McBain, J. *J. Am. Chem. Soc.* **1947**, *69*, 334.

19.   Sadaghiani, A. S.; Khan, A. *Langmuir*, **1991**, *7*, 898.

20.   Brown, W.; Rymden, R.; van Stam, J.; Almgren, M.; Svensk, G.; *J. Phys. Chem.*, **1989**, *93*, 2512.

21.   Kushner, L. M.; Hubbard, W.D. *J. Phys. Chem.* **1954**, *58*, 1163.

22.   Paradies, H. H.; *J. Phys. Chem.*, **1980**, *84*, 599.

23.   Robson, R. J.; Dennis, E.A. *J. Phys. Chem.* **1977**, *81*, 1075.

24.   Phillies, G. D. J.; Stott, J.; Ren, S.Z. *J. Phys. Chem.* **1993**, *97*, 11563.

25.   Streletzky, K.; Phillies, G. D. J., in preparation.

26.   Arauz-Lara, J. L., Medina-Noyola, M. *J. Phys. A: Math. Gen.* **1986** *19*, L117.

27.   Rehage, H., Hoffmann, H. *J. Phys. Chem.* **1988** *92*, 4712.

28.   Brown, W., Johannson, K., Almgren, M. *J. Phys. Chem.* **1989** *93*, 5888.

29.   Imae, T. *J. Phys. Chem.* **1990** *94*, 5953.

RECEIVED July 8, 1994

Chapter 17

Flexibility of Cetyltrimethylammonium 3,5-Dichlorobenzoate Micelles

P. D. Butler[1], L. J. Magid[1], and J. B. Hayter[2]

[1]Department of Chemistry, University of Tennessee,
Knoxville, TN 37996–1600
[2]Solid State Division, Oak Ridge National Laboratory,
Oak Ridge, TN 37831

While aqueous cetyltrimethylammonium 2,6-dichlorobenzoate (abbreviated CTA2,6ClBz) forms Newtonian fluids of spherical micelles, the CTA3,5ClBz analog exhibits viscoelastic behavior even at very low concentrations (three to four millimolar). In this brief report we present light scattering data for the CTA3,5ClBz that indicate a persistence length on the order of 500Å. We also discuss a novel use of the bending rod plot (I(Q)*Q vs. Q) as an important qualitative tool for examining the data prior to applying quantitative techniques.

CTAX (cetyltrimethylammonium) surfactants having certain aromatic counterions (X) produce viscoelastic solutions at quite low concentrations (on the order of 1 or 2 wt %) in water (1) . These systems are interesting technologically as well as from the perspective of basic science.

The CTA salicylate (o-hydroxybenzoate) is the most extensively studied of these surfactants (2-4). When the counterion is a mono- or dichlorobenzoate, para and/or meta substitution produces viscoelastic solutions, while CTA2ClBz and CTA2,6ClBz produce Newtonian aqueous solutions containing globular micelles. We present data here on the CTA 3,5-dichlorobenzoate system, where disubstitution has been used to enhance the effect (onset of viscoelasticity is well below 0.5%, with a cmc of 0.1mM).

Small angle neutron scattering (SANS)(5,6) and cryo-TEM (7) have been used to look at this system. Cryo-TEM pictures show that the solutions do indeed contain very long threadlike micelles that have considerable flexibility. The micelles are too long to make it possible to determine micellar lengths from micrographs and flexibility is hard to quantify. The radius of the micelles obtained by cryo-TEM is 25Å ± 5Å which is quite consistent with the SANS estimation of 23Å ± 3Å. Some estimates of length and persistence length from SANS have been attempted, but their validity is dubious at best. We have used light scattering (LS),

0097–6156/94/0578–0250$08.00/0

too often neglected if not deliberately ignored aspect of data reduction).

Figure 2 is a generic BR plot. The height of the peak between region I and region II relative to the height of the plateau of region III is directly related to the number of persistence lengths (or the ratio of the micelle length to the persistence length). If that number is small enough (less than one for example) the peak disappears altogether. The extent of the plateau region depends on the ratio of the persistence length to the radius of the micelle and would disappear if the persistence length were of the order of the radius. Thus one can judge qualitatively from such plots the extent of any flexibility and whether a simple rigid rod fit might be used on the data. One can of course also perform the analysis on such a plot and obtain the number of persistence lengths quantitatively.

Equally important is the fact that the Q axis on the generic plot has no scale, and is effectively infinite both in resolution and extent. In practice of course, any experiment has a very definite Q range and resolution. Therefore experiments can only see a portion of the full curve through the Q window. By varying techniques one can slide the window somewhat, but more importantly, as the parameters of the specific system change (length, persistence length, and radius) that portion of the curve visible in the window can shift dramatically.

Most of the BR plots have an upturn at high Q. This is due to the fact that reflection corrections necessitated by the large asymmetries involved are important and that at high angles the corrections could not be made. Upturns at low Q are undoubtedly due to the problems with sample filtration. The trends exhibited in these plots are nonetheless very clear. At a concentration of 0.3mM (roughly three times the CMC but well below the overlap concentration c^* of 2.5mM), the plot is a monotonically increasing function of Q. Thus only region I of Figure 2 is visible through the LS Q window indicating the rods are too short for this Q range to see any flexibility. Whether these are rigid rods or semiflexible rods cannot be determined from these data. In principle SANS could answer the question since its window effectively starts where the LS window ends and goes far enough to cover region IV (the rod radius). Unfortunately, the concentration is so low that such an experiment is impossible with any presently available neutron source and will have to await the construction of a next generation source such as the Advanced Neutron Source (ANS).

By 1mM (still well below c^*) the high Q portion of the curve is showing a marked turnover indicating the micelles have grown large enough for the finite size effects not to dominate completely at high Q any more. It is still not possible however to ascertain whether the rods are flexible, with the turnover being the top of the peak between region I and II, or whether they are rigid and the turnover is simply region I going smoothly into region III without any region II. Again SANS could in principle answer that question, but the concentration is still too low for presently available spectrometers.

At 2mM, still a bit below c^*, the plot looks almost like a mirror image of the one for the 0.3mM data: it is a monotonically decreasing function of Q. This indicates that the rods have grown so large that region I has moved completely to the left of the Q window. It is worth noting here that the implication of this is that while one can readily construct a Debye plot and extract an apparent R_g from these data, the size thus obtained must be considered completely unreliable as the data

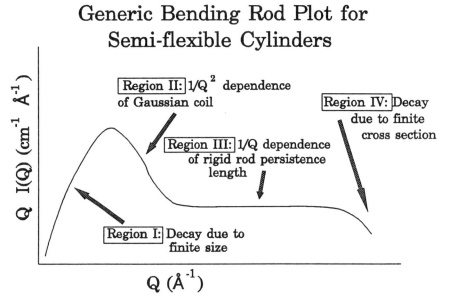

Figure 2.   Sketch of a generic BR plot delineating the various regions of interest.

are clearly not sensitive to finite size. Another piece of information comes from the fact that such a curve can only be interpreted as region II, and thus the micelles must be measurably flexible.

At 3.7mM, definitely above overlap, there is now a distinct turnover at low Q. This could be interpreted as a shrinking of the micelles causing region I to move right, back into the Q window. However, using the polymer analogy first proposed by Candau and Zana (*15*), this is a semi-dilute system and as de Gennes (*16*) has pointed out, the "finite" size in such a system is not the length of the polymer (or micelle), but the mesh size of the entangled network which will of course decrease with increasing concentration. In contrast to the 0.3mM case the main portion of the visible curve is definitely region II and hence the flexibility is still clear and presumably measurable despite the entanglements.

Finally, at 20mM, Region I is again prominent while region II seems to be disappearing into region III indicating that the mesh size is becoming of the order of the persistence length. In this case the persistence length can no longer be measured. Again Region I here represents a mesh size and not the micellar length.

Quantitative Estimates of Persistence Length. Unfortunately, there is no exact form factor for semiflexible rods, though several approximate ones exist. There also exist several methods besides a full fit of the data that yield a persistence length. Each approach has its own limitations. We have chosen four that are each likely to be applicable to at least a portion of our data. The first method was presented by ter Meer and Burchard (*17*) and is a Gaussian coil approach. The persistence length l_p is obtained from the expression:

$$l_p = 3 \frac{R_{g_{app}}^2}{M_{w_{app}}} M_L \tag{1}$$

where M_L is the mass per unit length of the micelle. Assuming that M_L does not change with length, we can calculate (*12*) a value from SANS data (*18*). We use a value of 1.6×10^{-13} g/cm. The second method is an application of the Benoit-Doty formula (*19*):

$$R_g^2 = L_c^2 \left[\frac{1}{3n} - \frac{1}{n^2} + \frac{2}{n^4}(n-1+e^{-n}) \right] + \frac{d^2}{8} \tag{2}$$

where $n = L_c/l_p$, $d = 2R_c$, L_c is the contour length of the micelle, and R_c is the cross sectional radius. For purposes of this calculation L_c was computed using M_w (*20*). The third method uses the ratio of peak height to plateau height in a BR plot as proposed by Denkinger et al (*12*), in which one uses the ratio to read off $n = L_c/l_p$ from the appropriate plot in their paper. In order to obtain the ratio, the plateau height of Region III was calculated from M_L (*12*). Finally the fourth and last method is to fit the data to the Peterlin form factor (*21*), which essentially treats rigidity as a perturbation to the Gaussian coil form factor.

The results from these various approaches are tabulated in Table I. The validity of each method is limited to some portion of the concentration range studied; therefore for a particular method, results are reported only for concentrations for which it is valid. We note that there is remarkable agreement between the values calculated from the different methods. The persistence length appears to actually increase rapidly from roughly 300Å to about 500Å where it remains until the concentration becomes too high for any of these methods to be applicable. The scatter in the fitted persistence length around its apparent equilibrium value leads to a relatively large error bar of 100Å. As regards the possible stiffening of the rods, we note that the BR plots of the lowest concentrations indicate that the data are rather insensitive to any flexibility, and the apparent increase in l_p must therefore be considered suspect. At the very least the error on the low concentration l_p must be as large or larger than that for the equilibrium value which, in itself, means the difference in persistence lengths is not significant. We must conclude therefore that, while the data show some evidence of an increase in l_p, they cannot confirm it.

Table I. l_p values (in Å) obtained from the different methods

conc. (mM)	method 1 Gaussian coil form	method 2 Benoit-Doty formulation	method 3 BR plot method	method 4 Peterlin's P(Q)
0.3	310	300		228
0.5	350	330		265
0.7	420	380	420	317
1.0	550	520	560	388
1.5	680	550	290	366
2.0				669
2.5				649
2.8				486
3.0				476
3.3				489
3.7				508
4.0				484
4.5				541
5				493
6				559

Conclusion

We have measured the persistence length of the CTA 3,5ClBz micelles to be 500Å $\pm$ 100Å. There may be a slight growth of the persistence length with concentration at very low concentrations, but it is too small to verify at the present time due to the rather large errors. This value is for solutions containing a fairly large amount of added electrolyte in the form of NaCl and it is reasonable to expect the persistence length of the micelles in an unsalted solution to be considerably larger. Unfortunately, present scattering techniques and theory do not permit such measurements.

A thorough examination of the data for semiflexible micelles in BR plot form can yield a wealth of qualitative information (besides the quantitative information). Gaining such an overview of the system before applying any quantitative technique is very important and can help avoid overinterpreting the data as was the case here in dealing with the apparent increasing persistence length at low concentrations. A more clear cut case deals with the size of the micelles at overlap which will be discussed in a future publication. In that case, use of the BR plots was the first strong indication that the sizes obtained were suspect, and indeed, subsequent application of more accurate, if much less precise, methods lead to the conclusion that the rods are much larger than initially thought.

Acknowledgments

This research was funded by the National Science Foundation (Grant CHE-9008589). P.B. extends his thanks and appreciation to Dr. J. Shibata for numerous stimulating and helpful conversations. We would also like to acknowledge DOE, Office of Energy Research, Division of Materials Science (under contract DE-AC05-84OR21400 with Martin Marietta Energy Systems) for financial support.

Literature Cited

1. Gravsholt, S. *J. Colloid Interface Sci.* **1976**, *57*, 575.
2. Gravsholt, S. in *Proceedings of the VIIIth International Congress on Rheology*, Naples, September 1-5 1980; p 629 (and references therein).
3. Imae, T. *J. Phys. Chem.* **1990**, *94*, 5953.
4. Rehage, H.; and Hoffmann, H. *Faraday Discuss. Chem. Soc.* **1983**, *76*, 363.
5. Gee, J. C. Ph.D. dissertation, University of Tennessee, 1991.
6. Magid, L. J. in *Ordering and Organization in Ionic Solutions*; Ise, N. and Sogani, I, Eds.; Proceedings of the Yamada Conference XIX, Kyoto Japan, November 9-12, 1987; World Scientific Publishing Co. Pte Ltd.: Singapore; pp. 288-301.
7. Magid, L. J.; Gee, J. C.; and Talmon, Y. *Langmuir* **1990**, *6*, 1609.
8. Quirion, F.; and Magid, L. J. *J. Phys. Chem.* **1986**, *90*, 5435.
9. Chu, B. *Laser Light Scattering*; Academic Press: New York, NY, 1974.
10. Brown, W.; and Rymden, R. *Macromolecules* **1986**, *19*, 2942.
11. Russo, P. S.; and Karasz, F. B. *J. Chem. Phys.* **1984**, *80*, 5313.
12. Denkinger, P.; Burchard, W.; *J. Polym. Sci., Part B* **1991**, *29*, 589.
14. Manfred, S.; Paradossi, G; and Burchard, W. *Makromol. Chem., Rapid Commun.* **1985**, *6*, 767.
14. Denkinger. P.; Burchard. W.; and Kunz, M. *J. Phys. Chem.* **1989**, *93*, 1428.

15. Candau, S. J.; Hirsch, E.; and Zana, R. *J. Phys. (FR)* **1984,** *45,* 1263.
16. de Gennes, P. G. *Scaling Concepts in Polymer Physics*; Cornell University Press: Ithaca, NY, 1979.
17. ter Meer, H.; Burchard, W. *Polym. Commun.* **1985,** *26,* 273.
18. Gee, J. C. *unpublished observations.*
19. Benoit, H.; and Doty, P. *J. Phys. Chem.* **1953,** *57,* 958.
20. The study of the size and polydispersity of these systems will be published at a later date.
21. Peterlin, A. *J. Chim. Phys.* **1950,** *47,* 669.

RECEIVED July 8, 1994

SHEAR EFFECTS

Chapter 18

Shear Effects in Surfactant Solutions

Rhyta S. Rounds

Becton Dickinson, Vacutainer Systems, 1 Becton Drive,
Franklin Lakes, NJ 07417-1885

Surfactants exhibit a broad spectrum of rheological properties dependent upon the internal structure of the surfactant micelles within the fluid media. Newtonian or highly non-linear viscoelastic behavior can occur as a function of concentration, temperature and solution environment, reflective of the complex phase behavior of these materials. Even dilute isotropic solutions can manifest very significant time-dependent non-linear behavior in simple shearing flows as the applied external flow field induces change in the morphology of the discrete surfactant units. Examples are provided of nonionic, cationic, anionic and mixed surfactant systems in various aqueous environments and phases exhibiting shear rate and time dependent rheological behavior in both steady shear and oscillatory measurements.

Surfactants are remarkable in the complex rheological behavior they manifest throughout the various regions of the phase diagram and within each phase as well. Even within the isotropic phase, Newtonian and non-Newtonian behavior can occur with dramatic shear effects such as rheopexy, thixotropy, dilatancy and pseudoplasticity. This unusual phenomenon has been well documented for specific cationic surfactant systems (1-6). For surfactants exhibiting shear effects, rheological behavior within critical shear rate ranges can become strong functions of the applied shear rate, duration of the shearing interval and surfactant/solvent composition. Behavior of concentrated mesomorphic surfactant systems is also frequently punctuated with dramatic time and shear dependent rheological functions within regions of the liquid crystalline phase with shear effects occurring in both steady shear rotational and oscillatory measurements.

   Rheological characterization of time and shear rate dependent material functions for surfactant systems can be a difficult and time intensive process. However, shear effects are readily observed in simple shear stress growth measurements, $\sigma^+ = f(\dot{\gamma}, t)$, spanning an appropriate shear rate range. Under these experimental conditions, a fluid is subjected to a step change in shear rate at $t=0$, as shown in Figure 1. In response to the applied shear rate, the shear stress, σ, is

0097–6156/94/0578–0260$08.00/0

measured as a function of time. These measurements can directly identify and quantify any time dependent viscometric behavior such as rheopexy or thixotropy and strain dependent functions such as pseuoplasticity or dilatancy. Such measurements at constant shear rates can also provide insight into the kinetics of any time dependent shear induced structural transitions occurring within the surfactant matrix. Several examples stress growth characteristics are provided in Figure 2.

"Thixotropic loop" measurements are commonly used for qualitative evaluations of shear effects. During a thixotropic loop evaluation, a time dependent ramped shear rate is applied to the experimental material in acceleration and deceleration as shown in Figure 3. As in stress growth measurements, the shear stress is commonly the measured response variable. Such measurements are difficult to interpret, however, due to the coupled time and shear rate independent variables. As such, stress growth measurements are preferred to thixotropic loop measurements to quantify and isolate the independent effects of shear rate and the duration of the applied flow field. Ramped shear rate measurements do provide, however, a preliminary assessment of shear effects although the absence of a "thixotropic loop" does not preclude the existence of time and shear dependent shear effects since the fluid "memory" or kinetic processes can coincide with the time scale of the shear rate ramp. Figure 4 is a graphical presentation of several common flow curves that can be obtained in these types of qualitative measurements.

Dilute Solutions

Cationic Surfactants. Cationic surfactants are singular in the complex rheological properties they can exhibit. This is especially true of a select subset of this class of surfactants containing mainly pyridinium or trimethylammonium headgroups and strongly binding counterions. While most surfactant solutions in the isotropic solution phase are Newtonian fluids, cationic detergents at dilute concentrations near the cmc can exhibit dramatic non-linear viscoelastic behavior and shear effects. It has been documented that changes in micellar assembly occurring in simple shearing flow produce corresponding changes in rheological behavior.

Research attributes the unusual viscometric behavior of these cationic systems to the shear induced formation of rod-like micellar chains, micellar "polymerization" (*7*), resulting in a dense three dimensional network structure (*8-10*). When the length of the surfactant polymer-like segment, L, equals the mean separation distance, D, viscoelastic and rheopectic effects can be observed. The supramicellar structure formed from individual rods, entangled and connected by temporal linkages, result in systems exhibiting rheological behavior typical of both linear and crosslinked polymers. There are, however, noteworthy differences exhibited by these specific cationic systems and polymers due primarily to the life-times or temporary nature of the "crosslinkages" or entanglement modes of micellar systems (*11-12*). In this respect, the entangled rod-like cationic surfactant network is truly unique.

Cationic surfactants are also unusual in the breadth of rheological behavior that can occur within a relatively narrow concentration and shear rate range in the dilute regime. For example, within the concentration range of 3-5 mmol/dm^3 of N-cetyl-N,N,N-trimethylammonium bromide and sodium salicylate, rheopectic, thixotropic, non-Newtonian Power Law and Newtonian behavior has been documented (*13,14*).

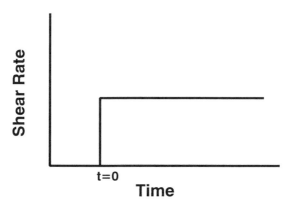

Figure 1. Step change in shear rate in stress growth measurements.

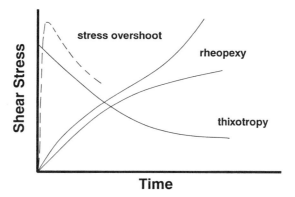

Figure 2. Several examples of stress growth behavior observed with materials exhibiting time dependent shear effects.

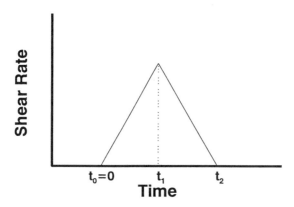

Figure 3. Shear rate ramp functions in thixotropic loop measurements.

Figure 5 summarizes the stress growth behavior, $\sigma^+ = f(\dot{\gamma}, t)$, of the CTA-Sal complex, at 253 ppm (.6 mmol dm^{-3}) within the shear rate range of .59 - 1.35 sec^{-1} using a Contraves viscometer with couette fixture as determined by Gravsholt (*13*) at 25°C. Stress relaxation following steady shear, $\sigma^- = f(\dot{\gamma}, t)$, is also provided. At shear rates < .99 sec^{-1}, Newtonian behavior is demonstrated through a shearing time interval of 3600 seconds, although slight shear thickening is also evident at .99 sec^{-1}. With a slight shear rate increment to 1.35 sec^{-1}, significant rheopectic behavior occurs following a lengthy induction interval. Approximately 450 seconds of shearing at 1.35 sec^{-1}, or a strain of approximately 600, was required to induce the observed rheopectic shear effect. The time scale or strain needed to induce the rheopectic response caused by the change in micellar morphology is most noteworthy.

In surfactant solutions, as in all complex fluids, rheological behavior is intimately linked to internal microstructure and micellar architecture. If a perturbation of the equilibrium thermodynamic state of the self-assembly is induced by shear, the transitional metastable nonequilibrium state will exhibit a modified rheological response. The rheopectic behavior observed in Figure 5 is a dramatic example of such a shear effect. It is proposed that the shear thickening behavior is due to a transition in micellar order within the isotropic phase to long rod-like micellar chains. The solutions will revert to the thermodynamic equilibrium conformation, however, following cessation of shear. From the length of the induction periods, the kinetics for this chain formation process and ensuing structural transition are quite slow.

Stress growth measurement also obtained by Gravsholt (*13*) for 5 mmol dm^{-3} CTA-Sal at .046 s^{-1} (25°C) is provided in Figure 6 showing a significant stress overshoot followed by thixotropic behavior. This stress growth behavior is frequently observed within the anisotropic surfactant phase and across various classes of materials. Such a response is dependent on the Weissenberg number which is a dimensionless ratio of fluid relaxation time to the applied shear rate. A point of distinction is that an equilibrium shear stress is not achieved following a shearing time interval of 2400 seconds. Stress relaxation is also significantly retarded indicating multiple relaxation times. At this surfactant concentration, pseudoplasticity is also observed in the shear rate range of .0006 - .01 sec^{-1}, while pseudoplasticity and thixotropic behavior is evident at higher shear rates in the range of .0135 - .085 sec^{-1}. This dilute cationic surfactant system demonstrates complex non-Newtonian behavior and significant shear effects in the dilute solution isotropic phase within the low shear rate regime.

The viscoelastic behavior demonstrated by the dilute cationic surfactants is most unusual. In the dilute cationic solutions of the type CTA-X, a strongly binding counterion is critical in the development of viscoelasticity. It has been demonstrated that substituted benzoic acid, salicylate, m-chlorobenzoate and p-chlorobenzoate induce viscoelasticity while o-chlorobenzoate, m-hydroxybenzoate, and p-hydroxybenzoate apparently do not alter the micellar structure. This phenomenon is known to be pH dependent (*15,16*) and a function of the solvent ionic strength. It is also known that oppositely charged surfactants, organic counterions and uncharged esters and aromatic hydrocarbons can induce viscoelastic effects in shear (*17-18*).

Dynamic mechanical studies of CTAB as a function of sodium salicylate (and salicylic acid) concentration have revealed three types of rheological behavior, dependent upon the molar concentration of salicylate (*19,20*). From the rheological

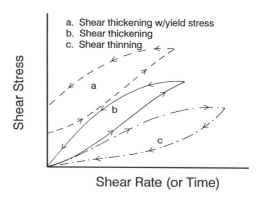

Figure 4. Examples of thixotropic loop shear stress patterns.

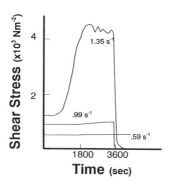

Figure 5. Stress growth behavior of .6 mmol dm^{-3} CTA-Sal at 25°C at various shear rates (Reproduced with permission from Ref 13. Copyright 1980 Plenum Press).

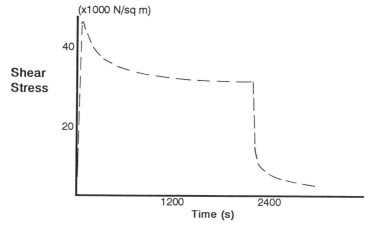

Figure 6. Stress growth function of 5 mmol dm^{-3} CTA-Sal at 25°C (Reproduced with permission from Ref 13. Copyright 1980 Plenum Press).

and electron microscopic studies, it is suspected at a 1:1 molar complex of NaSal and CTAB ($C_s/C_D=1$), thread-like micellar structures of uniform radius occur. At $C_s < < C_D$, dilute flexible polymer-type viscoelasticity occurs without entanglement. Various researchers have shown that complexation occurs between Sal⁻ and CTA⁺ leading to the formation of thread-like micellar chains. As C_S approaches C_D, behavior similar to that of concentrated entangled linear polymers is observed. In excess C_S, a single relaxation time can be observed by the Cole-Cole plots, indicating Maxwell model behavior and it has also been determined that relaxation does not appear to be a function of temperature or CTAB concentration but highly dependent upon free salicylate concentration. A hypothesis is that the free salicylate counterion promotes reorganization of the micelles as a virtual catalyst at entanglement sites and relaxation does not occur via reptation. Results with salicylic acid reflect similar transitions occurring as a function of detergent/acid ratios from dilute polymer Rouse-like behavior to entanglement to Maxwell behavior.

For the same surfactant system, at a concentration of 2 mM of CTAB and sodium salicylate, Figure 7 summarizes the steady state viscosity function as a function of shear rate within the temperature range of 10.1 - 35.9°C (*21*). Both pseudoplasticity and dilatancy are observed within this temperature range. Although a decrease in viscosity with increasing shear rate is apparent initially, following a critical shear rate, an abrupt increase in viscosity occurs, a dilatant shear effect, with pseuoplasticity accompanying further increases in shear rate. The critical shear rate at which shear thickening occurs increases with increasing temperature. For example, at 10.1°C, the viscosity transition occurs at approximately 50 sec⁻¹, whereas the critical shear rate is extended beyond 300 sec⁻¹ at 35.9°C It is also reported that cessation of flow during the induction period, prior to the development of a steady shear stress, results in instantaneous stress relaxation.

Anionic Surfactants. Dilute anionic surfactants are not known to exhibit time-dependent and non-linear viscosity behavior in the isotropic phase. In combination with cationic surfactants in the appropriate ionic environment, however, rheopectic and dilatant behavior can occur following the formation of shear induced rod-like micellar structures (*22,23*). One example of such a mixed surfactant system is tetradecyldimethyl aminoxide (TDMAO), sodium dodecylsulfate (SDS) with sodium chloride. Stress growth behavior for a 10 mM solution of TDMAO and SDS at a mole ratio of 6:4 including 24 mM NaCl is provided in Figure 8 at 50 sec⁻¹ and 25°C using the Rheometrics RFS 8500 and cone and plate tooling. Rheopectic behavior is observed similar in form to the CTA-Sal cationic surfactant system in Figure 5, however, there are notable differences. The experimental shear rate is significantly higher for the mixed anionic/cationic surfactant pair and the induction time intervals markedly shorter than the single cationic surfactant. The induction period for the TDMAO/SDS mixture is approximately 20 seconds. At a shear rate of 50 sec⁻¹, the strain for the development of rheopexy is approximately 1000. For this surfactant pair, induction time appears to be inversely proportional to shear rate in the experimental conditions investigated in this study.

The TDMAO/SDS solution clearly exhibits dilatant behavior. Shear thickening with increasing shear rate is summarized in Figure 9 for a shear rate range spanning two orders of magnitude, representing the steady state viscosity as a function of shear

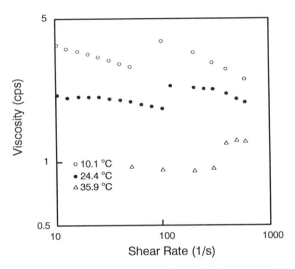

Figure 7.   Steady shear viscosity as a function of shear rate (Reproduced with permission from Ref 21. Copyright 1993 Society of Rheology).

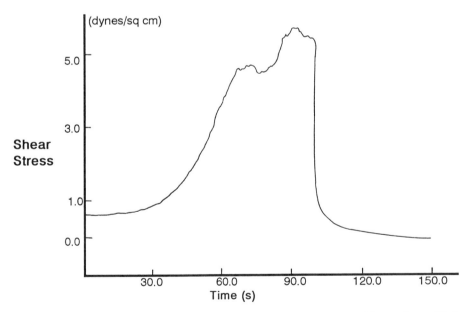

Figure 8.   Stress growth behavior of TDMAO/SDS/NaCl at 50 sec^{-1} at 25°C (Reproduced with permission from Ref 22. Copyright 1993 Academic Press).

rate. A hydrodynamic diameter of 22 nm for this micellar solution was obtained using dynamic light scattering, with micelle length estimated to be 50 nm and number density of 1.92×10^3 m^{-3}.

Nonionic Surfactants. It has been reported that polyethylene oxide can experience degradation in shear complicating even dilute solution viscosity measurements (*24,25*). Chain scission has been found to be a function of the applied shear gradients, the shear duration, and the measured viscosity. No dependence on solution surfactant concentration has been observed, and degradation or chain scission found to follow simple first order kinetics. Figure 10 summarizes viscometric measurements for two PEO samples differing only in molecular weight within the concentration range of 1-4% within a shear rate range of 0 to 1000 sec^{-1}. The data is summarized as a function of parameter A defined as $A = \eta\dot\gamma^2 t$ where t = time duration of applied flow field.

This is an example of a shear effect that is irreversible. Previously discussed shear induced structural transitions were only reflective of a temporary nonequilibrium metastable shear induced structural transition.

Liquid Crystals

The rheology of lyotropic and thermotropic surfactant liquid crystals can be punctuated with dramatic shear effects throughout the various mesomorphic phases and within each phase as well. A single surfactant liquid crystal can exhibit a broad spectrum of rheological properties within the boundaries of a single phase including both time and shear rate dependent rheological functions. Within each phase, rheological response is dependent on composition and solution environment, and marked differences in rheological behavior can occur within narrow composition variations. Rheological characterization of surfactant liquid crystals with the accompanying shear effects that can be encountered is also encumbered with experimental difficulties.

The complications encountered in the rheological characterization of surfactant liquid crystals include slippage within the liquid crystal during measurements, orientation during shear and at tooling fixtures, defects within the ordered phase and sample handling and pretreatment prior to characterization. Each of these can influence the magnitude of the rheological properties and reproducibility of test results. The presence of defect structures within the liquid crystal geometry is perhaps the most difficult to properly resolve.

Cationic Surfactants. Cationic surfactants are well known for their pronounced dilute solution shear effects. Equally dramatic, however, are the anomalous flow properties at higher concentrations within liquid crystal phases. This is well demonstrated by the unusual flow properties of a ternary aqueous solution of hexadecyltrimethyl ammonium bromide (CTAB), hexanol and water in the lamellar liquid crystal phase (*26*). A lamellar liquid crystal solution containing 60% water and a molar ratio of 1.26:3 CTAB/hexanol exhibits pronounced rheopexy when subjected to a series of periodic pulses of 77.5 sec^{-1} of varying pulse width at 25°C in couette flow, as shown in Figure 11. The effect of each shear rate pulse appears to be additive and extremely long relaxation times have been determined for re-equilibration

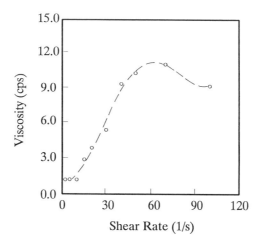

Figure 9. Steady shear viscosity as a function of shear rate at 25°C (Reproduced with permission from Ref 22. Copyright 1993 Academic Press).

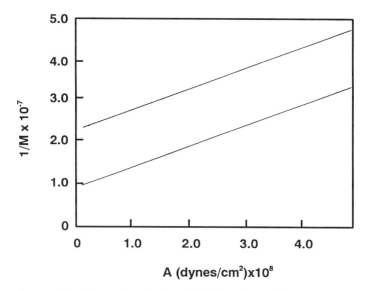

Figure 10. Shear degradation of PEO (adapted from Ref 24).

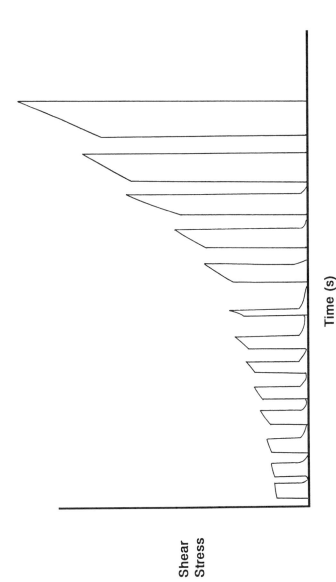

Figure 11. Thixotropic loop of CTAB/Hexanol liquid crystal (Reproduced with permission from Ref 26. Copyright 1978 Academic Press).

to the initial structure. Bohlin and Fontell have quoted intervals in excess of 6 months for re-equilibration at room temperature following cessation of the shearing flow. X-ray diffraction of samples subjected to shear do not reveal any changes in the repeat distance of the lamellar configuration. The formation of a metastable micro-structure is proposed, however, involving large oriented domains based on optical polarized microscopy prior to and following shear.

It is of further interest that shear effects are not uniformly distributed throughout a single phase of the phase diagram. Figure 12 shows that not all samples within the lamellar liquid crystalline region of the CTAB/hexanol/water system respond identically but that rheological behavior and shear effects are a strong function of water content. Further, significant hysteresis is observed for the higher water content ($>60\%$) liquid crystals. The hysteresis loop, shown in Figure 13A, for the system containing 60% water at the same molar ratio, was obtained at one minute applications at each shear rate. The influence of water is summarized in Figure 13B, representing an equilibrium viscometric flow curve at each shear rate. It is mentioned that steady state shear stress values were not obtained at select shear rates rapidly but required shearing for long durations extending to hours.

Anionic Surfactants. Viscometric "thixotropic loop" measurements of a sodium soap, sodium caprylate/decanol/water system, within the various mesophase regions, identify qualitatively a broad range of rheological responses (27). Figure 13 summarizes results of these measurements for the lamellar phase containing a decanol/sodium caprylate molar ratio of 2.45 and 40% vs 50% water at 20°C. Shear rate is arbitrarily represented in fixture rotation speed and data was obtained using the Ferranti-Shirley viscometer, cone and plate fixture, and 60 second acceleration/deceleration time intervals. Although slight deviation from non-Newtonian behavior is evident at 40% water, significant shear effects are observed for the lamellar liquid crystal phase prepared with 50% water. A shear effect is prevalent throughout the acceleration segment of the thixotropic loop measurement. During deceleration of the shear rate ramp, no significant effects are observed and results are similar to the total response obtained for the liquid crystal containing 40% water. This effect begins to occur at 42% water for rotation speeds in excess of 50 rpm using the same shear rate ramp time span, as shown in Figure 14. The shear effects at 42% continues in both acceleration and deceleration for rotation speeds in excess of 60 rpm for the selected ramp conditions. At this liquid crystal composition, dilatancy, rheopexy or both shear effects are in place. No significant effect is observed at low rotation speeds. The overall response is strongly dependent upon molar ratio and water content within this liquid crystal phase.

Stress growth and thixotropic loop measurements for a water/sodium caprylate hexagonal phase composition at 52/48 wt % (0% decanol) produce interesting results. Although the thixotropic loop measurement, obtained at ramped experimental conditions of 100 rpm/60 sec, indicates no apparent shear effects, the stress growth measurements at 100 rpm shows significant thixotropy with a lengthy shear stress equilibration, in excess of 300 seconds, at 100 rpm. A comparison of these two measurements demonstrates the deficiency of thixotropic loop measurements in capturing shear effects and the importance of the Weissenberg Number in designing rheological experiments. Stress growth measurements at constant shear rate or

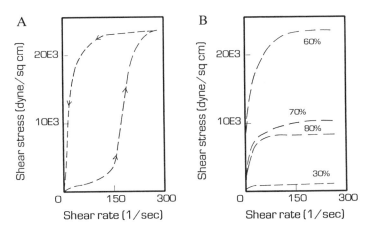

Figure 12. Thixotropic loop (A) and steady shear flow curve (B) of CTAB/Hexanol liquid crystal (Reproduced with permission from Ref 26. Copyright 1978 Academic Press).

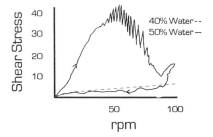

Figure 13. Thixotropic loop measurement at 20°C of a sodium caprylate lamellar liquid crystal (adapted from Ref 27).

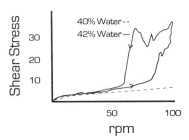

Figure 14. Thixotropic loop measurement at 20°C of a sodium caprylate lamellar liquic crystal (adapted from Ref 27).

rotation speeds were not completed at each experimental condition to discern the individual effects of shear rate and duration of shear. It is difficult to extrapolate from thixotropic loop measurements the nature of such dramatic shear effects. Nevertheless, there are distinct differences in rheological behavior and complex shear effects occurring within the liquid crystal phases.

Stress growth measurements for an aqueous sodium alkylethersulfate lamellar liquid crystal containing 69.1% (wt%) surfactants obtained with a Rheometrics RFSII and parallel plate geometry (25 mm) at 25°C is provided in Figure 15 for a shear rate of .05 sec^{-1}. A significant stress overshoot is observed with slow equilibration to a steady shearing stress. A steady state shear stress is not achieved within the measurement timescale. Similar behavior is obtained within the shear range of .05 - 50 sec^{-1}. Following the stress overshoot, thixotropic behavior is clearly demonstrated. The molecular weight of the surfactant used in this is 430 with a homolog distribution within the range of C_{12}-C_{15}. Liquid crystals were prepared with deionized water and equilibrated for one month prior to rheological measurement. Measurements repeated following an additional six month equilibration demonstrated no change in rheology. The liquid crystalline phase was identified through polarized light microscopy.

The stress growth behavior of the anionic lamellar liquid crystal surfactant, represented as $\eta^+ = f(t)$, is similar in form to the 5 mmol dm^{-3} CTA-Sal solution behavior shown in Figure 6. The anionic liquid crystal, however, demonstrates a greater degree of thixotropy in comparison to the dilute cationic solution. Following the stress overshoot, the viscosity decrease is in excess of 60% of the maximum value at the overshoot within a shearing time interval of 400 seconds. For the cationic solution, the decrease is not as dramatic. This is not surprising as the mechanism controlling rheological behavior is quite different for these two surfactant systems.

One interpretation of the stress growth function shown in Figure 15, consistent with the findings of Horn and Kleman (28), is that the liquid crystal is undergoing reorientation or shear facilitated alignment of the surfactant bilayers parallel to the shearing plane producing the thixotropic effect. This is not a structural degradation frequently encountered with highly interactive dispersions; in fact, the reverse may occur. With small molecular surfactant liquid crystals, containing structural defects within the liquid crystalline two dimensional matrix, simple shear may be affecting an order transition from a multiple defect crystallite "dispersion" to a highly ordered lamellar liquid crystal depending on the strain magnitude and complexity of the applied flow field. Higher shear rates, however, can induce disruption of the lamellar bilayers distorting rheological behavior. Interpretation of test results is further complicated by the presence of slip planes, complex micellar configuration and interaction/coordination along sensor boundaries and the need to define quiescent condition of the specimen prior to testing.

Dynamic mechanical characterization of the alkylethersulfate lamellar liquid crystal is provided in Figures 16 and 17. The strain dependence of the storage and loss moduli indicates non-linear viscoelastic behavior at low strains and the mechanical frequency spectrum further indicates non-terminal behavior. The frequency dependence of the complex moduli is indicative of a surfactant gel. This lamellar liquid crystal composition is highly elastic and strain sensitive in the quiescent state with linear viscoelastic behavior occurring at strains <.001. No pretreatment was

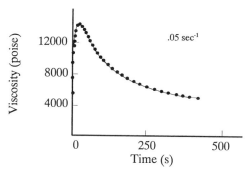

Figure 15. Stress growth function at 25°C and .05 sec⁻¹.

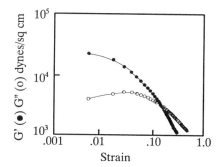

Figure 16. Storage and loss modulus as a function of strain at 10 radians/second and 20°C.

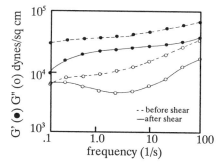

Figure 17. Frequency dependence of storage and loss modulus as a function of frequency before and after shear alignment at a strain of .01 and 25°C.

provided to the sample prior to testing for the measurement indicated as "before shear" in Figure 17 and identical results were obtained with sample equilibration periods up to 3 hours following sample loading. Additional measurements were completed to insure that the no-slip boundary condition was not violated during flow. Excellent reproducibility was obtained throughout all measurements with less than 10% variance.

Prealignment of the lamellar liquid crystal in oscillatory strain results in a change in the frequency dependence of the storage and loss moduli, also shown in Figure 17. The lamellar liquid crystal was subjected to a strain of .5 at a frequency of 10 radians/second for a 20 minute period and dynamic mechanical properties determined before and after the conditioning interval. The effect of the oscillatory strain is a decrease in magnitude of the two moduli and an increase in frequency functionality. It would appear that the frequency dependence of the complex moduli are significantly influenced by the presence of defects within the liquid crystal matrix.

It has been observed that the rheology of the anionic lamellar liquid crystal surfactant determined in simple uni-directional shearing flows can be quite different from that experienced in three-dimensional complex flow fields. This is likely to be due to the disruption of the liquid crystal architecture and catastrophic disruption of the ordered micellar planar matrix. In complex flow fields, such as experienced in mixing, dramatic shear thickening rheopectic behavior can be observed with evidence of excessive normal stresses. Mixing results in the Weissenberg effect at moderately low agitation levels. The corresponding increase in sample turbidity would suggest formation of an intermediate liquid crystallite "dispersion" causing the increase in viscosity and apparent elasticity. This is demonstrated in Figure 18 showing the torque output as a function of mixing time at 75 rpm using a single turbine agitator in a 1 liter vessel at 25°C for the sodium alkylether sulfate previously discussed. Significant rheopectic behavior is observed and due to the high normal stresses occurring during mixing, mixing could not be maintained for longer time intervals. Although this surfactant exhibits pseudoplastic and thixotropic effects in simple shearing fields, rheopexy and dilatancy are experienced in more complex flow fields in which the liquid crystalline matrix is not preserved.

Nonionic Surfactants. Larson et al have recently shown n-nonyl-1-β-D-glycopyranoside (Mw = 322) in the smectic phase exhibits terminal behavior in the viscoelastic spectrum as a function of frequency (29). Non-terminal behavior is attributed to the presence of defects in the macromolecular structure. Progressive ordering through large strain amplitude global realignment can be achieved through preshearing with an oscillatory strain amplitude of 1 and a frequency of 10 radians/sec. The effect of shear alignment is self-evident. The rheological measurements completed with this thermotropic liquid crystal demonstrate the need for clarification of liquid crystal order prior to characterization. Non-terminal behavior of lyotropic surfactant and thermotropic liquid crystals is frequently attributed to gel behavior and not the presence of defects within the lamellar liquid crystal structure.

Horn and Kleman also evaluated the effect of liquid crystal defects on the viscometric behavior of a non-ionic surfactant (n-octyl-cyanobiphenyl) coupling viscometry with optical microscopy. Their experimental results indicate that if liquid crystal defects occur, viscosity will be significantly influenced. Experimental results are provided in Figure 19 showing that defects result in increased flow resistance. These effects contribute complexity to the interpretation of rheological measurements of surfactant liquid crystals as it depends upon the ordered state of the liquid crystal matrix and laminar flow throughout the fluid continuum.

Lamellar liquid crystals formed from alklyl polyethylene glycol ethers exhibit shear effects as well (*30*). Flow curves of for $C_{12}4EO$ containing 60 wt% water are provided under thixotropic loop experimental conditions for consecutive measurements. This lamellar liquid crystalline surfactant system demonstrates typical thixotropic shear effects with decreasing flow resistance with repetitive shearing. Estimates of yield stresses or "yield values" are frequently extrapolated from thixotropic loop measurements from the intercept of the flow curve with the shear stress axis in both acceleration and deceleration. The measurements provided indicate the presence of a "yield value" which depends strongly on shear history. A summary of yield values was obtained under these experimental conditions as a function of ethylene oxide composition and carbon chain length for systems containing 10 mole% surfactant. Unfortunately, experimental shear ramp conditions are not specified. For oriented lamellar liquid crystals, oriented in shear, yield values show little dependence on surfactant structure further demonstrating the importance of liquid crystalline order on the rheological behavior. Yield values appear to decrease following shear, a further indication of the importance of defect structures in shaping rheological behavior.

Lyotropic liquid crystals formed with phosphated polyoxyethylene, hexane and water also exhibit complex shear effects when subjected to rapidly accelerated/decelerated high shear flow fields within the lamellar and hexagonal liquid crystal phases (*31*). Under the selected experimental conditions, using a Ferranti-Shirley cone and plate viscometer, at shear rates in excess of 1800 sec^{-1} obtained within a 10 second interval, lamellar liquid crystals appear more viscous than the hexagonal phase with rheopectic hysteresis observed for the lamellar phase and thixotropy evident for the hexagonal phase. Rheological response for the lamellar liquid crystalline phase again appears to be highly dependent upon water concentration. A model of the flow behavior is characterized by a single Maxwell element with two Voigt units. The relative viscosity differences between the two liquid crystalline phases is attributed to the slippage of lamellar bimolecular layers. The rheopectic behavior is suspected to be due to the strong hydrogen bonding and dipolar interactions within the phosphated nonylphenol ethoxylate and alcohol ethoxylate. This may, however, be due to the disruption and formation of a high degree of disorder in the lamellar liquid crystal matrix as previously discussed.

Summary

There has been a growing interest in the rheology of surfactant systems during the past few decades leading to a much broader understanding of the interdependence of surfactant morphology and mechanical properties. Rheological studies in tandem with comprehensive analytical characterizations are needed to define the structures responsible for the various shear effects just described, which are not yet clearly understood. Although there has been marked progress in this area recently, the rheology of surfactant systems is still a relatively young research area.

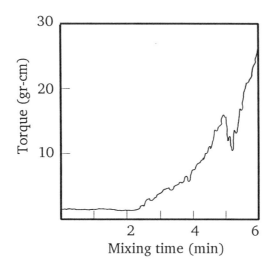

Figure 18. Rheopectic behavior in mixing at 25°C.

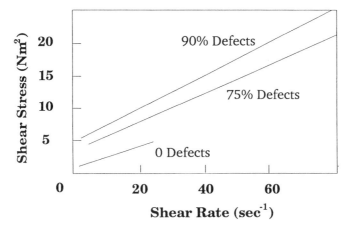

Figure 19. Flow curve as a function of defect structure (adapted from Ref 28).

Surfactants can exhibit shear effects and flow anomalies and the experimental time scales and total energy input is critical in the resulting rheological responses. In addition to the novelty and curiosity of such behavior, the rheology is indeed very important as well. This is especially true if rheological characterizations are required for process specification and design where various shear conditions are present for extended time intervals with accelerated/decelerated flow fields. An excellent example is the anomalous thixotropic/pseudoplastic behavior noted for surfactant mesophases in uni-directional flow fields attributed to simple alignment versus flow behavior in more complex flow fields.

Literature Cited

1. Hoffman, H.; Hoebl, H.; Rehage, H.; Wunderlich, I. *Tenside Detergents.* **1985**,*22*,6.
2. Hoffman, H.; Loebl, M.; Rehage, H. *Proc Int Sch Phys "Enrico Fermi"*, 1990.
3. Hoffman, H.; Rehage, H. *Surfactant Science Series.* Volume 22, (1987).
4. Wunderlich, I.; Hoffmann, H.; Rehage, H. *Rheol Acta* **1987**,*26*,6.
5. Hoffman, H. *Adv Colloid Interface Sci.* **1990**,*32*,2-3.
6. Clausen, T.M.; Vinson, P.K.; Minter, J.R.; Davis, H.T.; Talmon, Y.; and Miller, W.G. *J Phys Chem* **1992**,*96*,1.
7. Nash, T. *Journal of Applied Chemistry.* **1956**, December.
8. Rehage, H.; Hoffman, H. *Faraday Discuss. Chem. Soc.* **1983,** 76.
9. Hoffman, H.; Rehage, H.; Schorr, W. *Surfactants Solutions* (Proc Int Symp). Plenum, N.Y. 1982.
10. Rehage, H.; Hoffman, H. *Molecular Physics.* **1991**,*74*,5.
11. Cates, M.E. *J Phys Chem* **1990**,*91*,1.
12. Drye, T.J.; Cates, M.E. *J Chem Phys.* **1992**,*96*,2.
13. Gravsholt, S. *Rheol (Proc Int Congr)* **1980**,*8*,3.
14. Gravsholt, S. *J Colloid Interface Sci.* **1976**,*57*,3.
15. Wan, L.S.C. *J Pharm Sci.* **1966**,*55*,1395.
16. Hyde, A.J.; Johnstone, D.; *J Colloid and Interf Sci.* **1975**,53.
17. Gravsholt, S. *Proc Int Congr Surf Activ.* **1973**.
18. Rehage, H.; Hoffman, H. *Mol Phys.* **1991**,*74*,5.
19. Shikata, T.; Hirata, H. *J Non-Newtonian Fluid Mech.* **1988**,*28*,22.
20. Shikata, T.; Kotaka, T. *J Non-Cryst Solids.* **1991**,*131-133*, Pt 2.
21. Hu, Y., Wang; S.Q., Jamieson, A.M. *J of Rheology.* **1993**,*37*,3.
22. Hu, Y., Wang; S.Q., Jamieson, A. *J of Colloid and Interf Sci.* **1993**,*156*,31-37.
23. Rehage, H.; Hoffmann, H. *Mol Phys.* **1991**,*74*,12.
24. Asbeck, W.; Baxter, M. *ACS Polymer Division. Annual Meeting.* September, 1958.
25. Vink, M. *Makromol Chem.* **1960**,*46*,51.
26. Bohlin, L.; Fontell, K. *J of Colloid and Interface Sci.* **1978**,*67*,2.
27. Solyom, V. P.; Ekwall, P. *Rheol Acta.* **1969**,*8*,3.
28. Horn, F.J.; Kleman, M.N. *J Phys.* December, 1978.
29. Larson, R.G.; Winey, K.I.; Patel, S.S.; Watanabe, H.; Bruinsma, R. *Rheol Acta.* **1993**,*32*,3.
30. Paasch, S.; Schambil, F.; Schwuger, M.J. *Langmuir.* **1989**,*5*,6.
31. Groves, M.J.; Ahmad, A.B. *Rheol Acata.* **1976**,*15*,9.

RECEIVED September 23, 1994

Chapter 19

Formation of Nonequilibrium Micelles in Shear and Elongational Flow

Shi-Quing Wang, Y. Hu, and A. M. Jamieson

Department of Macromolecular Science,
Case Western Reserve University, Cleveland, OH 44106

Micelle formation in aqueous solutions in the presence of shear and elongational flows is studied by rheological measurements. Cetyltrimethylammonium bromide(CTAB)/sodium-salicylate (NaSal) in water is found to self-assemble into larger micellar structures in flow than in quiescence. The critical rate for shear thickening is inversely proportional to the micellar relaxation time or lifetime. We also present strong evidence that an elongational flow is more favorable to micellar growth than shear. By changing the molar ratio of NaSal to CTAB, we can control the extent of flow-induced micellar formation, as measured in terms of shear and elongational viscosities.

Cationic surfactants solutions often form very large micellar structures in solution and thus show unique viscoelastic properties. Linear and nonlinear viscoelastic behaviors of surfactant solutions have been extensively studied as a function of surfactant concentration (*1-4*). Rehage and Hoffmann (*2*) first reported that the viscosity of a 0.9 mM Cetylpyridinium-salicylate solution slowly increases with time (rheopexy) when subjected to shear flow at a sufficiently high shear rate, and that it takes an unexpectedly long time, several minutes, for the system to reach steady state. Later work on the rheological behavior of dilute micellar solutions has clearly established (*2,5*) that shear induced collisions between the initially short micelles lead to micellar growth beyond the equilibrium sizes. This statement is based on four sets of observations: (a) The stress grows over a period of time that is exceedingly long in comparison to the relaxation time of the equilibrium solution; (b) Above a critical shear rate, shear thickening occurs; (c) The relaxation times of the shear-thickened solutions are substantially longer than those of the equilibrium

0097–6156/94/0578–0278$08.00/0

solutions; (d) The induction period for stress growth is inversely proportional to the shear rate.

This chapter summarizes our current understanding of the mechanism for shear thickening and shear-induced micellar growth in dilute solutions of cetyltrimethyl-ammonium bromide (CTAB)/sodium-salicylate (NaSal). We are especially interested to know the rheological conditions under which shear thickening and micellar growth take place. For example, it remains unclear what factors determine the critical rate for shear thickening. Here we demonstrate that the micellar lifetime is a crucial parameter which controls the micellar growth kinetics in flow, and thus determines the critical shear rate.

There have been several studies of the effect of NaSal concentration on the equilibrium properties of CTAB solutions. NMR studies (6) indicate that NaSal plays an essential role in the formation of long wormlike micelles. Light scattering studies reveal a complex variation of micelle size as a function of NaSal concentration (7,8). Viscoelastic properties of micellar solutions also change significantly with NaSal concentration (9,10). Specifically, the viscoelasticity increase with NaSal concentration up to a certain level depending on the CTAB concentration and a decrease is observed upon a further increase of NaSal. This is interpreted as due to the growth of thread like micelles, as NaSal is incorporated. When the concentration of NaSal is larger than a certain value, the excess NaSal decreases micelle lifetime. In other words, micellar break-up rate increases when excess NaSal is present. It is therefore of interest to examine the effect of NaSal concentration on the shear-thickening and elongational-thickening characteristics.

In addition, we compare the rheological behavior of CTAB/NaSal solutions in shear and elongational flows. We find that the elongational viscosity is several hundred to several thousand times larger in magnitude than the shear viscosity at comparable rates of deformation, indicating more extensive micellar growth in an extensional flow. Formation of large micellar structures is further supported by stress growth and relaxation experiments. The mechanism for micellar growth in such an extensional flow field is discussed.

Experimental

Materials.   CTAB and NaSal were from Sigma Chemicals Co. and Aldrich Chemical Company, Inc. respectively. NaBr was from Fisher Scientific Company. All materials were used as received. The solvent was distilled water. CTAB concentration was fixed at 1 mM for shear measurements and 2 mM for elongational measurements. NaSal concentration was varied to obtain NaSal/CTAB ratios between 0.5 to 4.0. All solutions were equilibrated for at least two days before measurement. However, no filtration was used, in contrast to previous procedures (5). Filtering may remove not only large (≥ 0.5 μm) impurities but may also alter the concentration of filtered solutions.

Methods.   A Rheometrics Fluid Spectrometer RFS 8500, with cone and plate geometry (50 mm diameter, 0.02 radians cone angle), was applied to measure shear flow properties. A Rheometrics RFX rheometer was used to study elongational flow properties, in opposing jets geometry using the sucking mode. The diameter of the jet orifices and the distance between the two orifices are each 4 mm. Both stress growth upon start-up flow and stress relaxation upon cessation of flow were monitored as a function of time to gain insight into the kinetic processes leading to micellar growth.

Results and Discussions

Shear Thickening. Since the micellar break-up time is very sensitive to salt concentration, we can gain insight into the micellar growth kinetics and the accompanying build-up of the shear thickening state by varying the molar ratio of NaSal to CTAB. In Figure 1 the steady state viscosity at a shear rate of 100 sec^{-1} is plotted as a function of NaSal/CTAB molar ratio for 1 mM solution. The shear viscosity exhibits a peak at the unit ratio of NaSal/CTAB, indicating formation of the large micellar structures. We deduce from this observation that the flow-induced micelles are most stable when the molar ratio of NaSal/CTAB equals unity. Thus, at this ratio, the micellar lifetime is longest and decreases as the sodium salicylate concentration is varied away from 1 mM. We therefore expect to observe a maximum in the relaxation time τ measured by flow birefringence relaxation experiments. This is indeed the case as shown in Figure 2.

It is important to realize that the relaxation of a shear thickened state as observed in Figure 2 involves, in principle, two different processes. Upon cessation of shear, non-equilibrium micelles larger than the equilibrium ones tend to break up and return to smaller sizes on an average time scale referred to as the micellar lifetime τ_s. As the micelles undergo the fragmentation process, the relaxation of flow-induced birefringence accelerates, i.e., the fragmented micelles disorient faster because they are smaller. The other mode of relaxation is the usual conformational relaxation process. The micellar lifetime can be most straightforwardly probed by applying consecutive start-up shear with increasing lengths of intervening zero shear intervals, and measuring the residual stress. This has been discussed in some detail in our previous publications (5,12). From such experiments, it can be shown (12) that the micellar lifetime τ_s is at a maximum when the ratio NaSal/CTAB is unity, in agreement with Figure 2.

Furthermore, it can be illustrated by a simple theoretical analysis (13) that the magnitude of τ_s essentially determines the minimum shear rate required to produce a shear thickening state in a micellar solution. Shear thickening can take place if the micellar lifetime τ_s is long enough at a certain surfactant concentration, irrespective of how short is the relaxation time of the equilibrium solution. Actually at low concentrations (≤ 2 mM), the magnitude of τ_s cannot be derived from the apparent relaxation time τ_{eq} of an equilibrium solution, because micelles may be so short that their deformed configurations relax quickly without interference from the micelles break-up. Although τ_{eq} is very small, τ_s might be very long. The kinetic coagulation process leading to micellar growth and shear thickening does not depend on the magnitude of τ_{eq}, but on that of τ_s. Therefore it is understandable that the critical shear rate $\dot{\gamma}_c$ for shear thickening of an 1 mM CTAB solution with an equimolar of NaSal is as low as 10 sec^{-1}, since $\dot{\gamma}_c$ is inversely proportional to the

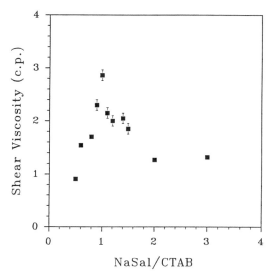

Figure 1. Steady state shear viscosity as a function of sodium salicylate to CTAB molar ratio at the CTAB concentration of 1 mM. A maximum is reached when the ratio is unity. (Reproduced with permission from ref. 12. Copyright 1994, ACS)

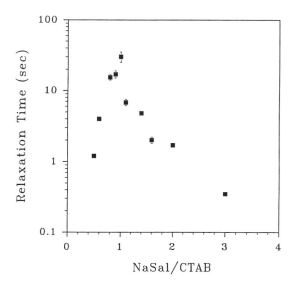

Figure 2. Relaxation times τ, measured from decaying flow birefringence upon cessation of shear, as a function of the NaSal to CTAB molar ratio at the CTAB concentration of 1 mM. Again there is a peak at the unit ratio. (Reproduced with permission from ref. 12. Copyright 1994, ACS)

comparatively long micellar lifetime τ_s $(13,14)$. Consistent with the above discussion, the critical shear rate is found to exhibit a minimum at the equimolar ratio of sodium salicylate to CTAB in 1 mM aqueous solution, as shown in Figure 3.

Elongational Thickening. The question of how micelles react to an extensional flow has been discussed theoretically $(15,16)$. In one study the effect of an elongational flow was incorporated in terms of an additional potential energy term, i.e. an elastic stretching energy, in a generalized thermodynamic formulation of the free energy for a dilute solution of rodlike micelles. It was concluded that the elongational flow should cause micelles to shrink in size at a high enough strain rate (15). This treatment completely ignores any kinetic effects due to the flow convection. A second analysis discusses re-organization of micelles in an elongational flow within a kinetic framework, assuming that only collinear collisions might occur between rodlike micelles (16). This theory not only predicts micellar growth in an elongational flow, but also deduces that the tendency for self-assembly of surfactant molecules into larger micelles is much stronger in an elongational flow than in shear. This conclusion is based on the assumption that rodlike micelles are more readily aligned in an extensional flow than in shear and therefore coagulation through collinear collisions would occur more easily. As a test of this collinear collision theory, Prud'homme and Warr (17) made elongational flow measurements of tetradecyltrimethyl-ammonium bromide/sodium salicylate micellar solutions and concluded that an elongational flow indeed appears to promote micellar growth more effectively than shear.

However, it appears difficult to rule out another interpretation of the experimental data presented in Ref. 17. That is, stretching of a micellar network beyond the flexible chain limit could also produce an elongational viscosity larger than three times the shear viscosity, which would not be inconsistent with the experimental measurements. Thus the question remains whether elongational-flow-enhanced micellar growth can take place. If flow-enhanced coagulation indeed occurs, it will not be realized by collinear collisions which are forbidden by the kinematics of an elongational flow, but it can occur through side-to-side collisions that may have an even larger cross-section than that encountered in a shear flow. This process is sketched in Figure 4 where only the velocity component in the radial direction is drawn and the effect of Brownian motion is not indicated.

When a start-up elongational flow of strain rate 2 sec^{-1} is applied to a 2 mM CTAB solution with 2 mM concentration of sodium salicylate at 24°C, one observes stress growth during a period of several minutes as shown in Figure 5. Stress relaxation measurement of the equilibrium solution indicates that the characteristic relaxation time τ_{eq} is only a fraction of a second. Thus it is difficult to attribute this rheopexy to chain stretching of micelles. In the plateau state where the stress grows no longer, the elongational viscosity is measured to have a magnitude many thousand times higher than the corresponding shear viscosity over a range of strain rates from 1 to 100 sec^{-1} as shown in Figure 6. Also in contrast, the solution in shear shows no rheopexy for shear rates below 100 sec^{-1} (5). Thus Figure 5 and Figure 6 indicate together that micelles may have reorganized into much larger structures in elongational flows at the specified strain rates. The micellar growth is truly remarkable as manifested in terms of the enormously high elongational

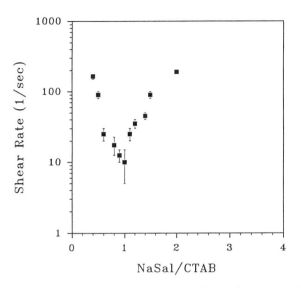

Figure 3. The critical shear rate for shear thickening at 1 mM CTAB concentration as a function of NaSal/CTAB ratio. A minimum critical shear rate is required for the equimolar NaSal/CTAB solution. (Reproduced with permission from ref. 12. Copyright 1994, ACS)

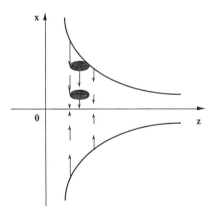

Figure 4. Schematic illustration of a uniaxial three dimensional elongational flow described by $v = \dot{\varepsilon}(-0.5x e_x - 0.5y e_y + z e_z)$, where only the radial component of the velocity field is indicated. The oval shaped objects are meant to represent elongated and aligned micelles in such an extensional flow. The micelle further away from the middle line has a larger radial velocity due to convection toward the middle line and will collide into the other micelle.

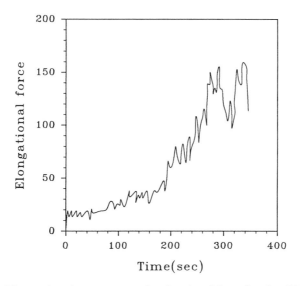

Figure 5. Elongational stress growth of a 2 mM equimolar NaSal/CTAB solution upon onset of a start-up flow of strain rate 2 sec^{-1}.

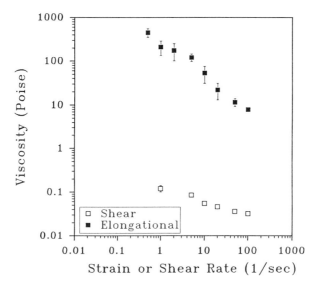

Figure 6. Elongational and shear viscosities of the 2 mM equimolar NaSal/CTAB solution as respective functions of strain or shear rate.

viscosity. This implies, in agreement with the results of Prud'homme and Warr (*17*) that an elongational flow is more effective than shear in causing coagulation between micelles. In an elongational flow such coagulation may take place through side-to-side collisions as sketched in Figure 4. Note that the induction time as shown in Figure 5 is a comparatively long period during which micelles can experience side-to-side encounters. Since there is a stagnation point in the opposing jets geometry, micelles near the stagnation point may grow over a long period, which can happen only if the micellar lifetime is comparably long. The following relaxation experiment confirms this.

Here we probe the re-dispersion of the elongational-flow-induced micelles upon flow cessation by a stress relaxation experiment. The decay of the elongational stress is found to be considerably slower than the relaxation of the shear stress. In Figure 7, the elongational stress relaxation process is shown after cessation of an elongational flow of strain rate 2 sec^{-1}. The behavior is very similar to that observed in shear stress relaxation of a shear-thickened state. A large-amplitude instantaneous decay is followed by a small-amplitude slow relaxation. The former represents the viscous stress contribution generated by flow, which drops to zero as soon as the flow stops. The latter is due to the decay of flow-induced structures. The slow elongational stress relaxation shown in Figure 7 may therefore be interpreted as evidence for the formation of large micellar structures prior to flow cessation and as the origin of the extremely high elongational viscosity shown in Figure 6. It is important to remark that stress relaxation upon cessation of a shear flow at any shear rate up to 100 sec^{-1} decays immediately, implying that micelles in shear do not grow larger and are much smaller than those formed in the elongational flow. Such a long relaxation time associated with the small decaying tail in Figure 7 implies a long micellar lifetime, and this is consistent with existence of the long induction time.

Thus our studies on micellar solutions serve to demonstrate most clearly that effects of even an extensional flow on a self-organizing thermodynamic system cannot be described within a generalized Gibbs statistical mechanical framework, because the effect of flow convection cannot be readily built into such a theoretical formulation. Thus a quantitative theoretical description of flow effects on self-assembling fluids such as micellar and block copolymer solutions remains a formidable theoretical problem.

Summary

We have studied reorganization of micelles through self-assembly in dilute aqueous solutions of CTAB/NaSal mixtures that are subjected to either shear or elongational flow. Our experiment to study effects of NaSal concentration supports the idea that the flow thickening phenomenon depends largely on the magnitude of the micellar lifetime. We show that the critical shear rate for the apparent shear thickening is approximately inversely proportional to the overall micellar relaxation time and micellar dissociation time. We report additional evidence that micelles recombine more effectively in an elongational flow, and the corresponding micellar growth is due to flow enhanced side-to-side coagulation. The speculation of micellar growth in elongational flow is supported by the long stress growth process, by the large

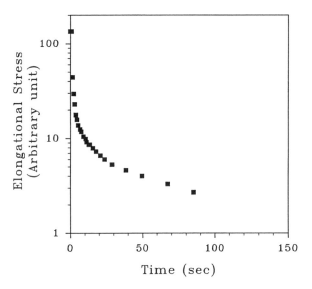

Figure 7. Elongational stress relaxation of the 2 mM equimolar NaSal/CTAB solution upon cessation of the flow of strain rate 2 sec^{-1}.

elongational viscosity, and by the presence of a slow relaxation component in the elongational stress upon cessation of flow.

Acknowledgments

This work was partly supported by the National Science Foundation through the Materials Research Group program (DMR 91-22227) at the Case Western Reserve University.

Literature Cited

1. See the review papers by H. Hoffmann and M. Cates in this book.
2. Rehage, H.; Hoffmann, H. *Rheol. Acta* **1982**, *21*, 561; Rehage, H.; Wunderlich, I.; Hoffmann, H. *Progr Colloid & Polymer Sci* **1986**, *72*, 51; Rehage, H.; Hoffmann, H. *J. Phys. Chem.* **1988**, *92*, 4712.
3. Candau, S. J.; Hirsch, E.; Zana, R. *J. Physique* **1984**, *45,,*1263.
4. Shikata, T.,; Hirata, H. *J. Non-Newtonian Fluid Mechanics* **1988**, 28, 171.
5. Hu, Y.; Wang, S. Q.; Jamieson, A. M. *J. Colloid Interface Sci.* **1993**, *156*, 31; Hu, Y.; Wang, S. Q.; Jamieson. A. M. *J. Rheol.* **1993**, *37*, 531.
6. Olsson, U.; Söderman, O.; Guéring, P. *J. Phys. Chem.* **1986**, *90*, 5223.
7. Ng, S. C.; Gan, L. M.; Chew, C. H. *Colloid Poly Sci* **1992**, *270*,64.
8. Imae, T. *J. Phys. Chem.* **1990**, *94*, 5953.
9. Shikata, T.; Hirata, H.; Kotaka, T. *Langmuir* **1987**, *3*, 1081.; Shikata, T.; Hirata, H. Kotaka, T. *Langmuir* **1988**, *4*, 354.
10. Imae, T.; Hashimoto, K.; Ikeda, S. *Colloid Polym Sci* **1990**, *268*, 460.
11. Wunderlich, I.; Hoffmann, H.; Rehage, H. *Rheol Acta* **1987**, *26*, 532.
12. Hu, Y.; Rajaram, C.V.; Wang, S. Q.; Jamieson. A. M. *Lamgmuir*, **1994**, *10*, 80.
13. Wang, S.Q. unpublished.
14. Wunderlich, I; Hoffmann, H.; Rehage, H. *Rheol. Acta* **1987**, 532.
15. Wang, S.Q.; Gelbart, W.M., and Ben-Shaul A. *J. Phys. Chem.* **1990**, *94*, 2219.
16. Cates, M.; Turner, M.S. *Europhys. Lett.* **1990**, *11*, 681.
17. Prud'homme, R.K.; Warr, G.G. see the paper in this book.

RECEIVED July 21, 1994

Chapter 20

Structure of Complex Fluids under Flow and Confinement

X-ray Couette Shear Cell and the X-ray Surface Forces Apparatus

Stefan H. J. Idziak[1], Cyrus R. Safinya[1,2], Eric B. Sirota[3],
Robijn F. Bruinsma[4], Keng S. Liang[3], and Jacob N. Israelachvili[1,5]

[1]Materials Department and Materials Research Laboratory,
University of California, Santa Barbara, CA 93106
[2]Physics Department, University of California, Santa Barbara, CA 93106
[3]Corporate Research Laboratories, Exxon Research and Engineering
Company, Annandale, NJ 08801
[4]Physics Department, University of California, Los Angeles, CA 90024
[5]Department of Chemical Engineering, University of California,
Santa Barbara, CA 93106

A review of recent synchrotron x-ray scattering studies of the non-equilibrium structure of flowing complex fluids using the X-Ray Couette Shear Cell (X-CSC) is presented and contrasted with studies of fluids under confinement with and without flow using the newly developed X-Ray Surface Forces Apparatus (X-SFA). This apparatus allows for simultaneous measurements of forces and structures of confined complex fluids under static and flow conditions. Under bulk flow, the non-equilibrium structures of the smectic-A phase of the liquid crystal (8CB), and the lyotropic liquid crystal lamellar L_α phases of surfactant membranes were investigated in the X-CSC apparatus. At high shear rates, the lamellar multilayer membrane phase orients primarily with the layers parallel to the shearing plates. At low shear rates a more complex phase behavior is observed with varying degrees of orientation. The liquid crystal layer orientation is strikingly different with the smectic-A phase material orienting with the layers perpendicular to the shearing plates at high flow rates. In thin 3900 Å films achieved in the X-SFA apparatus between two shearing mica surfaces, 8CB exhibits many distinct planar layer orientations; including a bulk flow-forbidden orientation. These experiments demonstrate the importance of using a direct structural probe such as synchrotron scattering in elucidating collective molecular structures under confinement and flow both in bulk and thin films.

At the macroscopic level, the distortion of complex fluids (*1-3*) such as colloidal suspensions, polymeric solutions, and liquid crystals, under flow is manifested as a breakdown of Newtonian hydrodynamics that requires the introduction of new normal stress terms in the hydrodynamic equations. In a landmark paper, Weissenberg first showed the necessity of considering normal stress differences in describing flowing surfactant solutions (*4*). The bulk macroscopic response of the material under stress results from the deformation of a large-scale interior structure with long orientational and translational relaxation times (*1-3*). This is in striking contrast to liquids comprised of small molecules (e.g. an oil molecule like decane), where, under bulk flow, the classical Newtonian theory of hydrodynamics gives an adequate description of the dynamics. Aside from the scientific interest, the response of complex fluids to flow fields is central to many technological processes; for example, in melt, gel, and solution spinning of polymer fibers and films (*5,6*). Therefore, in-situ bulk studies with a structural probe such as x-rays and neutrons are crucial since they allow for the imaging and subsequent correlation of the microstructure and the macroscopic mechanical properties.

In contrast with non-equilibrium flow induced structural arrangements, confinement may also lead to strong distortions of the collective structure of the trapped molecules of either simple or complex fluids, but now at equilibrium. For example, molecular dynamics simulations (*7-9*), indicate that increasing confinement may change the density and positional order, in addition to the molecular orientational order, of molecules and aggregates. This occurs especially as the gap size becomes comparable to an inherent length scale such as the diameter of suspended colloidal particles or the radius of gyration of dissolved polymer coils and, ultimately, the diameter of the trapped solvent molecules themselves.

In this work, we will review recent work in large gap flow in a specially designed X-Ray Couette Shear Cell (X-CSC) (*10-13*) and compare the results to small gap flow in the newly developed X-Ray Surface Forces Apparatus (X-SFA) (*14*). The bulk data was obtained on the dilute lamellar L_α phases of surfactant membranes (*12*) and on the smectic-A (SmA) phase of the liquid crystal 4-cyano-4'-octylbiphenyl (8CB) (*10-13*) composed of rodlike molecules (Figure 1). For these liquid crystalline materials, their bulk equilibrium structures have been previously investigated by x-ray scattering (*15-21*). The L_α phase is comprised of thin water layers coated with surfactant and cosurfactant molecules [sodium dodecyl sulfate (SDS) and pentanol] separated by dodecane (Figure 1C). The SmA phase we studied is thermotropic, and exhibits a temperature induced phase transition to the nematic phase at 32.5 °C. The nematic (N) phase is a liquid that has orientational order but no translational order (Figure 1A) (*22*). The critical temperature for the N-SmA transition we denote as T_{NA}. Both the L_α and SmA systems can be viewed as stacks of liquid layers within which the molecules are free to diffuse. We now discuss the experimental x-ray spectrometer before starting our discussion of the structure of bulk liquid crystals and membranes under bulk flow. This will be followed by the structure of the same liquid crystal material under confined conditions with and without flow.

Synchrotron X-Ray Scattering Spectrometers

X-Ray Couette Shear Cell (X-CSC). The experiments using the X-CSC were primarily carried out at the National Synchrotron Light Source (NSLS) on the Exxon beam line X-10A using 8 KeV radiation. Preliminary experiments were carried out at the Stanford Synchrotron Radiation Laboratory (SSRL) on beam line 6. Refering to

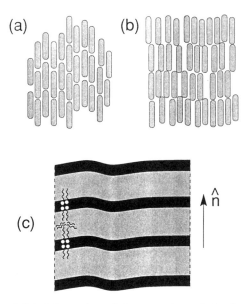

Figure 1. (a) and (b) Schematics of the nematic and (the layered) smectic-A phases of rod shaped liquid crystal molecules. (c) Sketch of the layered L_α phase of a surfactant membrane. (Reproduced with permission from ref. 13. Copyright 1993 AAAS.)

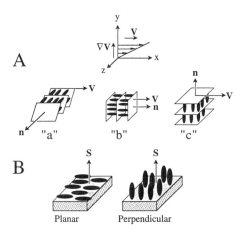

Figure 2. (A) Definition of real-space orientation directions used in the text: the "a," "b," and "c" layer orientations refer to the layer normal $\hat{n}$ pointing along the $\hat{z}$, the velocity ($\mathbf{v}$), and the velocity gradient ($\nabla\mathbf{v}$) directions, respectively. (B) The planar and perpendicular (homeotropic) orientations of elongated molecules on a surface are shown.

real space as defined in Figure 2A, three orthogonal directions can be defined where (v) and (∇v) are along the x- and y-axes, and the neutral direction parallel to the cylinder axis is along the z-axis. In reciprocal space, q_x, q_y, and q_z, are parallel to v, ∇v, and the neutral directions. Because of the small x-ray beam size of order 100 microns, the x-ray path may go through the X-CSC cell either through the center or the edge of the cell or any position in between these extreme positions. A Braun PSM-50 linear-detector placed 95 cm after the sample was positioned so that the scattering component collected in the channels was along q_x (q_y) parallel to v (∇v) when going through the center (edge). The setup gave a q_z resolution of 3×10^{-4} Å^{-1} (HWHM), with a linear-detector resolution of 1.2×10^{-3} Å (HWHM).

X-Ray Surface Forces Apparatus (X-SFA). The measurements using the X-SFA (*14*) were conducted on beamline 10-2 at the Stanford Synchrotron Radiation Laboratory, using the intense 31 pole wiggler x-ray source. A Si(111) monochromator was used at 8 keV with the beam focused at the sample position. A 125 μm pinhole was placed several centimeters before the sample in order to define the spacial extent of the beam. The diffraction patterns were measured using a Bicron scintillation detector. A combination of slits on the diffractometer yielded a longitudinal in-plane resolution of $\delta q_{\parallel} = 0.003$ Å^{-1} and an out-of-plane resolution of $\delta q_{\perp} = 0.02$ Å^{-1}.

X-Ray Couette Shear Cell Experiments

The complex fluid experiments were carried out between concentric cylinders transparent to x-rays, in a specially designed X-ray Couette Shear Cell (X-CSC) as discussed previously (*23*). In the experiments, the outer cylinder was rotating with the inner cylinder held fixed. This geometry results in an almost linear velocity profile with a constant shear rate $\dot{\gamma} = v/D$, where the gap size D varied between 250 to 2000 μm. The "a-", "b-", and "c-" layer orientations, originally introduced by Miesowicz (*24*) to describe flowing nematics, refer to cases with the layer normal n̂ pointing respectively along the ẑ direction, the velocity (v), and the velocity gradient (∇v) directions (see Figure 2A).

In the lamellar L_α phase, six mixtures were studied with the dodecane volume fraction Φ varying between 0.54 and 0.62, which corresponds to multilayer phases with the interlayer spacing increasing from about d= 118 Å to 160 Å (*12,13*). At high shear rates (*13*), the layers were oriented primarily parallel to the shearing plates as expected intuitively ("c" orientation of Figure 2A). The x-ray data in Figure 3A shows a three-dimensional plot of the scattering intensity for a typical L_α sample at Φ = 0.58 with d = 140 Å at high shear rates $\dot{\gamma} = 6000$ sec^{-1}, in the plane in reciprocal space containing $q_{\nabla v}$ (parallel to ∇v) and q_z (parallel to z). The peak maximum, which occurs at $q_{\nabla v} = 2\pi/d = 0.045$ Å^{-1} and $q_z = q_v = 0$, demonstrates that the layer normal n̂ is along the ∇v direction. For Φ ranging between 0.54 and 0.62, the shear rate where this main orientation was observed decreased from $\dot{\gamma} \approx 2200$ to 1100 sec^{-1}. At low shear rates, where the formation of spherical multilayer liposome structures has been observed (*25*), the scattering peak was isotropic (*12*), indicative of randomly oriented lamellae.

In the SmA phase, the corresponding behavior was different and quite unexpected. At low shear rates in the smectic phase, the layer normal was found to be distributed non-uniformly about an axis parallel to the flow direction, indicative of a heterogeneously mixed regime of SmA domains with layer orientations between that of the "a" and "c". At high shear rates, a pure orientation regime is observed where the layers orient perpendicular to the shearing plates ("a" orientation of Figure 2A). A data set is shown in this regime for a shear rate of $\dot{\gamma} = 300$ sec^{-1} at T = T_{NA} - 0.117 °C (Figure 3B). The peak position of the scattering maximum is rotated by $\pi/2$ with respect to that of the L_{α} data. The peak scattering intensity occurs in reciprocal space at a wave vector $q_z = q_0$ $\hat{n}$ with $q_0 = 2\pi/d = 0.198$ Å^{-1} (and $q_{\nabla_v} = q_v = 0$), where the SmA layer spacing is d = 31.73 Å. In other words, the SmA layers themselves are undergoing constant shear flow while the layer center of mass position does not change. The same orientation has recently also been reported for block-copolymer layered materials under shear flow (26,27).

The SmA data showing the (shear rate - temperature) phase diagram in the SmA phase in the vicinity of the transition temperature $T_{NA} = 33.58$ °C is summarized in Figure 4. A larger shear rate is required to remain in the pure "a" layer orientation state as the temperature is lowered below T_{NA} into the SmA phase. The detailed description of the orientational regimes of the N phase was reported in reference (10,11). While the experimentally observed "a" orientation for 8CB in the vicinity of the nematic to smectic-A phase transition regime is unexpected, a numerical study of the nonlinear dynamics of the liquid crystal nematic director shows that it results because of the flow distortion of thermal fluctuations (13). We now turn to a discussion of the structure of 8CB in thin films under flow.

X-Ray Surface Forces Apparatus Experiments

In this section, we consider very recent synchrotron x-ray scattering structural studies with the newly developed X-ray Surface Forces Apparatus (X-SFA) (14), in the "meso-scale" size range that spans the regime between about 500 and 10,000 Å. An understanding of the properties of such complex fluid systems is of fundamental scientific interest and also of importance to many technological fields such as lubrication, the flow of colloidal and biocolloidal particles through narrow membraneous or biological pores, and the processing of ceramic and polymer composite materials and films. The X-SFA is ideally suited for directly probing such meso-scale structures, under both stationary static and flowing non-equilibrium conditions. Significantly, the new technique is uniquely capable for separately probing the distinct effects of confinement and flow.

The effects of confinement on molecular conformations, dynamics (i.e. molecular transport), and thermodynamic phase behavior of liquids in porous media (with a pore size distribution) has been recently studied with spectroscopic (28,29), calorimetric (30), and neutron scattering techniques (31,32). At the true molecular level such studies have primarily been feasible with the Surface Force Apparatus (SFA) methodology, which allows for direct force, friction, and under flow, rheological measurements (33-36).

The X-Ray SFA (14) is based on a conventional miniature SFA (Mk III model) (37), which was modified for simultaneous on-line use with an intense synchrotron x-ray beam passing through the surfaces. In this device, the gap thickness between two atomically smooth surfaces can be adjusted from a few ångstroms to 10 μm, with control to ±1Å, by means of a four-stage coupled micrometer/differential micrometer/differential spring/piezoelectric crystal mechanism. The surfaces typically

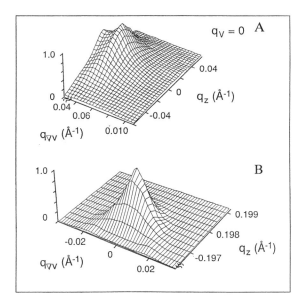

Figure 3. (A) A three-dimensional plot of the x-ray scattering intensity for an L_α membrane multilayer under high shear rates, plotted in reciprocal space in the $q_{\nabla v}$ - q_z plane, which shows a maximum at $q_{\nabla v} = 2\pi/d = 0.045$ Å^{-1} indicative of the intuitive "c" orientation with the layers sliding over each other. (B) Plot of the x-ray data in the smectic-A phase for a shear rate of $\dot\gamma = 300$ Hz at T = T_{NA} - 0.117 °C, where the position of the peak (at $q_z = q_0 \, \hat{n}$ with $q_0 = 2\pi/d = 0.198$ Å^{-1}, and $q_{\nabla v} = q_v = 0$, with d = 31.73 Å the layer spacing) scattering maximum is rotated by $\pi/2$ with respect to that of the L_α data indicating that the layers assume the unexpected "a" orientation in parts of the ($\dot\gamma$ - T) phase diagram shown in Figure 4. (Reproduced with permission from ref. 13. Copyright 1993 AAAS.)

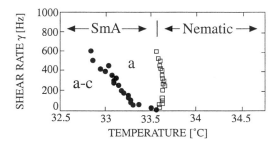

Figure 4. The (shear rate - temperature) phase diagram ($T_{NA} = 33.58$ °C at zero shear) in the smectic-A phase showing the pure "a" and the mixed "a-c" layer orientation regimes (shown schematically in Figure 2) as discussed in the text. (Reproduced with permission from ref. 13. Copyright 1993 AAAS.)

used are mica sheets in a crossed-cylinder geometry with diameters of order 2 to 5 cm, though in the experiments reported here much smaller diameters were used (described below). The X-SFA was mounted onto a Huber four-cycle diffractometer via an XYZ translation stage that allowed for the contact area between the two surfaces to be centered at the center of rotation of the four-cycle. The XYZ displacements have a range of ±1 cm, controllable to 3μm, in each direction.

Figure 5A shows a blow-up of the crossed cylinder geometry used in the X-SFA in which 2 mm diameter quartz capillary tubes with a wall thickness of 0.01 mm were used for supporting the mica sheets (14). Such thin-walled quartz glass tubes are necessary for performing x-ray diffraction experiments in the transmission mode where the beam must pass through the two glass supports as well as through the sample. The cylinders were 10 mm long and were flame sealed at both ends to improve rigidity. Back-silvered mica sheets were glued onto each cylinder to provide smooth surfaces for the measurements. The cylinders were then mounted in the X-SFA at 45° to the shear direction to enable x-ray access in reflection aside from transmission through the sample. The gap between the two cylinders was monitored by observing the Newton's rings pattern created by passing sodium light through the two cylinders. Horizontal shearing of the samples was performed through the use of a special piezoelectric bimorph lateral sliding device. A triangular wave voltage profile was applied to the device which created a tangential velocity of one cylinder with respect to the other.

A schematic of the two crossed cylinders is shown in Figure 5B. For the measurements presented below, the separation between the two surfaces was kept around 4000 Å. Note that this is the minimum spacing between the two surfaces; because of the curvature of the cylinders and the 125μm diameter of the incident x-ray beam, the actual spacing seen by the beam ranges from 4000Å to 10,000Å. Diffraction results from our first series of experiments are shown in Figure 6 for a shear rate, $\dot{\gamma}$, of 10 sec^{-1}. For these measurements, 8CB was chosen because, as described above, its bulk structure and orientation under flow has been characterized (10-13). It is also known that liquid crystals become oriented near surfaces or when confined between two surfaces. The two most common orientations are the "planar" and "homeotropic" orientations where the molecules align parallel or perpendicular to the surfaces, as illustrated in Figure 2B. Between mica surfaces, cyano biphenyl molecules such as 8CB are known to orient in the planar configuration, corresponding under flow, to the plane defined by the "a" and "b" directions. Thus, for 8CB between two shearing mica surfaces, flow and confinement are expected to have competing layer orientational effects, where the relative importance of the two effects is controlled by the experimental conditions of shear rate and gap thickness. 8CB was thus chosen since a direct comparison was desired to differentiate between flow and confinement effects.

To perform these measurements, the detector was set at a scattering angle $2\theta = 2.8°$, which corresponds to the 8CB layer spacing of 31.73Å. The incident x-ray beam was passed radially through both cylinders in the transmission mode. The X-SFA was then rotated about the incident beam through an angle χ (Figure 7A).

Figure 6 clearly shows that considerable alignment is occurring between the two crossed cylinders. No large qualitative change in the smectic layer orientation was observed between the case of no shear and the highest shear rate of 30 sec^{-1} studied. The peak at $\chi = -45°$ corresponds to the smectic layer normal $\hat{n}$ pointing along the

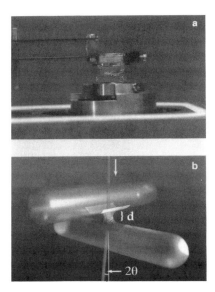

Figure 5. (a) Photograph of the two crossed capillary tubes used in the experiment and their mounts. (b) Schematic of the two crossed cylinders showing the gap (d) between the two confining surfaces. The x-ray beam, shown incident from the top in the transmission geometry, is scattered an angle 2θ by the sample.

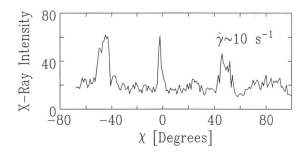

Figure 6. X-ray orientational scan showing the discrete planar molecular configurations at shear rate ($\dot{\gamma}$) equal to 10 sec^{-1}. The "a" orientation (see Figure 2) corresponds to χ being roughly $\pm 90°$; the "b" orientation to $\chi \approx 0°$. The shearing velocity was in the plane of scattering at $\chi \approx 0°$. These scans were taken on SSRL beamline 10-2. Each point was counted for approximately 17 seconds; the point spacing is $1°$.

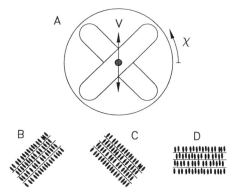

Figure 7. Orientation of the smectic layers of 8CB between the crossed cylinders of the XSFA as determined from the orientational scans shown in Figure 6. (A) Sketch of the two crossed cylinders, showing the shear direction as well as the orientation angle χ. The x-ray beam is shown passing through both cylinders. The peak at $\chi = -45°$ corresponds to the smectic layer normal lying parallel to the bottom cylinder (B), while the peak at $\chi = 45°$ corresponds to the layer normal being parallel to the top cylinder (C). The peak at $\chi = 0°$ is due to the layers taking up the "b" orientation (D).

lower cylinder axis shown schematically in Figure 7B. Likewise, the peak at $\chi = +45°$ corresponds to the layers lying normal to the top cylindrical axis (Figure 7C). The peak at $\chi = 0°$ is due to the layers orienting such that the layer normal is pointing along the shear direction, i.e.. in the "b" orientation (Figure 7D). This "b" orientation was not observed in the case of large gap flow studies in the smectic-A phase as it tends to tilt the layers, which is energetically costly (*12*). Clearly, the effects of confinement are dominating over flow-induced bulk layer orientational ordering. The "c" orientation was not observed.

There was no evidence of any smectic layer "a" orientation at zero shear. However, as the shear rate is increased, a weak broad peak does start to emerge at χ approximately 80°, indicating the onset of the "a" orientation seen in the bulk (*12*).

Higher resolution orientational scans around the peak at $\chi = 0°$ shown in Figure 6 reveal that the peak is not a single wide peak, but a collection of sharp independent peaks (*14*). This implies that the "b" orientation does not consist of a mosaic of small domains, centered about the expected orientation; instead, it is formed by *discrete domains*. This is a rather surprising result which we do not understand. However, the overall width of the "b" peak, which is 3.6 degrees, corresponds roughly to an angular variation in the orientation of the surfaces of the illuminated part of the cylinder (the 125 μm x-ray beam illuminates 1/50 of the cylinder circumference or 7.2 degrees).

Discussion and Conclusions

The majority of experimental work on flowing complex fluids has emphasized macroscopic rheological parameters, such as viscosity and normal stress measured either for large or small gaps. We have discussed two recently developed techniques, the X-Ray Couette Shear Cell and the X-Ray Surface Forces Apparatus, that are well suited for directly probing the structure of complex fluids under flow both for macroscopic fluids and in thin films. The data we showed emphasized the importance of a direct structural probe such as synchrotron x-ray scattering, in elucidating the micro-structure that underlies the macroscopic rheological shear and normal stress responses of flowing complex fluids.

The X-ray Couette Shear Cell (X-CSC) (*23*) enables one to probe the structural distortions in complex fluids such as liquid crystals, lyotropic surfactants and biomaterials, and polymeric fluids under non-equilibrium shear flow conditions. In particular, in contrast to existing neutron shear cells (*38-39*), the (X-CSC) is unique in that it allows the most relevant shear plane ($\mathbf{v}\text{-}\nabla\mathbf{v}$) to be accessed in addition to the other two planes containing the ($\mathbf{v} - \hat{\mathbf{z}}$) and the ($\nabla\mathbf{v} - \hat{\mathbf{z}}$) directions (which are the only planes currently accessible by neutron shear cells). We emphasize that the experiments are among the few microscopic studies, probing length scales < 100 Å and as large as a few microns, of the collective molecular structure under non-equilibrium steady-state conditions. Such studies have important bearings on our understanding of phases and phase transitions away from equilibrium and more generally of non-equilibrium statistical mechanical phenomena. More generally, the experiments (*10-13*) demonstrate that under flow, synchrotron x-ray diffraction techniques provide a powerful structural probe of steady-state dynamical behavior.

The Surface Forces Apparatus (SFA) is now generally accepted world wide as a powerful technique for measuring forces between surfaces in liquids (*33-36*). While researchers have studied the effects of confinement on molecular structure and dynamics in porous media with a large distribution of pore sizes (*28-32*), the SFA is the only technique that allows for such studies between two parallel surfaces with precisely controllable separations spanning a wide range from > 10,000 Å down to a few ångstroms.

Here, we have discussed a new technique (*14*) using an apparatus that we refer to as the X-ray Surface Forces Apparatus (X-SFA), that combines the force-measuring capabilities of the Surface Forces Apparatus (SFA) with the direct structural imaging capabilities of x-ray scattering. To illustrate the power of the technique, *in-situ* synchrotron x-ray scattering results of the structure and orientation of a confined thin liquid crystal film were presented which show how the molecules, which can be induced to order (differently) by flow, by surface fields and by confinement, respond to situations where all three competing constraints are operating at the same time. The new method opens the way for simultaneously measuring forces and micro-structure of fluids in thin films and confined geometries under static (equilibrium) but especially under dynamic (non-equilibrium) conditions.

Acknowledgments

(CRS) and (JNI) are grateful for partial support provided by the Office of Naval Research under grant N00014-93-1-0269. (CRS) is grateful for partial support of an Exxon Education Grant. The Couette Cell synchrotron x-ray scattering experiments described in this paper were carried out at on beam line X-10A at the National Synchrotron Light Source at the Brookhaven National Laboratories, which is supported by the U. S. A. Dept. of Energy. The X-SFA synchrotron x-ray scattering experiments were carried out on beam line 10-2 at the Stanford Synchrotron Radiation laboratory, which is supported by the U. S. A. Dept. of Energy. The Materials Research Laboratory at Santa Barbara is supported by NSF under grant No. DMR-9123048.

Literature Cited

1. See, for example, de Gennes, P.G. *Scaling Concepts in Polymer Physics* ; Cornell University Press: Ithaca, NY, 1979.
2. *Physics of Complex and Supermolecular Fluids*; Safran, S.A.; Clark, N.A., Eds.; John Wiley & Sons, Inc., 1987.
3. *Complex Fluids*; Sirota, E.B.; Weitz, D.; Witten, T.; Israelachvili, J.N. Eds.; Materials Research Society: Pittsburgh, PA, 1992, Vol. 248.
4. Weissenberg, K. *Nature* **1947**, 159, 310.
5. See e.g. Bastiaansen, C.; Schmidt, H.W.; Nishino, T.; Smith, P. *Polymer* **1993**, 34 (18), 3951.
6. Pearson, J.R.A. in *Mechanics of Polymer Processing*, Elsevier Applied Science: London, 1985.
7. Schoen, M; Diestler, D.J. and Cushman, J.H. *J. Chem. Phys.* **1987**, 87, 5464.
8. Rhykerd, C.L. Jr.; Schoen, M.; Diester, D.J.; and Cushman, J.H. *Nature* **1987**, 330, 461.

9. Thompson, P.A.; Robbins, M.O.; Grest, G.S. *Phys. Rev. Lett.* **1992**, 68, 3448.
10. Safinya, C.R.; Sirota, E.B.; Plano, R.; Bruinsma, R.F. *J. Phys. Condens. Matter* **1990**, 2, SA365.
11. Safinya, C.R.; Sirota, E.B.; Plano, R. *Phys. Rev. Lett.* **1991**, 66, 1986.
12. Sirota, E.B.; Safinya, C.R.; Plano, R.J.; Jeppesen, C.; Bruinsma, R.F.; *Mat. Res. Symp. Proc.* **1992**, 248, 169.
13. Safinya, C.R.; Sirota, E.B.; Plano, R.; Bruinsma, R.F.; Jeppesen, C.; Plano, R.J.; Wenzel, L.J. *Science* **1993**, 261, 588.
14. Idziak, S.H.J; Safinya, C.R.; Hill, R.S.; Ruth, M.; Warriner, H.E.; Kraiser, K.E.; Liang, K.S.; Israelachvili, J.N. *Science* **1994**, 264, 1915.
15. Litster, J.D.; Birgeneau, R.J. *Physics Today* **1982**, 26, May, 1982.
16. Pershan, P.S. *Structure of Liquid Crystal Phases* ; World Scientific: Singapore, 1988.
17. Davidov, D.; Safinya, C.R.; Kaplan, M.; Dana, S.S.; Schaetzing, R.; Birgeneau, R.J.; Litster, J.D. *Phys. Rev. B.* **1979**, 19, 1657.
18. Safinya, C.R. et al. *Phys. Rev. Lett.* **1986**, 57, 2718.
19. Safinya, C.R.; Sirota, E.B.; Roux, D.; Smith, G.S. *Phys. Rev. Lett.* **1989**, 62, 1134.
20. Smith, G.S.; Sirota, E.B.; Plano, R.J.; Clark, N.A. *J. Chem. Phys.* **1990**, 92, 4519.
21. Roux, D.; Safinya, C.R.; Nallet, F. in *Modern Amphiphilic Physics;* Editors, Ben-Shaul, A.; Gelbart, W.; Roux, D.; Springer: NY in press.
22. de Gennes, P.G. *The Physics of Liquid Crystals;* Oxford Univ. Press: London, 1974.
23. Plano, R.J.; Safinya, C.R.; Sirota, E.B.; Wenzel, L. *Rev. Sci. Instrum.* **1993**, 64, 1309.
24. Miecowicz, M. *Nature* **1946**, 158, 27.
25. Diat, O.; Roux, D. *J. Physique II* **1993**, 3(9), 1427; also see D. Roux this book.
26. Koppi, K.A.; Tirrell, M.; Bates, F.S.; Almdal, K.; Colby, R.H. *J. Phys.* **1992**, 2, 1941.
27. Koppi, K.A.; Tirrell, M.; Bates, F.S. *Phys. Rev. Lett.* **1993**, 70, 1449.
28. See e.g. Drake, J.M.; Klafter, J. *Physics Today* **1990**, 43 (5), 46 and references therein.
29. Drake, J.M.; Klafter, J.; Levitz, P. *Science* **1991**, 251, 1574.
30. Bellini, T. et al., *Phys. Rev. Lett.* **1992**, 69, 788.
31. Sinha, S.K. et al. to be published.
32. Clark, N.A. et al. to be published.
33. Israelachvili, J.N. *Intermolecular and Surface Forces;* Academic Press: London & New York, 1985 (1st edition), 1991 (2nd edition).
34. Van Alsten, J.; Granick, S. *Phys. Rev. Lett.* **1988**, 61, 2570.
35. Klein, J.; Perahia, D.; Warburg, S. *Nature* **1991**, 352, 143.
36. Israelachvili, J.N.; Homola, A.M.; McGuiggan, P.M. *Science* **1988**, 240, 189.
37. Israelachvili, J.N.; McGuiggan, P.M. *J. Mater. Res.* **1990**, 5(10), 2223.
38. Linder, P.; Oberthur, R. *Revue Phys. Appl.* **1984**, 19, 759.
39. Cummins, P.G.; Staples, E.; Millen, B.; Penfold, J. *Meas. Sci. Technol. (UK)* **1990**, 1, 179.

RECEIVED July 8, 1994

Chapter 21

Relation Between Rheology and Microstructure of Lyotropic Lamellar Phases

Didier Roux, Frederic Nallet, and Olivier Diat

Centre de Recherche Paul Pascal, Centre National de la Recherche Scientifique, Avenue Doctor Schweitzer, 33600 Pessac, France

Rheological behavior of lyotropic lamellar phases is studied as a function of the membrane repeat distance. The steady state rheology is described as a consequence of the so-called orientation diagram described previously. Three distinct regions of different orientations are described that are separated by two out-of-equilibrium transitions. We show that these transitions can be either discontinuous (subcritical) or continuous. In one of these transitions, one can go continuously from one regime to the other through a bifurcation point.

The effect of shear on systems having a large characteristic length allows us to describe the effect of shear on the microstructure. Typical systems studied are either near a second order phase transition *(1,2)*, or colloidal systems *(3,4)*. One basic issue of these works is to understand the viscoelastic behavior of fluids in terms of microstructure in the same way that statistical mechanics allows us to describe the stability and thermodynamics of equilibrium systems.

We have studied the effect of shear on a lyotropic lamellar phase. We have been able to show that the orientation taken by a lyotropic lamellar phase under shear can be described as steady states that are separated by transitions as a function of the characteristic distance between membranes and the shear rate *(5)*. Three different states of orientation of the lamellar phase have been described: an isotropic state, where the membranes form onion-like structures (multilayer spherical objects of size R much larger than the repeating distance d) exists at intermediate shear rates in between two other states (at either lower or higher shear rates) where layers are mainly parallel to the flow. The location of these regions in the shear rate/smectic-period plane has been called the orientation diagram *(5)*.

We also have studied the consequences of the so-called orientation diagram on the rheological properties of a lyotropic lamellar phase. Studying the shear rate as

0097–6156/94/0578–0300$08.00/0

a function of the stress, we have shown that the passage from one state to the other corresponds to out-of-equilibrium transitions. The first transition can be either discontinuous or continuous, depending upon the repeating distance, while the second transition seems to be always discontinuous. Upon approaching the bifurcation point where the first transition goes from continuous to discontinuous we show that oscillation in time may be observed. We interpret this behavior as due to a coupling between transitions and emphasize the fact that the rheological behavior has to be described within the framework of dynamic transition rather than static transitions.

A lyotropic lamellar phase made of water, Sodium Dodecyl Sulfate (SDS), pentanol and dodecane exhibiting a lamellar phase whose repeat distance lies between 60Å to 400Å *(6)* has been studied. This phase corresponds to layers of water surounded with surfactant separated with dodecane and is stabilized by undulation interactions *(7)*. A previous work, using Couette cells and different techniques, has shown that an orientation diagram can be described corresponding to different orientations of the smectic layers with respect to the flow field *(5)*. This diagram exhibits three states of orientation as shown in ref. 5. At very low shear rates ($\dot{\gamma}$ < 1 s^{-1}) and high surfactant concentrations, the membranes are mainly parallel to the flow with the smectic director parallel to the velocity gradient direction (region 1). In this state, many defects (probably dislocations) persist in the two directions perpendicular to the director and are presumably similar to the ones described by Oswald and Kléman for thermotropic systems *(8)*. At a higher shear rate or for more dilute systems, a new state appears where the smectic layers (membranes) form multilayer spherical droplets of a well defined size, controlled by the shear rate, ranging typically from 10 μm to less than 1 μm (region 2) *(5,9)*. At even higher shear rates, a state where the membranes are parallel to the flow with the smectic director parallel to the gradient of velocity direction is stable. This state has some similarities with the first one but no defects remains in the direction of the flow (region 3). When observations are made in a cell where the shear rate is fixed, regions 2 and 3 are separated with a region where the two states coexist (region 2+3).

In order to get rheological information, experiments have been carried out with a Rheometer Carrimed 100 that fixes the stress and measures the velocity (shear rate). We have used a Mooney cell corresponding to a Couette cell terminated by a cone/plate at the bottom in order that the shear rate is uniform throughout the cell. We have measured for different repeating distances the shear rate ($\dot{\gamma}$) as a function of the stress (S). Three regimes are described corresponding to 3 different power laws (S ∝ $\dot{\gamma}^x$). Regions 1 and 3 correspond to an exponent x = , and region 2 corresponds to x = 0.2. One sees that regions 1 and 3 correspond to a Newtonian behavior (S ∝ $\dot{\gamma}$, viscosity: η = constant, see Figure 1) but exhibit very different viscosities. Region 2 corresponds to a continuous shear thinning (viscosity decreasing with the shear rate η ∝ $\dot{\gamma}^{-0.8}$). We also see that the passage from regions 2 to 3 corresponds to a jump

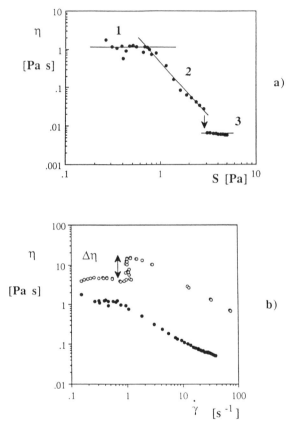

Figure 1. Viscosity as a function of the stress for different samples (71 % of oil volume fraction, 1a) and as a function of the shear rate (1b samples 50% and 69%).

in the shear rate for a given value of the stress, while the passage from region 1 to 2 corresponds to either a jump in the stress (or viscosity) at constant shear rate or to a continuous process depending upon the dilution.

Let us first study the second transition between regions 2 and 3. When the stress at which the transition occurs is reached (around 4 Pa for the 69% of oil sample and 50 Pa for the 50 %), a very small increase of the stress leads to a jump in the shear rate value. For the 69% sample the jump corresponds to an evolution of the shear rate from typically 200 s^{-1} to 900 s^{-1}. For the 50% sample the jump starts at 1000 s^{-1} and ends above the maximum shear rate that the apparatus can measure (1200 s^{-1}). If measurements are made relatively rapidly (not waiting for the complete steady state equilibrium) one observes an hysteresis cycle.

The first transition is more complex (between states I and II). Indeed, for concentrated samples ($\phi_{oil} < 68\%$) the transition is discontinuous in stres, but it becomes continuous for $\phi_{oil} > 68\%$. The transition between discontinuous ($\Delta\eta \neq 0$) to continuous ($\Delta\eta = 0$) transition is a bifurcation point *(10)*.

The fact that these transitions are out-of-equilibrium in nature leads to a behavior that is expected only from a description of stationary states. Indeed, in these transitions, either feedback effects or coupling between transitions can lead to time evolutions that are no longer stationary but oscillations or chaotic behavior as a function of time are expected in certain cases. We may expect to find such states in the rheological behavior of these systems *(11)*.

Let us come now to a microscopic explanation of the observed phenomenon. As already pointed out *(5-9)*, under shear the preferred orientation is given by the velocity in the plane of the layer and the gradient of velocity perpendicular to the layers, as observed at very low and high shear rates. Within this orientation the gap in which the lamellar phase is moving is of the order of 1mm with a precision that is at best of the order of 1/100 mm. Consequently, the lamellar phase develops defects that are only slightly anisotropic at a low shear rate *(13)*. In this observed state, the system flows by moving dislocations *(14)* and is very viscous but Newtonian ($\eta \approx$ 1000 mPa s) . For intermediate shear rates, the sample is forced to move faster and the dislocations cannot follow. The fact that the plate movements move faster than the rate of displacement of the dislocations creates a pressure perpendicular to the layer and the smectics develop an undulation instability corresponding to a lattice of dislocations at a length L=$(\lambda D)^{1/2}$ (D being the distance between the plates and λ the penetration length of the smectic phase which is of the order of the d-spacing) *(8)*. These dislocations forbid the system to flow and instead, the system bifurcates to another orientation which consists of small spheres rolling on each other to allow the flow to proceed. At a higher shear rate the lattice of dislocations corresponding to an undulation instability is so anisotropic *(14)* that no dislocations are left in the direction of the velocity and the oriented state beciomes again the most stable phase with a viscosity of the order of a few mPa s ($\eta \approx$ 3-5 mPa s).

In the intermediate region, we can calculate the characteristic size of the onions balancing two forces: an elastic one f_{el} and the viscous force f_{vis} *(15)*. The elastic force needed to maintain a lamellar phase at a size R can be expressed by:

$$f_{el} = 4\pi(2\kappa + \overline{\kappa})/d \qquad (1)$$

where κ and $\overline{\kappa}$ are respectively the mean and Gaussian elastic constant of the membrane.

The viscous force that a droplet experiences in a flow is *(15)*:

$$f_{vis} = \eta R^2 \dot{\gamma} = R^2 S \qquad (2)$$

with $\dot{\gamma}$ being the shear rate, η the viscosity and R the onion size and S the stress. Balancing (1) and (2) we can calculate the equilibrium size for the steady state:

$$R = \sqrt{\frac{4\pi(2\kappa + \overline{\kappa})}{S\,d}} = \sqrt{\frac{4\pi(2\kappa + \overline{\kappa})}{\eta\,d\,\dot{\gamma}}} \qquad (3)$$

Equation 3 has been quantitatively checked *(5,9)*. A more microscopic model has been proposed leading to basically the same scaling *(16)*.

In order to decide whether the transition between the states 1 and 2 is continuous or not we have to compare two lengths, L (at which the dislocation array forms) and R. If L > 2R, we may understand that as soon as the dislocation array wants to form, the system has no problem to bifurcate to the onion state. However, if L < 2R, the viscous force is not large enough to form onions of size at least equal to L. The system then stays at the limiting velocity below which no dislocation array is developed and builds up stress until it reaches the value needed to form onions of size L. The system undergoes a discontinuous transition with a jump in stress. Since the elastic constants are known for this system, we can put in some quantitative numbers: we find that *(6)*:

$$L = (8/4\pi \; \kappa/kT \; d \; D)^{1/2} \qquad (4)$$

As expected, L is an increasing function of d and R a decreasing one. The dilution (corresponding to a value d* of the d-spacing) at which we expect the transition to go from continuous to discontinuous can be quantitatively estimated. With κ being of the order of 1 kT and $\overline{\kappa}$ = -1 kT estimated from the variation of the onion size with the shear rate *(7)*, we get:

$$d* = \pi kT \sqrt{\frac{2}{\kappa \eta \dot{\gamma} D}} \qquad (5)$$

and is of the order of 30 nm which is effectively very close to the experimentally observed value.

In conclusion, one would like to stress that the complicated behavior exhibited by the rheology of the lyotropic smectic phase can be interpreted at the microscopic level as different states of orientation separated by out-of-equilibrium transitions.

Literature cited

1. Beysens, D.; Gladamassi, M. *J. Phys. Lett.* **1979**, *40*, 565
2. Safinya, C. R.; Sirota, E. B.; Plano, R.J.; Bruinsma, R.
   J. Phys. **1991**, *C2*, 365
3. Pieranski, P. *Contemp. Phys.* **1983**, *24*, 25
4. Ackerson, B. J.; Pusey, P. N. *Phys. Rev. Lett.* **1988**, *61*, 1033
5. Diat, O.; Roux, D.; Nallet, F. *J. Physique II France* **1993**, *3*, 1427
6. Bellocq, A. M. ; Roux, D. In *microemulsions: structure and dynamics*,
   Friberg, S. E.; Bothorel, P. Eds
   CRC Press, Boca Raton, F. L., USA 1987, p. 33
7. Roux, D.; Safinya, C. R. *J. Phys. France.* **1988**, *49*, 307
8. Oswald, P. ; Kléman, M. *J. Physique Lett.* **1983**, *43*, L411-L415
9. Diat, O.; Roux, D. *J. de Phys. II* 1993, *3*, 9
   Roux, D.; Diat, O. French patent number 92-04108
   Roux, D., Diat, O. and Laversanne, R. PCT, FR 93-00335
10. Guggenheim, J.; Holmes P. In *Non linear oscillations dynamical systems
    and bifurcations of vector fields* (Springer, New York 1983).
    Nicolis, G.; Prigogine, I. In *Self-organization in nonequilibrium systems*
    (Wiley, New York 1977)
11. This will be discussed in a forthcoming publication.
12. de Gennes, P.G. In *The Physic of Liquid Crystals*, (Clarendon Oxford 1974)
13. Clark, N.; Meyer, R.; *Appl. Phys. Lett.* **1973**, *22*, 493
14. Oswald, P.; Ben-Abraham, S. I. *J. Physique* 1983, *43*, 1193-1197
15. Taylor, G. I. *Proc. Roy. Soc.* 1932 , *A 138*, 41-48
    and *Proc. Roy. Soc.* **193** , *A 146*, 501-523
16. Prost, J.; Leibler, L.; Roux, D. to be published

RECEIVED July 24, 1994

Chapter 22

Steady Shear Behavior of Ternary Bicontinuous Cubic Phases

Gregory G. Warr and Chih-Ming Chen

Physical and Theoretical Chemistry Division, University of Sydney,
New South Wales 2006, Australia

The steady shear rheology of bicontinuous cubic phases comprising didodecyldimethylammonium bromide, water and dodecane or cyclohexane has been studied. Yield stresses and non-ideal plastic flow were measured in all systems, with a transition to linear flow behavior at higher shear rates. The yield stress is found to depend on the composition of the cubic phase, but is less sensitive to details of the lattice structure. A fracture mechanism for the yielding is proposed. In the nonlinear flow region, slow relaxation of the fluid structure was observed, with a time constant of approximately 250s. Possible structures for the cubic phase under shear are discussed.

Cubic phases are among the least well-understood surfactant self-assembly phases. In the hierarchy of self-assembly structures they rank among the most complex, exhibiting crystalline long-range order in three dimensions. They are optically isotropic and hence often hard to identify by simple tests, unlike lamellar or hexagonal phases. Cubic phases are generally sufficiently stiff that they will not flow under their own weight, and have been referred to among other things as "viscous isotropic phases" (1), "viscoelastic gel" (2), and "ringing gels" (3). They are also known as microemulsion gels, and this can lead to confusion since highly viscous microemulsions are also known to form. However, these lack the order which is observed in small-angle scattering studies of cubic phases.

In general, little is still known about the structure of most cubic phases. Models for their structure fall broadly into two classes. In the first the cubic phase is comprised of discrete particles distributed in an ordered arrangement. These may be spherical micelles sitting on lattice points of a crystal array, or the micelles themselves may be anisotropic but may arrange themselves to form an isotropic unit cell. The second model is one in which the cubic phase is bicontinuous. This has recently been shown to be the case for systems containing the surfactant didodecyldimethylammonium bromide, which is the focus of this work.

0097–6156/94/0578–0306$08.00/0

The surfactant didodecyldimethylammonium bromide (DDAB) is known to form a wide variety of self-assembly phases in the presence of water and an immiscible oil (*4-6*). In water alone, DDAB has a solubility of around 0.1mM, and above this concentration forms two lamellar phases, one between 3 and 28 wt% surfactant (L_α) and the other above 75 wt% (L'_α) (*7*). At intervening compositions the systems are biphasic. Each of these lamellar phases can be swollen somewhat be the addition of a water-immiscible second solvent, and this leads also to the formation of microemulsion, cubic, and hexagonal phases in different regions of the ternary composition diagram. Partial phase diagrams of DDAB, water, and dodecane or cyclohexane showing several of the single phase regions are shown in Figure 1. The shaded areas correspond to two- or three-phase regions, or to single phases whose boundaries are not well known.

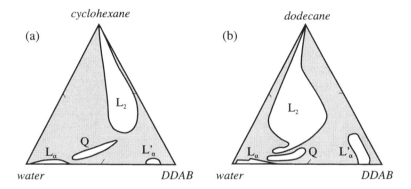

(a) cyclohexane — L_2, Q, L_α, L'_α — water, DDAB

(b) dodecane — L_2, Q, L_α, L'_α — water, DDAB

Figure 1   Phase diagrams of DDAB, water, and (a) cyclohexane or (b) dodecane systems showing cubic (Q), microemulsion (L_2), and two lamellar (L_α and L'_α) phases.

The phase diagram is dominated by the L_2 or microemulsion phase, which is bicontinuous throughout most of its range of stable compositions, and consists of an oriented monolayer of DDAB separating oil and water domains. The structure is easily visualised as an interconnected network of water tubes dispersed in oil (*6*). Each tube is coated with an oriented monolayer of surfactant "wrapped around" the water region. A wide range of compositions is accommodated by changing the number density of tube intersections, or nodes, and the number of tubes which intersect at each such node. As water is added to a particular microemulsion, diluting it towards the water corner of the ternary phase diagram, the connectivity of the microemulsion decreases and it eventually disconnects into discrete water droplets dispersed in oil. This structure occupies a thin wedge of the L_2 region at high water contents.

The cubic phase structure in these systems has been identified as a surfactant bilayer swollen with oil, and lying on a geometric surface of constant mean curvature (8-10). This surface divides the majority solvent, water, into two interpenetrating but unconnected labyrinths. On a local scale, the bilayer resembles that within a swollen lamellar phase, consisting of two oriented surfactant monolayers with oil between the alkyl tails. Like most surfactant self-assembly phases, the structure can be rationalised in terms of the packing of the surfactant molecules at the hydrophobic/hydrophilic interface. The optimal packing is governed by the anisotropy in the lateral interactions between adjacent head groups and those occurring between oil-swollen chains, and is encapsulated in the surfactant packing parameter, $v_{eff}/a_0 l_c$. Here v_{eff} is the volume of the surfactant alkyl tails, including penetration of dodecane or cyclohexane between the tails, a_0 is the molecular area occupied by the surfactant head group, and l_c the alkyl chain length of the surfactant. This parameter essentially measures whether the fully extended alkyl chains would occupy more or less area than the head group, and hence form structures with an interfacial film that is concave or convex with respect to the hydrophobic region. In DDAB systems $v_{eff}/a_0 l_c$ is very near to one, so that planar structures are favored. This is the case in the lamellar phases formed in aqueous solution. The presence of oil increases $v_{eff}/a_0 l_c$, so that inverted structures such as those found in the microemulsion phase arise.

As shown in Figure 1, the more dilute lamellar phase can only be swollen by a few percent by the addition of oil before becoming unstable. Between this phase boundary and the onset of the microemulsion phase, which has a $v_{eff}/a_0 l_c > 1$ (5, 6), lies the cubic phase, consistent with a surfactant film which is slightly concave towards water, but which still supports a bilayer structure.

The cubic phase is highly ordered, displaying sharp peaks in neutron and x-ray diffraction experiments. Hence the surfactant bilayer adopts a regular structure in its division of the sample space.

It has recently been demonstrated that the regions marked as cubic phases in Figure 1 are actually composed of a sequence of distinct phases, all of cubic symmetry, and with intervening two- and three-phase regions. The most extreme case discussed up to the present is that of DDAB with water and styrene (9), in which the cubic region contains five separate single phase regions, all with different structures and all bicontinuous. Cyclohexane has been shown to have at least two cubic phases (10), as have octane and dodecane (11). With other oils the situation is less clear.

This chapter describes a preliminary investigation of the shear rheology of ternary cubic phases of DDAB in the presence of water and either cyclohexane or dodecane as the immiscible oil. The behavior of samples spanning the single phase regions of both oils under a constant applied stress have been examined, and in the case of cyclohexane we have studied systems with compositions identical to those of Barois et al. (10), so that the structure is well characterised.

Experimental Section

Cubic phases were prepared by mixing didodecyldimethylammonium bromide, DDAB (Eastman), with water and cyclohexane (Aldrich, >99%) or dodecane (Fluka, AR grade). The DDAB was recrystallised from ether (dried over sodium) before use, and the water was doubly-distilled, the second stage from alkaline permanganate solution. Components were sealed in a stoppered glass vessel and warmed to 85°C until a transparent solution was formed. The samples studied are more fluid at this temperature and mixing is therefore facilitated. Each sample was then gently shaken to remove bubbles and ensure that any material adhering to the sides of the flask was incorporated into the homogeneous phase formed. The sample was then cooled over a period of approximately two hours to room temperature.

Four samples spanning the known cubic phase regions of the DDAB/water/dodecane and DDAB/water/cyclohexane systems were examined, and the compositions studied are listed in Table I.

Flow behavior was studied using a Deer constant shear stress rheometer in the Couette configuration. The gap between the inner and outer cylinders was 1.5 mm. This apparatus is particularly useful for examining systems with a yield stress, as the usual procedure of extrapolation to zero shear rate is avoided. The sample (approx. 14 ml) is thermostatted in a thermal jacket and the inner cylinder driven at constant torque by a motor operating on an essentially frictionless air bearing. The shear rate is measured from the angular velocity of the inner cylinder at a given applied stress, calculated from the known geometry of the experiment. Shear rate and shear stress were calibrated against a variety of aqueous solutions of glycerol. All calibration solutions were found to be Newtonian and no yield stresses were detected down to the sensitivity limit of the instrument, 5×10^{-3} Pa.

Results and Discussion

All cubic phases examined display a yield stress that depends strongly on composition and on the type of oil used. The measured yield stresses for each of the systems studied are listed together with their compositions in Table I. As can be seen, cubic phases containing cyclohexane have higher yield stresses than those with dodecane. 115 Pa is the maximum stress achievable with the geometry we

Table I. Compositions and yield stresses of DDAB cubic phases.

| | *Compositions* | | | | | |
	DDAB	water	oil	yield stress /Pa	lattice spacing /Å	space group[10,11]
cyclohexane						
a	50.5	36.8	12.8	>115.0	73.9[10]	*Pn3m*
b	47.0	44.6	8.5	>115.0	85.4[10]	*Pn3m*
c	40.6	52.1	7.4	101.2	126.7[10]	*Im3m*
d	32.6	60.2	7.2	78.2	155.3[10]	*Im3m*
dodecane						
e	37.6	56.0	6.4	87.4	~107[11]	*Pn3m*
f	32.2	63.0	4.8	52.9	~133[11]	*Pn3m*
g	24.8	70.0	5.2	36.8	~109[11]	*?C4*
h	22.5	73.5	4.0	34.5	~115[11]	*?C4*

employed, and thus represents a lower limit on the actual value.

The observation of yield stresses in surfactant systems is unusual, although foreshadowed by the elasticity which leads to their "ringing gel" appellation (3). DDAB-water-dodecane microemulsions, which are also bicontinuous over much of their composition range, have no measurable yield stress and are Newtonian at shear rates from 0.3 to at least 10^4 s^{-1} (12, 13). Although many micellar solutions display complex elastic behavior and form interpenetrating networks, there are to our knowledge no reports of micellar solutions which yield. Even the L_3 or sponge phase, which is a disordered bilayer network structure most like the cubic phase, has no yield stress and is Newtonian at shear rates up to 200 s^{-1} (14).

Because of high yield stresses of the cyclohexane systems, only the flow curves of dodecane-containing cubic phases could be examined. These were all characterised by non-ideal plastic flow. Above their yield stresses, all samples exhibited an initially nonlinear shear rate-shear stress response, displaying constant plastic viscosities only at higher shear. In the nonlinear region the flow displayed some hysteresis due to a slow relaxation of the structure, but at high shear rates this behavior ceased. Flow curves obtained after the achievement of steady state for DDAB/water/dodecane systems are shown in Figure 2.

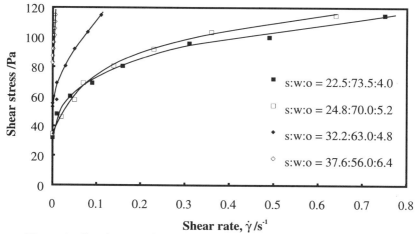

Figure 2   Steady state flow curves for DDAB/water/dodecane cubic phases.

Yield Stresses   Cubic phases of DDAB are triply-periodic bilayer networks separating two interpenetrating but distinct water domains. The midplanes of these bilayers are described by periodic minimal surfaces. An example of such a structure with body-centered cubic symmetry is depicted schematically in Figure 3a, showing a unit cell. Figure 3a shows the Schwartz P surface, which describes the midplane of an oil-swollen surfactant bilayer separating the interpenetrating water labyrinths. It should be noted that, although the surfactant film is drawn as a 2-dimensional surface, it actually has a finite thickness due to the surfactant chain lengths and the

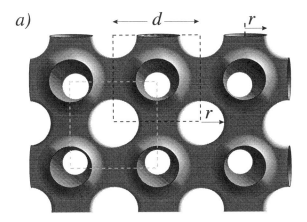

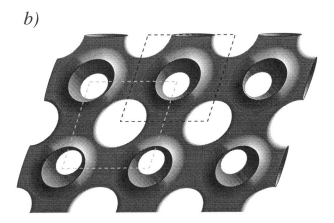

Figure 3 Schematic diagram of cubic phase with simple cubic structure showing lattice spacing, d, and neck radius, r.

interlayer oil. Each labyrinth forms a simple cubic lattice, but the two equivalent labyrinths are related by translation, giving rise to overall bcc symmetry.

From SAXS studies (*10*) we know that the two cyclohexane samples with measurable yield stresses (*c* and *d*), both belong to the *Im3m* space group as shown in Figure 3a (see Table I). Those outside the measuring range have a different structure, and Barois *et al.* attribute the diffraction pattern to a *Pn3m* space group, which has diamond symmetry and is described by the Schwartz D surface. Where the symmetries of the dodecane samples are known, they correspond to the *Pn3m* space group, however the structures of the two highest water content samples have not been resolved (*11*). For cyclohexane the compositions studied are identical to those of Barois *et al.* (*10*), and for dodecane we have listed the nearest compositions from the more recent work of Maddaford and Toprakcioglu (*11*).

A shear stress applied to such a structure translates one layer of unit cells with respect to other layers and also deforms the bilayer, transforming circular cross-sectional elements into ellipses, as shown in Figure 3b. This transformation conserves the volume fractions of the two water networks, but affects both local curvature and interfacial area of the bilayer. Apart from wall slip, the stress can only be relieved by the unit cells jumping into register above the next available lattice site.

Because of the multiply connected structure, this is impossible to achieve without breaking the bilayer, which forms the sample-spanning network. Hence yielding of the material must be associated with the destruction of the bilayer. It might therefore be reasonable to expect that the yield stress depend as a first approximation simply on the volume fraction of network-forming material. *i.e.* oil + surfactant. Yield stresses for both dodecane and cyclohexane systems are plotted together in this way in Figure 4. A monotonic increase in yield stress with increasing bilayer volume fraction is evident, and results for both systems seem to lie on a single curve.

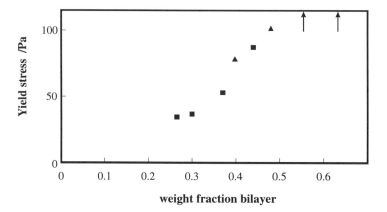

Figure 4 Yield stress dependence on composition of (■) dodecane and (▲) cyclohexane cubic phases. Arrows indicate out of range samples.

The lattice parameters, which correspond to the dimensions of the unit cells for these systems, are listed in Table I. There is no obvious universal relationship between crystallographic structure and rheology at this level, although it is noteworthy that for samples within the same symmetry group, the yield stress is smaller for a larger unit cell. To describe the yield stresses we therefore look to the global properties of bicontinuous cubic phases.

Although not extensively documented in self-assembly systems, yield stresses are well known in concentrated colloidal dispersions where an ordered phase is formed. The order in colloidal crystals gives a superficial similarity to bicontinuous cubic phases, and the relief of a shear stress is believed to occur by jumping of particles from lattice site to lattice site, as was suggested above for cubic systems. The difference arises in the topology: there are no physical connections between layers in a colloidal crystal, so that rupture of an actual surface is not needed for flow, which occurs by motion of discrete particles. Shear forces in such systems have been modelled with considerable success using a sinusoidal interaction potential, U, between sliding layers of particles of the form (*15*)

$$U = U_0 \left\{ \cos\left(\frac{2\pi x}{\lambda} - \pi \right) + 1 \right\} \qquad (1)$$

where x is the direction of shear, λ is the distance between particles, and U_0 the well depth. In the interconnected, triply periodic structures that constitute cubic phases, a layer sliding model, thus, is also a likely candidate to describe the flow. However, the nature of the interactions between layers is quite different from that for particles.

Referring again to the section of a deformed bcc cubic phase depicted schematically in Figure 3b, upon shearing the film is expected to rupture around the shortest perimeters. That is, at the narrowest necks of the conduits where the smallest cross-sectional area of membrane will be exposed. In a surface of zero mean curvature there are two such symmetry-related necks, one for each of the interpenetrating labyrinths.

The yield stress for a given surface geometry should therefore depend on the total area of hydrophobic material exposed to water at such an annulus, and the hydrophobe/water interfacial tension, $\gamma_{o/w}$. If the radius of the neck is r, and the bilayer thickness t, then the yield stress is given by the work of exposing two annuli of hydrophobe per rupture

$$\sigma_y = 4\pi r t n_\perp \gamma_{o/w} \qquad (2)$$

where $n_\perp$ is the number of channels perpendicular to the plane of shear ruptured per unit volume. The difference between physically connected and particulate systems is that once the surfactant film is ruptured the interlayer interaction is severed and hence there is no periodicity in the interaction.

As the cubic phases lie along lines of approximately constant surfactant/oil ratio, the thickness of the bilayers ought not vary with water content, so only r and $n_\perp$ will vary with composition. For a bcc lattice with a surface of zero mean curvature, symmetry dictates that the radius of curvature must be $d/4$, where d is the dimension of the unit cell. The number density of break points is simply proportional to the number density of unit cells, $1/d^3$, giving

$$\sigma_y \propto 1/d^2 \qquad (3)$$

For other symmetries a similar simple relationship will exist between the curvature of the film and the lattice spacing, so that equation 3 should still apply, although the constant of proportionality may differ. The only two samples for which both yield stress and lattice spacing are known do indeed scale as indicated, although this is hardly a rigorous test and much further work is needed.

In order to calculate the yield stress explicitly, we must estimate the number of necks which must be ruptured to permit flow. Using equation 2 for a bcc lattice with t = 28 Å (10) and $\gamma_{o/w}$= 15 mN m^{-1}, and *assuming* that every neck in the shear gradient direction must be ruptured (this is equivalent to breaking the cubic phase into layers of thickness *d*), the calculated yield stresses of samples *c* and *d* would be 5.5 x 10^5 and 8.3 x 10^5 Pa, respectively. This is clearly beyond reasonable bounds. However, using the experimental yield stresses suggests that the number of necks ruptured in each sample in the 1.5 mm gap between inner and outer cylinder in our apparatus is 12. That is to say, for both samples the equivalent of twelve layers need to be ruptured before the system can flow. The numerical value is not in itself significant, and is changed if another value for *t* and $\gamma_{o/w}$ is selected, however it would remain the same for both systems. The main conclusion we can draw from this is that the cubic phase probably fractures along a small number of planes when the yield stress is exceeded.

Of course only *one* layer need actually be cleaved fully in order that flow ensue in this system. Ruptured necks in the sample play the part of defects in the crystal lattice, and the simple model developed above cannot address interactions between such defects, nor how stress is transmitted through the sample. One can easily imagine how the fracturing of connections would weaken the surrounding structure, leading to a concentration of defects and possibly an "unzipping" of connections leading to yielding of the material.

Flow Curves and the Dynamic Structure of Cubic Phases The yield mechanism leads us to consider the structure of the sample under steady shear and the shape of the flow curves. As Figure 2 indicates, the shear rate-shear stress curves are initially nonlinear, but subsequently become linear further above the yield stress. In the nonlinear region the samples exhibit some flow hysteresis, with a higher shear rate measured for a given applied stress upon increasing stress than is measured when stress is decreased to the same value. The hysteresis loop for sample *h* is shown in Figure 5. Note that the observed hysteresis is confined to a narrow range of shear stresses, or shear rates. At higher shear stresses, in the linear part of the flow curve, no hysteresis was observed.

All samples also recovered their initial yield stress when the shear stress was slowly decreased towards σ_y. This is due to the dynamic nature and resilience of surfactant self-assembly structures, which distinguishes them so markedly from many analogous polymeric structures. As with micellar solutions under various flow conditions (16, 17), destruction of the self-assembly structure is not irreversible, and reduction of the applied stress allows the cubic phase to re-form its equilibrium structure.

The flow hysteresis indicates a slow relaxation of the structure, the kinetic characteristics of which are discussed below. The dynamic structure formed at higher shear rates reorganises rapidly on the time scale of experimental observation, a few seconds, but at lower shears this is not so. Any proposed structure under

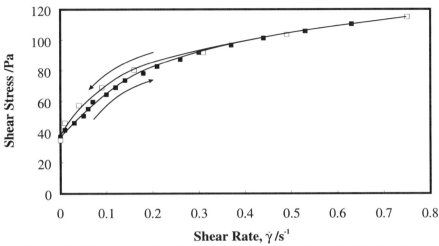

Figure 5 Flow curve for dodecane cubic phase (sample *h*), showing hysteresis loop obtained in intermediate shear region.

shear must accommodate this observation.

At first glance, the obvious structure for cubic phases to form under shear would be sliding layers of a "holey" lamellar phase. With increasing shear rate the structure could be fractured into thinner layers, which would slide past each other more and more easily, giving the observed shear thinning behavior. The hysteresis might then be due to the achievement of a steady state thickness of the layers from a history- dependent initial fracture pattern. At higher shear rates the number of layers is greater, so that a changing distribution is not apparent in the flow curves. However, this neglects the fluctuations that might occur in the layers and the torque that is applied to a sample in simple shear flow. As Roux has recently demonstrated for a range of lamellar phases, the application of shear destroys the long-range order in the system and leads to the formation of multilamellar vesicles (*18*). In fact they observe a lamellar phase at shear rates below 0.5 s^{-1}, vesicles between 0.5 and several hundred reciprocal seconds, and the recovery of an annealed or defect-free lamellar phase at even higher shear rates.

This raises at least the possibility of similar behavior in cubic phases once they are fragmented into layers. At present we cannot distinguish whether a stable layered phase is formed or not under steady shear, however a light scattering and optical birefringence apparatus currently under construction in our laboratory may enable us to do so.

Stress Relaxation The flow hysteresis observed in these systems relaxes with a time constant of about 300 s, which we measured by monitoring the relaxation of the shear rate after the abrupt application of a constant shear stress. Upon increasing the applied stress, the system jumps to a new shear rate which then decreases slowly to a final, steady-state value. This time dependence persists over a period of 5-10 minutes.

Shear rates as a function of time for dodecane systems (sample h) held at a constant shear stress are plotted in Figure 6. The relaxation of the shear rate at constant applied stress can be described by an exponential decay

$$\dot{\gamma}(t) = \dot{\gamma}_{\infty} + (\dot{\gamma}_0 - \dot{\gamma}_{\infty}) e^{-t/\tau} \tag{4}$$

where $\dot{\gamma}_0$ is the zero time shear rate and τ is a time constant describing the decay of the shear rate. $\dot{\gamma}_{\infty}$ is the ultimate steady state shear rate shown in Figure 2, and is very close to the value obtained on the downward arm of the hysteresis loop in Figure 5. Solid lines in Figure 6 are fits to the data according to equation 4, and the scatter indicates the shear rate resolution of the instrument.

The relaxation of the shear rates is first order for all of the samples. Although the amplitudes differ, the time constant for all four curves shown is the same within experimental uncertainty and equal to 250 ± 50 s. This slow relaxation time is somewhat unexpected if one compares it with other self-assembly phases. DDAB microemulsions have longest relaxation times of around 10^{-4}s, and sponge phases have no relaxation times longer than 10^{-2}s, except in exceptional circumstances (*13, 14, 18*). It is therefore unlikely that the rate-limiting process is that of membrane fusion, which occurs also in disordered phases. The solid character of cubic phases distinguishes them from these other structures, and the slow relaxation must be related to the recovery of long-range order and return of the layers into register with respect to one another. We cannot, however, distinguish at present whether this occurs in an anisotropically layered structure, or one which is fragmented as are lamellar phases.

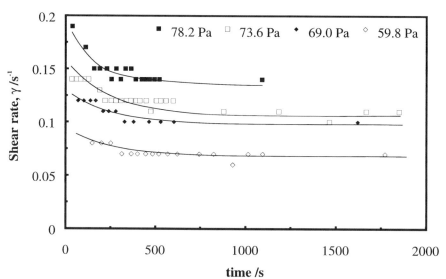

Figure 6   Shear rate relaxation curves for cubic phase sample h held at constant shear stresses. Stresses were increased abruptly from zero to obtain initial readings.

Conclusions

The relationship of the network structure, lattice spacing and symmetry, to the yield behavior of cubic phases has been demonstrated, and contrasted with the rheology of microemulsion and sponge phases. We have described the yield behavior in terms of the fracture of one or more layers, and speculated on the structures formed under steady shear. Near the yield stress a slow structural relaxation was observed, but at higher shears any such effect is too fast to be observed in our experiments.

Acknowledgments

We wish to acknowledge valuable discussions with Dr Peter Harrowell.

Literature Cited

1.   Rosevear, F.B. *J. Soc. Cosmet. Chem.* **1968**, 19, 581.
2.   *Microemulsions, Theory and Practice*; Prince, L.M. Ed.; Academic Press: NY, 1977.
3.   Gradzielski, M; Hoffmann, H. In *The Structure, Dynamics and Equilibrium Properties of Colloidal Systems*; Bloor, D.M.; Wyn-Jones, E., Eds.; Kluwer Academic Publishers: 1990, p. 427.
4.   Chen, S.J.; Evans, D.F.; Ninham, B.W. *J. Phys. Chem.* **1984**, 88, 1631.
5.   Fontell, K.; Ceglie, A.; Lindman, B.; Ninham, B.W. *Acta Chimica Scand. Ser. A.* **1986**, A40, 247.
6.   Zemb, T.N.; Hyde, S.T.; Derian, P.-J.; Barnes, I.S.; Ninham, B.W. *J. Phys. Chem.* **1987**, 91, 3814.
7.   Dubois, M.; Zemb, T.N.; Belloni, L.; Delville, A.; Levitz, P.; Setton, R. *J. Chem. Phys.* **1992**, 96, 2278.
8.   Hyde, S.T. *J. Phys. Chem.* **1989**, 93, 1458.
9.   Ström, P; Anderson, D.M. *Langmuir* **1992**, 8, 691.
10.   Barois, P.; Hyde, S.; Ninham, B.; Dowling, T. *Langmuir* **1990**, 6, 1136.
11.   Maddaford, P.J.; Toprakcioglu, C. *Langmuir* **1993**, 9, 2868.
12.   Chen, C.-M.; Warr, G.G. *J. Phys. Chem.* **1992**, 96, 9492.
13.   Anklam, M.R.; Prud'homme, R.K.; Warr, G.G. *AIChE Journal* in press.
14.   Snabre, P; Porte, G. *Europhys. Lett.* **1990**, 13, 641.
15.   Chen, L.B.; Zukoski, C.F. *Phys. Rev. Lett.*, **1990**, 65, 44.
16.   Rehage, H.; Hoffmann, H. *Faraday Disc. Chem. Soc.* **1983**, 76, 363.
17.   Prud'homme, R.K.; Warr, G.G., *Langmuir* in press.
18.   Diat, O.; Roux, D. *J. Phys. II*, **1993**, 3, 9.

RECEIVED July 8, 1994

POLYMER–MICELLE INTERACTIONS

Chapter 23

Polymer–Surfactant Complexes

Yingjie Li and Paul L. Dubin

Department of Chemistry, Indiana University–Purdue University
at Indianapolis, 402 North Blackford Street, Indianapolis, IN 46202

This paper reviews research on the interaction between polymers and
surfactants in dilute aqueous solution. Recent work on the
quantitative characterization of polymer-surfactant complexes by
using light scattering and fluorescence is emphasized. The paper
concludes by indicating some of the unanswered questions in this
field.

Complex formation between surfactants and polymers (including polyelectrolytes)
has been the subject of intense research effort, because of both fundamental and
technological interests (*1-12*). From a practical point of view, knowledge acquired
from the study of polymer-surfactant interactions can be applied to important
industrial and biological processes. Some industrial situations in which polymers and
surfactants are used conjointly include enhanced oil recovery by polymer-micelle
flooding (*13*), rheology control (*14*), drug release from pharmaceutical tablets (*15*),
and various applications in cosmetic formulations (*16*). Polymer-micelle association
may, in addition, be viewed as a type of polymer-colloid interaction. In particular,
the interaction of polyelectrolytes with oppositely charged micelles may prove to be
a useful model system for polyion-colloid systems. Such systems are represented in
a wide range of situations including water purification and the precipitation of
bacterial cells with polycations (*17*), and the stabilization of preceramic suspensions
(*18*). In the biological realm, they are central to the immobilization of enzymes in
polyelectrolyte complexes (*19*) and the purification of proteins by selective
precipitation and coacervation (*20-23*). In fact, the fundamental interactions that
govern the nonspecific association of DNA with basic proteins (*24,25*) must also be
identical to the ones that control the binding of charged colloids to oppositely
charged polymers.

There has been some significant progress toward a further understanding of
polymer-surfactant complex formation since 1986 when two significant review

0097–6156/94/0578–0320$08.00/0

articles were published (*4,5*). One obvious feature of the recent research in this field is the presentation of more direct and quantitative results by techniques such as dynamic and static light scattering (*26-28*), steady-state and time-resolved fluorescence (*29*), electrophoretic light scattering (*30*), and NMR. This review tends to emphasize the more recent progress in studies of the structure of polymer-surfactant complex based on such techniques. Attempts have been made to minimize duplication with earlier reviews, although some overlap cannot be avoided. The review is mainly concerned with so-called dilute solutions; phase behavior in more concentrated systems, a renewed interest, has been covered in an excellent review by Lindman and Thalberg (*10*). For convenience of presentation the research in this field will be divided into the following topics: (a) nonionic polymer-anionic surfactant; (b) nonionic polymer-cationic surfactant; (c) nonionic polymer-nonionic surfactant; (d) polyelectrolyte-oppositely charged surfactant; (e) polyanion (or polycation)-cationic (or anionic) surfactant.

Nonionic polymer-anionic surfactant complex. Nonionic polymer and anionic surfactant constitute the most investigated polymer-surfactant complex system. The association between nonionic polymers such as poly(ethylene oxide) (PEO) and poly(vinylpyrrolidone) (PVP) with sodium dodecyl sulfate (SDS) has been verified by a variety of techniques (*31-41*). With increasing surfactant concentration, no interaction between the polymers and the surfactants is detected until a critical aggregation concentration, known as the CAC,[*] is reached (*10,31,35*). It has to be noted that the CAC can be detected by a variety of techniques such as surface tension, conductivity, and fluorescence. The CAC is usually lower than the normal CMC in the absence of polymers (*10,31,35*). This is a clear indication of interaction between polymer and surfactant. Above the CAC, surfactants form aggregates[**] bound to polymer. The polymer chains become saturated with surfactant aggregates at a concentration C_2, above which polymer-surfactant complexes coexist with free micelles (*31,35*).

Using small-angle neutron scattering (SANS) to study complexation between PEO and SDS, Cabane and Duplessix (*33*) identified an intrapolymer complex, which contains one PEO macromolecule whose radius of gyration (R_g) is comparable to that of the free PEO coil. Therefore, the interaction between PEO and SDS does not produce a change in size of the PEO coils. By using the contrast variation method, they also showed from the SANS results at the Guinier region that the SDS within the complex (the overall macroaggregate of SDS, not the SDS micelles) has the same R_g as that of the PEO. Therefore, the SDS must be distributed throughout the polymer coil. The SDS molecules in the complex form micelles of ~2 nm in radius. The SANS experiment also showed that the complex is stoichiometric.

The exact nature of the interaction between PEO and SDS is open to debate. At issue are both the location of polymer segments with regard to the micelle and the nature of the attractive force. Cabane (*32*) concluded from NMR that the PEO chain is wrapped around the SDS micelle with some of the segments of PEO adsorbed at the hydrocarbon/water interface, while most of them form loops in the surrounding water. The PEO residues interact with the hydrated methylene group

---

*Please see Note, page 332.

**Please see Note, page 333.

and with polar groups of the SDS but do not penetrate the hydrocarbon core. Moroi *et al.* (*33*) concluded that the complex arises mainly from the interaction between the ionic head group and the -CH_2O- of PEO and is hardly affected by hydrophobic interaction between the surfactant tail and the -CH_2O- group. The results by Brackman and Engberts (*42*) imply that head group hydration inhibits the binding of polymer to micelles. Nagarajan (*43,44*) and Ruckenstein *et al.* (*45*) suggested that the driving force for complex formation is the reduction in interfacial energy between the hydrocarbon core and the local solvent medium. François *et al.* (*41*) also stated that -CH_2- groups of PEO tend to associate with the aliphatic part of the micelles. Kwak and coworkers (*46*) determined the apparent distribution coefficient of PEO in SDS micelles by using the NMR paramagnetic relaxation method. Their results showed significantly higher degree of solubilization of PEO in the SDS micelles than that concluded by Cabane (*32*). Dubin and coworkers (*47*) examined the influence of micelle counterions (Na^+, Li^+, NH_4^+) of dodecyl sulfate micelles on the interaction between micelles and PEO by using dye solubilization and dynamic light scattering. Their results suggested that the cation plays a role in this interaction, by simultaneously coordinating with the polymer oxygens while being electrostatically bound to the micelle. This argument regarding the nature of interaction in the PEO-SDS system, one of the most thoroughly investigated systems, clearly shows the difficulty of research on polymer-micelle complexes. The variability of conditions, such as polymer molecular weight (MW), concentration of polymer and surfactant, and presence or absence of supporting electrolyte, all make comparisons of these results difficult. More systematic investigation, on a wider range of systems, is obviously necessary.

In recent years steady-state and time-resolved fluorescence have been used to investigate the microenviroment in the polymer-micelle system and to determine the aggregation number of micelles bound to polymers (N). It has been found in several systems that N is smaller than that in free micelles (N_0) because polymer stabilizes the micelles. Such systems include PEO-SDS (*36,39,41,48-52*) PVP-SDS (*36,48,51,52*), and (poly(propylene oxide) (PPO))-SDS (*52,53*). Lissi and Abuin (*51*) found that N was independent of the polymer MW (from 6k to 100k for PEO and 40K to 400K for PVP). N increases in the presence of added salt. Zana *et al.* (*48*) found that N increased with increasing SDS concentration and with decreasing polymer concentration. van Stam *et al.* (*54*) also concluded that SDS micelles bound to PEO grow in size rather than number with increasing SDS concentration.

Maltesh and Somasundaran (*49*) studied the conformation change of PEO upon its association with SDS using fluorescence of either pyrene end-labeled PEO or pyrene probe. Initial binding (at [SDS] ~ 4×10^{-5} kmol/m^3) of SDS causes PEO to contract whereas saturation of the polymer by SDS causes the coiled PEO to stretch out. They also found that the stronger the binding of the cation to PEO the less the interaction between SDS and PEO, and the smaller N. From fluorescence measurements of SDS in the presence of pyrene end-labeled PEO, both Quina *et al.* (*55*) and Winnik and co-workers (*66*) showed lower CAC compared to SDS-PEO. It should thus be noted that the pyrene-labeled PEO represents a hydrophobically modified system and is no longer the same as PEO.

Usually it is more difficult to study the binding of monomeric surfactant molecules, rather than micelles, to nonionic polymers (*10*). Schild and Tirrell (*57*) showed that this difficulty could be overcome by using a surfactant with a fluorophore, sodium 2-(*N*-dodecylamino)naphthalene-6-sulfonate ($C_{12}NS$). They could observe *monomeric* surfactant bound to poly(*N*-isopropylacrylamide) (PNIPAAM), as well as micelles bound to PNIPAAM at higher $C_{12}NS$ concentration.

Kido and co-workers (*58*) studied the interactions between SDS and poly(*N,N*-dimethylacrylamide) (PDMA), poly(*N*-isopropylacrylamide) (PIPA), poly(*N,N*-acrylopyrrolidone) (PAPR), and PVP using a solvatochromic probe. Their results indicated that the complexation between polymer and surfactant is influenced not only by the alkyl chain length, but also by the polymer side groups.

It is very surprising that laser light scattering has not been used often in the study of polymer-surfactant complex. Gilányi and Wolfram (*59*) used static light scattering to study the MW and apparent R_g of PVP-SDS (in 0.1 M $NaNO_3$) with increasing SDS concentration from 0 to 3.70 mmol SDS per gram of PVP. It is interesting that while the MW of the complex showed a steady increase with increasing SDS concentration, the R_g of the complex decreased first and then increased, in agreement with their viscosity measurement. Dubin *et al.* (*60*) studied the effects of ionic strength and PEO MW on the complex of PEO-LDS (lithium dodecyl sulfate) by using static light scattering, dynamic light scattering, electrophoretic light scattering, and dialysis equilibrium. With increasing ionic strength, the number of LDS micelles bound per chain increases strongly, and the electrophoretic mobility of the complex becomes more negative. Dubin *et al.* suggested two effects of the simple salt: (a) it weakens the simultaneous binding of the counterion to oxygen and LDS, and (b) it also diminishes intermicellar repulsion allowing more micelles to bind. The diffusion coefficient obtained from dynamic light scattering versus PEO molecular weight showed the same scaling relation as that of the LDS-free PEO. For PEO-LDS complex in 0.4 M LiCl, $D_o = k_1 M^{-0.52}$; for PEO in 0.4 M LiCl, $D_o = k_2 M^{-0.55}$. This suggests that the PEO-LDS complex behaves like a polymer coil in good solvent. They also showed that the electrophoretic mobility of the complex is independent of the PEO MW from 25k to 996k. Static and dynamic light scattering were also used by Brown *et al.* (*61*) to study the PEO-SDS complex. The complex expands with increasing SDS concentration and reaches a maximum at the saturation point of weight concentration ratio $C_{SDS}/C_{PEO} = 5$ and decreases with further increase in the SDS concentration. Therefore, the work on PEO-SDS by Brown *et al.* (*61*) shows a dependence of R_g and viscosity upon increasing SDS concentration opposite to that found by Gilányi and Wolfram (*59*) on PVP-SDS. The dependence of polymer conformation upon addition of surfactant is a topic with contradictory reports, as indicated by Goddard (*4*) and by Lindman and Thalberg (*10*). It is noted that most of the results are obtained from viscosity measurements. In our opinion, it is difficult to define the intrinsic viscosity in such a system because it may not be clear whether the solute is defined as the polymer or the complexes formed therefrom. Consequently, it is not easy to relate this viscosity to the polymer conformation. We believe that more systematic work, preferably using light scattering, is needed to clarify the conformation change of the polymers.

Nonionic polymer-cationic surfactant complex. In general, nonionic polymers interact strongly with anionic surfactants, but weakly with cationic surfactants, if at all. But why this is so is not clear. For example, this behavior has been explained as due to (a) the bulkiness of the cationic head group (*43-45*), (b) the electrostatic repulsion between polymer and surfactant due to the possible positive charge of polymer upon protonation (*37*), and (c) possible difference in interaction of anions and cations with the hydration shell of the polymer which favors interaction with anionic surfactants (*53*). However, alkylammonium halides also interact more weakly with nonionic polymers than do comparable anionic surfactants (*10*). Since the choice of cationic surfactant head group is rather limited, it would be interesting to see how the change of the bulkiness of anionic surfactants will change the interaction between the surfactant and the polymer. Also, the second explanation (positive charge of the polymer chain) should be tested by pH-dependence studies.

Recently, interactions between nonionic polymers and cationic surfactants have been observed when hydrophobic polymers are used. In this case, the association of such hydrophobic polymers with surfactants may stabilize the polymers in aqueous solution. For example, Winnik *et al.* (*62*) showed complex formation between hydroxypropylcellulose (HPC) and hexadecyltrimethylammonium chloride (HTAC) micelles, using fluorescence probe and fluorescence label experiments. Shirahama and co-workers (*63*) even found binding of hexadecyltrimethylammonium bromide (HTAB) and hexadecyldimethyl-2-hydroxyethylammonium bromide (BINA) to poly(vinyl alcohol) (PVA) and PEO, which are hydrophilic polymers. PVP does not bind HTAB and BINA. On the other hand, if the hydrophobicity of surfactant is increased, interactions between nonionic polymers and cationic surfactants may also be enhanced (*64*).

Carlsson *et al.* (*65,66*) observed complex formation of ethyl(hydroxyethyl)cellulose (EHEC) with N-tetradecylpyridinium bromide (TDPB) or dodecyltrimethylammonium ions (DTA$^+$). Both an increase in ionic strength and a rise of temperature led to a more pronounced binding. The temperature dependence was explained as increased hydrophobicity of EHEC with increasing temperature. The interaction between EHEC and HTAC or HTAB has been investigated by Zana *et al.* (*67*) by means of conductivity, self-diffusion and time-resolved fluorescence quenching. In the presence of EHEC, they observed a decrease in CMC, an increase in micelle ionization degree, and a decrease in N with increasing temperature, in agreement with the observation by Carlsson *et al.* (*65,66*).

Winnik *et al.* (*68,69*) showed complex formation between hydrophobically modified PNIPAAM and HTAC. They performed parallel fluorescence studies with both labeled and unlabeled polymer, and found that the perturbation due to pyrene labeling was negligible in this case. In comparison with the results in Ref. 55 and 56, it seems that this is because the polymer samples used in Ref. 68 and 69 are already hydrophobic even apart from the pyrene labeling. Detailed structural changes of polymers due to surfactant binding could be thus investigated from fluorescence techniques.

Complex formation of PPO-HTAX (X$^-$ for Cl$^-$, Br$^-$, ClO$_3^-$, and NO$_3^-$) has been reported by Witte and Engberts (*52*), Brackman and Engberts (*70*), Sierra and Rodenas (*71*), and De Schryver and co-workers (*72*), using mainly conductivity, viscosity, and fluorescence. Brackman and Engberts (*70*) also observed complex formation between HTAB and poly(vinylmethylether) (PVME). The presence of polymer gives rise to a CAC lower than the CMC, a higher micellar ionization degree, and a smaller N for the bound micelle in the presence of PPO. De Schryver and co-workers (*72*) also suggest that the polymer chain is wrapped around several micelles and disturbs only the Stern layer of the micelles by means of its penetrating segments, which renders the polymer-bound micelle more hydrophobic and a better solubilizing agent, even though the surfactant aggregate is smaller. The viscosity results also indicate expansion of the polymer coil due to HTAX micelle binding to poly(vinyl alcohol)-poly(vinyl acetate) (PVOH-Ac). It is interesting that Brackman and Engberts (*70*) found from their viscosity measurements that in the presence of PVME, rodlike micelles of hexadecyltrimethylammonium salicylate (HTASal) and tosylate (HTATs) change to spherical polymer-bound micelles. N is decreased for PPO and PVME but unchanged for PEO and PVP. The change of rodlike micelles to spherical micelles in the presence of PVME has also been reported by using ^{13}C NMR (*73*) and diffusion-ordered 2D NMR (*74*).

Nonionic polymer-nonionic surfactant complex. The interaction between nonionic polymers and nonionic surfactants is usually very weak. However, since the driving force for polymer-micelle interaction is the reduction in Gibbs energy of the total system, interaction between nonionic polymer and nonionic surfactant could occur if a sufficiently hydrophobic polymer is used. Brackman *et al.* (*75*) have shown evidence of complex formation between PPO and *n*-octyl thioglucoside (OTG) by microcalorimetry. However it is not easy to understand why the CMC, as determined by microcalorimetry and bromophenol blue absorption, is not affected by the presence of PPO. Winnik (*76*) also showed, by fluorescence, an interaction between OTG and pyrene-labeled HPC, again without a decrease in CMC. Obviously more systematic investigation is needed to look further into this interesting behavior.

Polyelectrolyte-oppositely charged surfactant complex. In contrast to the interaction between nonionic polymers and ionic surfactants, the interaction between polyelectrolytes and oppositely charged surfactants is dominated by electrostatic forces, although hydrophobic interactions may play a secondary role (*5,8,10*). It is useful, at the outset, to distinguish between "strong polyelectrolytes", such as poly(styrenesulfonate) (PSS), which are always 100% ionized, and "weak polyelectrolytes", such as poly(acrylic acid) (PAA), whose charge depends upon, e.g. pH. In general, the interaction between polyelectrolytes and oppositely charged surfactants starts at a very low surfactant concentration (CAC), usually a few orders of magnitude lower than the CMC of the free surfactant. Unlike the nonionic polymer-ionic surfactant system, the complex usually can not coexist with free micelles because precipitation is observed as the addition of ionic surfactant brings the polyelectrolyte close to charge neutralization. In some cases, however, further addition of ionic surfactant may resolubilize the precipitate (*5*). In any event, phase separation effects result in the restriction of most studies of strong polyelectrolytes with oppositely charged surfactants to surfactant concentrations well below the CMC.

If the hydrocarbon tail of the surfactant is increased, the interaction between the polyelectrolyte and the surfactant is enhanced (5, 10). On the other hand, the interaction is also enhanced if the hydrophobicity of the polymer is increased (77). the CAC is increased when simple salt is added (5, 8, 10, 78), indicating a reduced interaction between polyelectrolyte and surfactant upon addition of salt. This is opposite to the effects of simple salt on the micelle formation in the absence of polyelectrolyte, where salt causes a decrease in the CMC. The effect of polyelectrolyte concentration is, to some extent, similar to that of a simple salt (10, 79).

If the binding of surfactant ions to polymers is viewed according to one-dimensional lattice models, statistical mechanical treatments developed in the context of biophysical chemistry can be utilized. Schwarz (80) and Satake and Yang (81) developed binding models based on the treatment of the helix-coil transition by Zimm and Bragg (82). The following relationships were derived:

$$Ku = (C^f_D)_{0.5}^{-1} \qquad (1)$$

and

$$(d\beta/d\ln C^f_D)_{0.5} = u^{0.5}/4 \qquad (2)$$

where K is the binding constant for the binding of a single surfactant ion onto an isolated site on the polyelectrolyte; u is a cooperativity parameter characterizing the interaction between adjacent bound surfactants; $(C^f_D)_{0.5}$ is the equilibrium concentration of free surfactant at $\beta = 0.5$ and β is the degree of binding (ratio of the number of bound surfactant to the number of ionic groups on the polyelectrolyte). Although this treatment could offer a good way to describe the intrinsic binding strength and binding cooperativity (8), it does not provide much structural information.

Škerjanc et al. (83) theoretically explored the reasons for the low CAC in the presence of polyelectrolyte. From solutions of the Poisson-Boltzmann equation for the cell model of a polyelectrolyte solution with two kinds of monovalent counterions of different size, they calculated the counterion distribution around the polyelectrolyte. The calculation shows that the local concentration of surfactant at the surface of polyelectrolyte is much higher than the total or bulk concentration. Therefore if such local concentration reaches the CMC (this CMC value might not equal exactly to the CMC of free micelles), the corresponding total surfactant concentration should be the CAC. This explains the low CAC compared to the CMC.

The properties of polyelectrolytes play a major role in complex formation. Perhaps the most fundamental parameter characterizing the polyelectrolyte is the linear charge density, ξ, defined as

$$\xi = e^2/(4\pi\epsilon bkT) \qquad (3)$$

where e is the protonic charge, ϵ the dielectric constant, b the distance between two adjacent charges on the polymer chain, k the Boltzmann constant, and T the absolute temperature. Generally an increase in ξ causes a stronger interaction (*10*). However, ξ is not the only factor. For example, the following order of binding strength with DTA^+ is observed (*8*): PSS > dextran sulfate (DexS) > sodium poly(acrylate) (NaPAA) (pH unknown), although they all have a similar ξ (2.8). Comparison between polymers of such radically different structures as DexS versus NaPAA may complicate analysis in terms of ξ. Furthermore, one must be careful with weak polyelectrolytes like NaPAA because pH is a determining factor, so that ξ is also pH dependent. Hydrophobicity of the polymer is another factor. Thus, Shimizu *et al.* (*77*) have shown that the polymer-surfactant interaction is enhanced with increasing hydrophobicity of the polymer chain. The flexibility of the polyelectrolyte chain is believed to be another factor. We would like to point out that the interaction may also depend upon how strong the polyelectrolyte is. For example, polymeric sulfonate groups and carboxylate groups may have very different effects on complex formation. Such dependence becomes obvious in the studies of N.

Abuin and Scaiano (*84*) investigated the interaction between PSS, a strong polyelectrolyte, and DTAB, by using fluorescence. They found $N \sim 7\text{-}10$ (at [PSS] = 28 mM and [DTAB] = 14 mM, $\beta \sim 50\%$, with CMC of DTAB 16 mM). The interaction also induces contraction of the polyelectrolyte chain, as deduced from viscosity measurement.

Almgren and co-workers (*85*) also investigated the interaction between PSS and DTAB by using time-resolved fluorescence quenching. They obtained $N \sim 30\text{-}40$ (at [PSS] = 5 mM and [DTAB] from 1.27 to 3.61 mM), only about half of N_0 of DTAB. N does not change much with increasing β (from 25% to 52%), but the number of surfactant aggregates increases with increasing β. This is different from the PEO-SDS system where van Stam *et al.* (*54*) found that the bound surfactant aggregates grow in size with increasing surfactant concentration. N is also insensitive to the type of counterion (Cl⁻ versus Br⁻) in the presence of PSS, but is sensitive in the absence of PSS. This implies that the surfactant counterions may be expelled by the polyelectrolyte. Almgren *et al.* (*85*) believe that the hydrophobic backbone of PSS strongly influences the surfactant aggregate formation.

Chu and Thomas (*86*) investigated the interaction between sodium poly(methacrylate) (NaPMA), a weak polyelectrolyte, and decyltrimethylammonium bromide (DeTAB) at pH=8. The N value at pH=8 is ~ 100 (at [NaPMA] = 1 g/L and [DeTAB] = 8 mM, with CMC of DeTAB 65 mM). This number is reduced to ~ 65 if the concentration of free DeTAB monomers is taken into account, still larger than N_0 of DeTAB. This is in direct contrast to the PSS-DTAB systems.

Thalberg *et al.* (*87*) investigated the interaction of DeTAB and DTAB with another weak polyelectrolyte, sodium hyaluronate (NaHy) at pH=7. N is relatively unaffected by the presence of NaHy (at [DeTAB] from 80 to ~ 200 mM and [DTAB] from 12 to 30 mM). They also found that a certain amount of counterion Br⁻ still remains at the surface of the surfactant aggregates, in contrast to the PSS-DTAB system (*85*).

To our knowledge, only Ref. 84-87 reported studies of N in the presence of polyelectrolytes. It is still not known why PSS gives rise to larger N values. We would like to point out that PSS is a strong polyelectrolyte while NaPMA and NaHy are weak ones. Kwak and co-workers have indicated that the binding strength of PSS is stronger than NaPAA (8). It would be interesting to study the effect of sulfonate group versus carboxylate group (with the same polymer backbone) on N to see whether the hydrophobicity of the polystyrene backbone or strength of the polyelectrolyte is the determining factor. In particular, it is important to look for the dependence of N upon degree of binding and the concentration of surfactant, relative to the values of free micelles.

The polymer conformation also undergoes a change upon binding by surfactants (10,84,88,89). Usually a contraction upon binding of surfactant is observed. However, it is again surprising that most of the information in this aspect is ambiguous, based on viscosity. Laser light scattering has not been used often.

Generally the results discussed above involve systems with surfactant concentration below or not much higher than CMC. This is understandable because the strong electrostatic interaction causes precipitation when the concentration of surfactants is increased. Such a situation greatly complicates research on polyelectrolyte-surfactant systems.

It is of interest to study the system where the surfactant concentration is well above the CMC, and the size of the surfactant micelle may be close to or even larger than the dimension of the polyelectrolyte. Thus, the description of the interaction between polyelectrolyte and oppositely charged surfactant could be divided into two separate disciplines. In the first case, the size of the surfactant aggregates is much smaller than the dimensions of the polymer, or a single surfactant molecule is involved, and the situation could be viewed as a kind of polymer-ligand binding, as discussed above. On the other hand, if the surfactant molecules form micelles with dimensions larger than that of polymers, their interaction can be viewed as a kind of adsorption of polymer onto a colloidal particle. The polymer could be taken as adsorbing species and the micelles as the adsorbing surface. In this case, some theoretical work has been done for polyelectrolytes. The adsorption of polyelectrolytes on planar charged surfaces has been treated by Wiegel (90), by Scheutjens, Fleer, and co-workers (91,92), and by Muthukumar (93). Odijk (94) considered the adsorption of a flexible polycation to a rigid cylinder. While differing in approach and in the details of the findings, all theories yield some common results. (a) The interaction energy increases with surface charge density σ, the polymer linear charge density ξ, and the Debye length κ^{-1} ($\propto I^{-1/2}$ where I is the ionic strength). (b) The interaction resembles a phase transition.

Most of Dubin's work on polyelectrolyte-surfactant complexes probably falls into the second category (95-108). It is also noted that most of the work from Dubin's group deals with systems in which the surfactant concentration is well above the CMC and the amount of micelles are much in excess, in contrast to other literature research.

In order to avoid precipitation in mixtures of strong polyelectrolytes with oppositely charged micelles, the binding strength (and extent of ion exchange) must be reduced. Practically, several ways could be used to attenuate the strong electrostatic interaction between polyelectrolyte and oppositely charged surfactant, such as use of mixed micelles of ionic and nonionic surfactant and addition of salt. (It would be interesting to see whether a polyelectrolyte with a low ξ could form a soluble complex with surfactant micelles at above the CMC). Dubin and Oteri (*95*) showed that the interaction between poly(dimethyldiallylammonium chloride) (PDMDAAC) and mixed micelles of SDS and Triton X-100 (TX100) takes place only when a critical molar ratio Y_c of SDS to TX100 has been reached, where Y is defined as

$$Y = [SDS]/([SDS]+[TX100]) \tag{4}$$

and is proportional to the average mixed micelle surface charge density.

In addition to the phase boundary Y_c corresponding to the reversible formation of soluble polyelectrolyte-mixed micelle complexes, Dubin and Oteri (93) observed a second phase separation, corresponding to irreversible precipitation, at a higher Y value, denoted by Y_p. Y_p usually exceeds Y_c by about 50% to 100%. The region between Y_c and Y_p represents a range of micelle surface charge density where soluble, reversible complexes could form. At high polyelectrolyte concentration, a maximum in the turbidity versus Y curve is observed in between Y_c and Y_p. However, it is noted that such a wide range of soluble complex formation is not observed in other systems. One speculation is that this wide range might be due to the compositional polydispersity of TX100 used.

After such observation, Dubin and co-workers have used a variety of techniques to investigate (a) the dependence of the phase separation points (Y_c and Y_p) upon variables such as ionic strength, polyelectrolyte concentration, polyelectrolyte molecular weight, surfactant concentration, and temperature (*95-98,102,104,105,108*), and (b) the structure of such soluble complexes (*96,98,99,101,104,105,108*). The major techniques used include turbidimetry (*95,97,98,102,104,108*), dialysis equilibrium (*108*), viscosity (*95,108*), static light scattering (*105,108*), dynamic light scattering (*96,98,99,104,105,108*), and electrophoretic light scattering (*108*).

The phase boundary Y_c has been found to be independent of polymer concentration, surfactant concentration, and polyelectrolyte molecular weight (at least to the first order). Therefore, the interaction involves a relatively short sequence of chain segments with the micelles. The phase boundary Y_c decreases slightly with increasing temperature.

The effect of I on the phase transition is significant. Addition of salt can screen the charge and attenuate the interaction. It was found that Y_c is proportional to $I^{1/2}$, where $I^{-1/2}$ is proportional to κ^{-1}, the Debye length. Therefore, the authors suggested that the diminution of micelles' electrostatic domain with linear increase in $I^{1/2}$ is exactly compensated for by a parallel linear increase in the micelle surface charge density, which is presumed to vary linearly with Y.

Dubin and co-workers (*98,100,102,103,106,107*) also extended their investigation to other systems, such as, PDMDAAC-(*n*-dodecyl hexaoxyethylene glycol monoether $(C_{12}E_6)$)-SDS,   PDMDAAC-(*N*-dodecyl-*N,N*-dimethyl-3-ammonio-1-propanesulfate)-SDS, (quaternized poly(4-vinylpyridine) (QPVP))-TX100-SDS, PSS-(*N,N*-dimethyldodecylamine oxide (DMDAO)), PSS-TX100-

DTAB, PAA-TX100-DTAB, and (poly(2-acrylamido-2-methylpropanesulfonate) (PAMPS))-DMDAO. The $Y_c \sim I^{1/2}$ relation seems universal.

Dubin and co-workers (103) studied the effects of linear charge density ξ on the phase boundary Y_c by using poly(ethyleneimine) (PEI) with TX100 and SDS. The results suggest a relation $Y_c \sim \xi^{-1}$.

In summary the work by Dubin and co-workers on the energetics shows an empirical relation by the surfactant charge density σ, I, and ξ as

$$\sigma \xi / \kappa = \text{constant} \tag{5}$$

This relation has been theoretically rationalized by Odijk (109).

Structural studies using mainly light scattering revealed that the hydrodynamic size of the complex versus Y showed a similar curve to the turbidity versus Y (96,104). The maximum is concentration dependent. Dubin and co-workers (104) suggested that the maximum corresponds to the formation of higher-order complex. Electrophoretic light scattering indicated that the complex underwent a charge reversal, i.e., from positive to negative, with increasing Y (108).

Static and dynamic light scattering were used by Dubin and co-workers (105,108) to investigate the weight-average MW (M_w), R_g, and hydrodynamic radius (R_h) of PDMDAAC, TX100-SDS mixed micelles, and the corresponding complexes. At Y=0.32 (soluble complex), the R_g value of the complex at 0.4 M NaCl and in excess micelles is 51 nm, with the R_g values for PDMDAAC and TX100-SDS being 28 nm and 21 nm, respectively. The R_h values are 25 nm, 15 nm, and 9 nm for complex, PDMDAAC, and TX100-SDS, respectively. Therefore, the PDMDAAC chains extended upon binding of TX100-SDS mixed micelles. On the other hand, the M_w value of complex was estimated to be 1×10^7 g/mol, compared to 1.5×10^5 g/mol and 1.9×10^5 g/mol for PDMDAAC and TX100-SDS mixed micelles, respectively. These values indicate the high density of the complex.

Dubin and co-workers (105,108) also found that whether inter-polymer complex forms depends upon the polyelectrolyte concentration. At below ~ 0.1 g/L, intra-polymer complex dominates, while above ~ 0.1 g/L, inter-polymer complexes are observed.

Polyanion-anionic surfactant complex. Binana-Limbelé and Zana (110) have investigated the effects of a polyanion, NaPAA, on the CMC and N of an anionic surfactant, SDS, using conductivity and time-resolved fluorescence probing. While no complex formation was observed, NaPAA decreases the CMC and increases N. Therefore, NaPAA behaves like a small electrolyte. However, if the polymer is hydrophobically modified, some interactions between polyanion and anionic surfactant could be observed. Thus, McGlade and co-workers (111) observed interactions between SDS and poly(1-octadecene-co-maleic acid) (POMA), but no interactions between SDS and poly(1-ethylene-co-maleic acid) (PEMA). Interaction with SDS was also observed by Iliopoulos et al. (112) using hydrophobically modified NaPAA. Maltesh and Somasundaran (113) clearly showed the effect of pH on the PAA-SDS interaction. PAA and SDS interact under acidic pH conditions, when the PAA concentration is low. Under conditions when the PAA is ionized, there are no interactions between PAA and SDS.

Concluding remarks. Although polymer-surfactant complexes have been the subject of extensive research efforts, a clear picture of their structure and energetics still has not emerged. At least to some extent, some basic questions still remain. The first question regards the thermodynamics of complex formation, i.e., what is the driving force for the complex formation for different polymers and different surfactants. A related question is the nature of the interaction. For example, why cationic surfactants interact weakly with nonionic polymers is still unexplained. The second question is about the mechanism of complex formation. For example, it is still not known whether surfactants aggregate first and then bind to the polymers, or aggregate around one or more *already* bound surfactant molecules, or whether these processes take place simultaneously. It is important to know whether the binding is cooperative or not because these factors may determine the final structure of the complexes. The third question regards the structure of the complex. For example, it is not known whether the polymer chains reside inside the surfactant aggregates or wrap around the surface of the surfactant aggregates. It is also not clear whether the surfactant aggregates restructure upon binding, and how the conformation of polymer changes upon binding. The fourth question regards the dependence of the nature of the interaction, the binding mechanism, and the structure of complex upon variables such as the chemical structure of polymers and surfactants, the ionic strength, and the temperature. The dependence of complex structure upon the changes of the length, the size, and the concentration of polymers and surfactants is also important. Systematic investigation of such dependence might be the key to the first three questions.

Such an unresolved situation occurs for several reasons. Firstly, complex formation between polymer and surfactant may be a consequence of several types of interactions, such as hydrophobic interaction, electrostatic attraction, as well as electrostatic and steric repulsion. Secondly, numerous experimental variables are involved, such as the nature and concentration of polymers and surfactants, the ionic strength, and the temperature. Systematic investigations are still lacking. For example, the choice of polymers and surfactants is somewhat limited, with the combination of nonionic polymers (e.g., PEO and PVP) with anionic surfactants (e.g., SDS) receiving most attention. For studies involving polyelectrolytes with oppositely charged surfactants, precipitation usually takes place even before the concentration of the surfactants reach the CMC, due to the strong electrostatic interaction. Thirdly, most of the techniques used, such as viscosity, surface tension, and surfactant electrode give indirect information. Lastly, there are few theories available. However, with systematic investigation involving a combination of techniques such as light scattering, fluorescence, NMR, and SANS, we should gain a better picture of the interaction in this complicated system. The phenomenology may also guide the development of theories.

Glossary
polymers
PEO   poly(ethylene oxide)

PVP	poly(vinylpyrrolidone)
PNIPAAM	poly(N-isopropylacrylamide)
PDMA	poly(N,N-dimethylacrylamide)
PIPA	poly(N-isopropylacrylamide)
PAPR	poly(N,N-acrylopyrrolidone)
HPC	hydroxypropylcellulose
PVA	poly(vinyl alcohol)
EHEC	ethyl(hydroxyethyl)cellulose
PSS	poly(styrenesulfonate)
NaPMA	sodium poly(methacrylate)
NaHy	sodium hyaluronate
PDMDAAC	poly(dimethyldiallylammonium chloride)
QPVP	quaternized poly(4-vinylpyridine)
PAMPS	poly(2-acrylamido-2-methylpropanesulfonate
POMA	poly(1-octadecene-co-maleic acid)
PEMA	poly(1-ethylene-co-maleic acid)
PVOH-Ac	poly(vinyl alcohol)-poly(vinyl acetate) (PVOH-Ac)
PAA	poly(acrylic acid)
NaPAA	sodium polyacrylate
DexS	dextran sulfate
surfactants	
SDS	sodium dodecyl sulfate
$(C_{12}NS)$	sodium 2-(N-dodecylamino)naphthalene-6-sulfonate
LDS	lithium dodecyl sulfate
HTAC	hexadecyltrimethylammonium chloride
HTAB	hexadecyltrimethylammonium bromide
BINA	hexadecyldimethyl-2-hydroxyethylammonium bromide
TDPB	N-tetradecylpyridinium bromide (TDPB)
DTA^+	dodecyltrimethylammonium ion
OTG	n-octyl thioglucoside
DTAC	dodecyltrimethylammonium chloride
DTAB	dodecyltrimethylammonium bromide
DeTAB	decyltrimethylammonium bromide
TX100	Triton X-100
$C_{12}E_6$	n-dodecyl hexaoxyethylene glycol monoether
DMDAO	N,N-dimethyldodecylamine oxide

Note

*, The term CAC has been used frequently in literature. We would like to point out that the CAC may not necessarily correspond to a surfactant concentration at which *surfactant aggregates* start to bind to polymers. It is possible that at this surfactant concentration *monomeric surfactant* starts to bind to polymer chain. Therefore, the CAC should be understood as a surfactant concentration at which interaction between polymers and surfactants (either surfactant aggregates or single surfactant molecules) takes place and complex starts to form. It should also be noted that CAC values from different techniques could be different.

**,   We have noticed that different names have been used for polymer-bound surfactant aggregate, such as clusters, hemimicelles, and micelles. The value of N of such moieties might range from 1 to N_0, depending upon the surfactant concentration. In this review, we try to use "aggregate" to describe those species formed below or not much higher than the CMC, and "micelle" to describe bound species formed well above the CMC. However, we found that it is not easy to distinguish these in some cases due to the lack of knowledge about the structure of complexes.

Acknowledgement. The support of grant DMR9311433 from the National Science Foundation is gratefully acknowledged.

Literature Cited

1.   Breuer, M. M.; Robb, I. D. *Chem. Ind.* **1972**, 530.
2.   Robb, I. D. In *Anionic Surfactants, Physical Chemistry of Surfactant Action;* Lucassen-Reynders, E. H. Ed.; Marcel Dekker: New York, 1981; p.109.
3.   Tsuchida, E.; Abe, K. *Adv. Polym. Sci.* **1982**, 45.
4.   Goddard, E. D. *Colloids Surf.* **1986**, *19*, 255.
5.   Goddard, E. D. *Colloids Surf.* **1986**, *19*, 301.
6.   Saito, S. In *Nonionic Surfactant, Physical Chemistry*; Schick, M. J. Ed.; Surfactant Science Series 23; Marcel Dekker: New York, 1987; Chapter 15.
7.   Smid, J.; Fish, D. *Encyclopedia of Polymer Science and Technology*; Wiley-Interscience: New York, 1988; Vol 11, p.720.
8.   Hayakawa, K.; Kwak, J. C. T. In *Cationic Surfactants. Physical Chemistry*; Rubingh, D. N.; Holland, P. M. Eds; Marcel Dekker: New York, 1991; Chapter 5, p.189.
9.   Piculell, B.; Lindman, B. *Adv. Colloid Interface Sci.* **1992**, *41*, 149.
10.   Lindman, B.; Thalberg, K. In *Interactions of Surfactants with Polymers and Proteins*; Goddard, E. D.; Ananthapadmanabhan, K. P. Eds.; CRC Press: Boca Raton, 1993; Chapter 5.
11.   Lindman, B.; Khan, A.; Marques, E.; Miquel, M.; Piculell, L.; Thalberg, K. *Pure Appl. Chem.* **1993**, *65*, 953.
12.   Brackman, J. C.; Engberts, J. B. F. N. *Chem. Soc. Rev.* **1993**, *22,* 85.
13.   Desai, N. N.; Shah, D. O. *Polym. Prepr.* **1981**, *22(2)*, 39.
14.   Brackman, J. C. *Langmuir* **1991**, *7*, 469.
15.   Alli, D.; Bolton, S.; Gaylord, N. S. *J. Appl. Polym. Sci.* **1991**, *42,* 947.
16.   Goddard, E. D. *J. Soc. Cosmet. Chem.* **1990**, *41*, 23.
17.   Kawabata, N.; Hayashi, T.; Nishikawa, M. *Bull. Chem. Soc. Jpn* **1986**, *59*, 2861.
18.   Cesarano, III, J.; Aksay, I. A. *J. Amer. Ceram. Soc.* **1988**, *71*, 1062.
19.   Margolin, A.; Sherstyuk, S. F.; Izumdov, V. A.; Zezin, A. B.; Kabanov, V. A. *Eur. J. Biochem.* **1985**, *146*, 625.
20.   Clark, K. M.; Glatz, C. E. *Biotechnol. Prog.* **1987**, *3*, 241.

21. Fisher, R. R.; Glatz, C. E. *Biotechnol. Bioeng.* **1988,** *32,* 777.
22. Bozzano, A. G.; Andrea, G.; Glatz, C. E. *J. Membr. Sci.* **1991,** *55,* 181.
23. Dubin, P. L.; Strege, M. A.; West, J. In *Large Scale Protein Purification* Ladish, M. Ed.; American Chemical Society: Washington D. C., 1990; Chapter 5.
24. Shaner, S. L.; Melancon, P.; Lee, K. S.; Burgess, M. T.; Record, Jr., M. T. *Cold Spring Harbor Symp. Quant. Biol.* **1983,** *47,* 463.
25. von Hippel, P. H.; Bear, D. G.; Morgan, W. D.; McSwiggen, J. A. *Annu. Rev. Biochem.* **1984,** *53,* 389.
26. Chu, B. *Laser Light Scattering;* Academic Press: New York, 1974.
27. Berne, B. J.; Pecora, R. *Dynamic Light Scattering*; Wiley: New York, 1976.
28. Schmitz, K. S. *An Introduction to Dynamic Light Scattering by Macromolecules;* Academic Press: Boston, 1990.
29. Zana, R. In *Surfactant Solutions. New Methods of Investigation;* Zana, R. Ed.; Surfactant Science Series, Vol 22; Marcel Dekker: New York, 1987.
30. Ware, B. R.; Haas, D. D. In *Fast Methods in Physical Biochemistry and Cell Biology;* Sha'afi, R. I.; Fernadez, S. M. Eds.; Elsevier: Amsterdam, 1983.
31. Jones, M. N. *J. Colloid Interface Sci.* **1967,** *23,* 36.
32. Cabane, B. *J. Phys. Chem.* **1977,** *81,* 1639.
33. Cabane, B.; Duplessix, R. *J. Phys. (Paris)* **1982,** *43,* 1529.
34. Wan-Badhi, W. A.; Wan-Yunus, W. M. Z.; Bloor, D. M.; Hall, D. G.; Wyn-Jones, E. *J. Chem. Soc. Faraday Trans* **1993,** *89,* 2737.
35. Fishman, M. L.; Eirich, F. R. *J. Phys. Chem.* **1971,** *75,* 3135.
36. Gilányi, T.; Wolfram, E. *Colloids Surf.* **1981,** *3,* 181.
37. Moroi, Y.; Akisida, H.; Saito, M.; Matuura, R. *J. Colloid Interface Sci.* **1977,** *61,* 233.
38. Shirahama, K.; Tohdo, M.; Murahashi, M. *J. Colloid Interface Sci.* **1982,** *86,* 282.
39. Turro, N. J.; Baretz, B. H.; Kuo, P. L. *Macromolecules* **1984,** *17,* 1321.
40. Schwuger, M. J. *J. Colloid Interface Sci.* **1973,** *43,* 491.
41. François, J.; Dayantis, J.; Sabbadin, J. *Eur. Polym. J.* **1985,** *43,* 491.
42. Brackman, J. C.; Engberts, J. B. F. N. *J. Colloid Interface Sci.* **1989,** *132,* 250.
43. Nagarajan, R. *Colloids Surf.* **1985,** *13,* 1.
44. Nagarajan, R. *J. Chem. Phys.* **1989,** *90,* 1980.
45. Ruckenstein, E.; Huber, G.; Hoffmann, H. *Langmuir* **1987,** *3,* 382.
46. Gao, Z.; Wasylishen, R. E.; Kwak, J. C. T. *J. Phys. Chem.* **1991,** *95,* 462.
47. Dubin, P. L.; Gruber, J. H.; Xia, J.; Zhang, H. *J. Colloid Interface Sci.* **1992,** *148,* 35.
48. Zana, R.; Lang, J.; Lianos, P. In *Microdomains in Polymer Solutions*; Dubin, P. L., Ed.; Plenum: New York, 1985; p.357.
49. Maltesh, C.; Somasundaran, P. *Langmuir* **1992,** 8, 1926; *J. Colloid Interface Sci.* **1993,** *157,* 14.
50. Zana, R.; Lianos, P.; Lang, J. *J. Phys. Chem.* **1985,** *89,* 41.
51. Lissi, E. A.; Abuin, E. *J. Colloid Interface Sci.* **1985,** *105,* 1.

52. Witte, F. M.; Engberts, J. B. F. N. *J. Org. Chem.* **1987**, *52*, 4767.
53. Witte, F. M.; Engberts, J. B. F. N. *Colloids Surf.* **1989**, *36*, 417.
54. van Stam, J.; Almgren, M.; Lindblad, C. *Prog. Colloid Polym. Sci.* **1991**, *84*, 13.
55. Quina, F.; Abuin, E.; Lissi, E. *Macromolecules* **1990**, *23*, 5173.
56. Hu, Y.; Zhao, C.; Winnik, M. A.; Sundararajan, P. R. *Langmuir* **1990**, *6*, 880.
57. Schild, H. G.; Tirrell, D. A. *Langmuir* **1990**, *6*, 1676.
58. Kido, J.; Imamura, Y.; Kuramoto, N.; Nagai, K. *J. Colloid Interface Sci.* **1992**, *150*, 338.
59. Gilányi, T.; Wolfram, E. In *Microdomains in Polymer Solutions;* Dubin, P. L., Ed.; Plenum: New York, 1985; p.383.
60. Xia, J.; Dubin, P. L.; Kim, Y. *J. Phys. Chem.* **1992**, *96*, 6805.
61. Brown, W.; Fundin, J.; Miguel, M. *Macromolecules* **1992**, *25*, 7192.
62. Winnik, F. M.; Winnik, M. A.; Tazuke, S. *J. Phys. Chem.* **1987**, *91*, 594.
63. Shirahama, K.; Himuro, A.; Takisawa, N. *Colloid Polym. Sci.* **1987**, *265*, 96.
64. Ullmann, E.; Mansel, L.; Mohrshulz, P. *APV-Inf.* **1981**, *20*, 1 (quoted in Ref. 8, p.225 as Ref. 114).
65. Carlsson, A.; Lindman, B.; Watanabe, T.; Shirahama, K.; *Langmuir* **1989**, *5*, 1250.
66. Carlsson, A.; Karlström, G.; Lindman, B. *J. Phys. Chem.* **1989**, *93*, 3673.
67. Zana, R.; Binana-Limbelé, W.; Kamenka, N. Lindman, B. *J. Phys. Chem.* **1992**, *96*, 5461.
68. Winnik, F. M.; Ringsdorf, H.; Venzmer, J. *Langmuir* **1991**, *7*, 905.
69. Winnik, F. M.; Ringsdorf, H.; Venzmer, J. *Langmuir* **1991**, *7*, 912.
70. Brackman, J. C.; Engberts, J. B. F. N. *Langmuir* **1991**, *7*, 2097.
71. Sierra, M. L.; Rodenas, E. *J. Phys. Chem.* **1993**, *97*, 12387.
72. Reekman, S.; Gehlen, M.; De Schryver, F. C.; Boens, N.; Van der Auweraer, M. *Macromolecules* **1993**, *26*, 687.
73. Wong, T. C.; Liu, C.; Poon, C.-D.; Kwoh, D. *Langmuir* **1992**, *8*, 460.
74. Morris, K. F.; Johnson, Jr., C. S.; Wong, T. C. *J. Phys. Chem.* **1994**, *98*, 603.
75. Brackman, J. C.; van Os, N. M.; Engberts, J. B. F. N. *Langmuir* **1988**, *4*, 1266.
76. Winnik, F. M. *Langmuir* **1990**, *6*, 522.
77. Shimizu, T. Seki, M.; Kwak, J. C. T. *Colloids Surf.* **1986**, *20*, 289.
78. Hayakawa, K.; Kwak, J. C. T. *J. Phys. Chem.* **1982**, *86*, 3866.
79. Thalberg, K.; Lindman, B.; Bergfeldt, K. *Langmuir* **1991**, *7*, 2893.
80. Schwarz, G. *Eur. J. Biochem.* **1970**, *12*, 442.
81. Satake, I.; Yang, J. T. *Biopolymers* **1976**, *15*, 2263.
82. Zimm, B. H.; Bragg, J. K. *J. Chem. Phys.* **1959**, *31*, 526.
83. Škerjanc, J.; Kogej, K. *J. Phys. Chem.* **1988**, *92*, 6382.
84. Abuin, E. B.; Scaiano, J. Ć. *J. Amer. Chem. Soc.* **1984**, *106*, 6274.
85. Almgren, M.; Hansson, P.; Mukhtar, E.; van Stam, J. *Langmuir* **1992**, *8*, 2405.

86. Chu, D.-Y.; Thomas, J. K. *J. Amer. Chem. Soc.* **1986**, *108*, 6270.
87. Thalberg, K.; van Stam, J.; Lindblad, C.; Almgren, M.; Lindman, B. *J. Phys. Chem.* **1991**, *95*, 8975.
88. Chandar, P.; Somasundaran, P.; Turro, N. J. *Macromolecules* **1988**, *21*, 950.
89. Herslöf, Å.; Sundelöf, L.; Edsman, K. *J. Phys. Chem.* **1992**, *96*, 2345.
90. Wiegel, F. W. *J. Phys. A: Math. Gen* **1977**, *10*, 299.
91. Evers, O. A.; Fleer, G. J.; Scheutjens, J. M. H. M.; Lyklema, J. *J. Colloid Interface Sci.* **1986**, *111*, 446.
92. Blaakmeer, J.; Bökmer, M. R.; Cohen Stuart, M. A.; Fleer, G. J. *Macromolecules* **1990**, *23*, 2223.
93. Muthukumer, M. *J. Chem. Phys.* **1987**, *86*, 7230.
94. Odijk, T. *Macromolecules* **1980**, *13*, 1542.
95. Dubin, P. L.; Oteri, R. *J. Colloid Interface Sci.* **1983**, *95*, 453.
96. Dubin, P. L.; Davis, D. D. *Macromolecules* **1984**, *17*, 1294.
97. Dubin, P. L.; Davis, D. D. *Colloids Surf.* **1985**, *13*, 113.
98. Dubin, P. L.; Rigsbee, D. R.; McQuigg, D. W. *J. Colloid Interface Sci.* **1985**, *105*, 509.
99. Dubin, P. L.; Rigsbee, D. R.; Gan, L.-M.; Fallon, M. A. *Macromolecules* **1988**, *21*, 2555.
100. Dubin, P. L.; Chew, C. H.; Gan, L.-M. *J. Colloid Interface Sci.* **1989**, *128*, 566.
101. Sudbeck, E. A.; Dubin, P. L.; Curran, M. E.; Skelton, J. *J. Colloid Interface Sci.* **1991**, *142*, 512.
102. Dubin, P. L.; Thé, S. S.; McQuigg, D. W.; Chew, C. H.; Gan, L.-M. *Langmuir* **1989**, *5*, 89.
103. Dubin, P. L.; Curran, M. E.; Hua, J. *Langmuir* **1990**, *6*, 707.
104. Dubin, P. L.; Vea, M. E.; Fallon, M. A.; Thé, S. S.; Rigsbee, D. R.; Gan, L.-M. *Langmuir* **1990**, *6*, 1422.
105. Dubin, P. L.; Thé, S. S.; McQuigg, D. W.; Gan, L.-M.; Chew, C. H. *Macromolecules* **1990**, *23*, 2500.
106. Davis, R. M.; Zhang, H.; Dubin, P. L.; McQuigg, D. W. *Polym. Prepr.* **1991**, *32(1)*, 292.
107. McQuigg, D. W.; Kaplan, J. I.; Dubin, P. L. *J. Phys. Chem.* **1992**, *96*, 1973.
108. Xia, J.; Zhang, H.; Rigsbee, D. R.; Dubin, P. L.; Shaikh, T. *Macromolecules* **1993**, *26*, 2759.
109. Odijk, T. *Langmuir* **1991**, *7,* 1991.
110. Binana-Limbelé, W.; Zana, R. *Colloids Surf.* **1986**, *21*, 483.
111. McGlade, M. J.; Randall, F. J.; Tcheurekdjian, N. *Macromolecules* **1987**, *20*, 1782.
112. Iliopoulos, I.; Wang, T. K.; Audebert, R. *Langmuir* **1991**, *7,* 617.
113. Maltesh, C.; Somasundaran, P. *Colloids Surf.* **1992**, *69*, 167.

RECEIVED July 8, 1994

Chapter 24

Interactions Between Water-Soluble Nonionic Polymers and Surfactant Aggregates

Josephine C. Brackman[1] and Jan B. F. N. Engberts

Department of Organic and Molecular Inorganic Chemistry, University of Groningen, Nijenborgh 4, 9747 AG Groningen, Netherlands

The influence of the chemical structure and charge of the surfactant on polymer-micelle interaction has been studied using various ionic and non-ionic surfactants, and surfactants of which the charge can be varied by (de)protonation. Octylthioglucoside, dodecyldimethylamine-oxide, alkylphosphates, alkylmalonates and cetyltrimethylammonium salts are the surfactants, and PEO, PVME and PPO are the polymers that were used. The results show how small changes in chemical structure or charge may greatly alter the propensity for association with polymers. Furthermore, the break-down of rod-like micelles of CTASal and NMe_4^+-dodecylmalonate to polymer-bound spherical micelles is described. This phenomenon illustrates how the accumulation of relatively weak forces that play a role in polymer-micelle interaction may severely change the morphology of the system and thereby the rheological characteristics of the solution.

Interactions between surfactant micelles and nonionic water-soluble polymers have been studied for several decades now (*1*). At an early stage of the research in this field, it was recognized that in the polymer-micelle complex the properties of both the micelles and the polymers are mutually modified. To mention the most important aspects in view of industrial applications, the solubilization power as well as the viscosity of an aqueous solution of polymer-bound micelles is often higher than that of the separate surfactant and polymer solutions (*2*). The commercial interest is reflected in the fact that many of the early reports on polymer-micelle interaction originated from industrial research laboratories (*3-7*).

The morphology of the complex formed between a nonionic water-soluble polymer and surfactant molecules has puzzled chemists from the first study of Saito (*2*) in 1955 onwards. Cabane in 1977 (*8*) established the model that is now quite generally accepted. According to this model segments of the polymer bind to the surface region

[1]Current address: DSM Research, P.O. Box 18, 6160 MD Geleen, Netherlands

of the surfactant micelles. The core of a micelle like SDS is shielded by the headgroups from the surrounding water for only about a third of its surface. Stabilization of the interface between the hydrophobic core and water is considered to be a major driving force for polymer-micelle interaction.

Although, the morphology of polymer-micelle complexes thus seems to be known in some detail, the subtle way in which the chemical structure of the surfactant and the morphology of the unperturbed micelle influence the tendency for association with nonionic polymers remains a challenging problem. Some generalizations concerning the influence of charge and headgroup volume on the association tendency have been proposed (*1*). However, our work on polymer-micelle interaction using a wider than conventional choice of surfactants revealed that one has to be extremely careful with generalizations in this field. In the case of charged polymers with charged surfactants electrostatic interactions clearly dominate. But in the case of water-soluble nonionic polymers with medium chain-length (C10-C16) surfactants several modes of interaction (hydrophobic interaction, dispersion forces, hydrogen bonding, hydration of the polymer and head groups, steric repulsion, and electrostatic interactions between surfactant headgroups, etc.) may contribute in the same order of magnitude to the total Gibbs energy of the system. The formation of polymer-bound micelles in this case results from an accumulation of relatively weak forces. Seemingly small changes in chemical structure of the surfactant or polymer may shift the balance of forces and may change which mode(s) of interaction dominate(s), without big changes in the overall Gibbs energy for aggregation. It is typical for these kinds of systems that small changes in Gibbs energy may result in large variations in morphology.

The effect of chemical structure of the surfactant and polymer on the propensity for mutual interaction has been one of the topics of our research in this field (*9-17*). Furthermore, we have initiated research on the interaction of polymers with rod-like micelles. Until recently, the study of polymer-micelle interaction concentrated on spherical micelles, though some studies on the interaction of vesicles with PEO (*18*) have been reported. Our work on rod-like micelles nicely illustrates how weak interaction forces may dramatically alter the morphology and concomittantly the rheological characteristics of the system.

Experimental

Experimental details may be found elsewhere (*9-15*).

The Influence of the Surfactant Structure on Polymer-Micelle Interaction

Nonionic Micelles. For a study of the effect of the surfactant structure on the propensity for interaction with polymers, the use of slightly hydrophobic water-soluble polymers, like poly(propylene oxide) (PPO) and poly(vinylmethylether) (PVME), is more convenient than the use of poly(ethylene oxide) (PEO) only. PPO and PVME will interact with a wider spectrum of surfactants since the Gibbs energy loss associated with the transfer of portions of the polymer from water to the micellar surface will contribute favorably to the association process.

With PPO (MW 1,000) and octylthioglucoside (OTG), for instance, we have shown (*10*) that association between this water-soluble polymer (in fact an oligomer) and nonionic micelles does occur contrary to expectations. A more cooperative clouding behavior of PPO (at 30°C) and lower Krafft temperature of OTG in D_2O in the presence of 0.5 g.dL^{-1} PPO (from 30°C to < 20°C) gave the first indications, which were confirmed by microcalorimetric measurements. However, the cmc of OTG (8.1 mM) is not noticeably affected by the presence of PPO. Association with nonionic micelles was hitherto thought not to take place, although only a limited set of studies showing mainly the absence of interaction of ethoxylated surfactants with PEO or poly(viny pyrrolidone) (PVP) had been reported. The principal driving force for polymer-micelle association was thought to be stabilization of the micellar Stern layer, and thereby the micelles proper. The core region of the common nonionic micelles is already totally shielded by the bulky nonionic headgroups, so polymer-micelle interaction was not expected. The absence of a reduction of the cmc in the presence of polymers seemed to confirm the absence of interaction with nonionic micelles. However, the occurence of polymer-micelle association has been confirmed with various techniques not only for OTG/PPO but also for OTG/hydoxypropylcellulose (*19*) and dodecyl dimethyl amine oxide (DDAO)/ PPO or PVME (MW 27,000) (*13*). We think that upon association two factors, namely the unfavorable steric repulsions between hydrated headgroups and polymer segments and the favorable transfer of the hydrophobic polymer to the micellar environment, compensate each other resulting in a negligible change in total Gibbs energy but appreciable change in morphology. Concomittantly, the cmc, which is determined by the total Gibbs energy for the formation of polymer-bound micelles, is not noticeably affected.

Nonionic and Cationic Micelles of Dodecyldimethylamine-oxide. The formation of polymer-bound micelles between the zwitterionic surfactant DDAO (*20-22*) at zero net charge and PPO or PVME was confirmed (*13*) by the reduction in aggregation number, measured by static fluorescence quenching of Ru(II)(bipy)$_2$(2,2'-didecyl-bipy) by 9-methylanthracene, from 76 in H_2O (at 20 mM) to 55 and 57 in a 0.5 g.dL^{-1} PPO or PVME solution respectively. The cloud point of PVME (34 °C) is not changed upon association with DDAO micelles at zero charge. However upon increasing the charge on the micelles by protonation the cloud point is raised in proportion to the degree of protonation and the effect is stronger with increasing DDAO concentration (Figure 1). This effect stems from the intermicellar repulsion of ionic polymer-bound micelles, causing coil expansion and repulsion between polymer coils. The same mechanism causes a rise in the viscosity and viscoelasticity of a PEO solution upon addition of SDS (*16*).

    In contrast to DDAO micelles at zero charge, the cmc of the cationic micelles formed from protonated DDAO is reduced in the presence of PPO and PVME (PEO has no effect). The reduction in electrostatic repulsion between headgroups upon formation of smaller polymer-bound micelles evidently results in a seizable change in Gibbs energy of micellization (*13*).

Spherical Micelles of n-Alkylphosphates. The use of surfactants, such as DDAO, for which the charge can be varied without drastically changing the chemical structure, has proven to be a valuable tool for separating the effects of chemical structure and charge. The charge of n-alkylphospates can be varied from -1 for n-$C_nH_{2n+1}PO_4H^-$ to

-2 for n-$C_nH_{2n+1}PO_4^{2-}$ by deprotonation, with pK_1 ca. 2 and pK_2 is ca. 7 (23). Hydrolysis of the monoanion is negligible at room temperature (for n-decylphosphate at pH 4.5 the rate constant is 8.2×10^{-6} s^{-1} at 100 °C, corresponding to a half life time of approximately 24 hours), whereas the dianion is totally unreactive (24). Both the monoanion and the dianion behave as surfactants (25). We emphasize that the actual charge of the surfactant molecules in the corresponding micelles will be somewhat lower than the structural charge (Z_0), which is the average charge of the alkylphosphate anions in the undissolved state (23,26). The aggregation behavior of the monoanion is quite comparable to that of SDS (23,25). Increasing the structural charge to $Z_0 = -2$, results in a substantial increase in headgroup repulsion, as evident from the larger headgroup area, lower aggregation numbers, and higher cmc and the larger area of contact between alkyl chain segments and water (23,25).

Cmc values for the two n-alkylphosphates, in water and in the presence of PPO, PEO 10k, PEO 20k and PVME have been measured using the pH method (14). For the ionic surfactants a reduction of the cmc is taken as evidence for polymer-micelle interaction. However, a quantitative comparison of polymer-micelle interaction necessitates an analysis of the Gibbs energy changes for micellization (14) in the presence and absence of the polymer. As a first approximation, the change in standard Gibbs energy of the micelle due to association with a polymer is given by (6),

$$\Delta G_{mic-pol} - \Delta G_{mic} = RT \ln (cmc_p/cmc)$$

where cmc_p is the cmc in the presence of polymer and cmc the corresponding value for the unperturbed micelle. These standard Gibbs energy changes are given in Table I and denote the changes in standard Gibbs energy when one mole of surfactant is transferred from normal to polymer-bound micelles plus the change in Gibbs energy of the polymer induced by this process. For comparison, ΔG_{mic} values are also listed for sodium n-decylsulfate (SDeS), and sodium n-dodecylsulfate (SDS) (27). Obviously, PPO and PVME interact with the surfactants listed in Table I. For PEO 10k or 20k only interaction with SDeS, SDS and n-decylphosphate at low structural charge is revealed.

A relevant comparison of headgroup effects on polymer-micelle interaction should be confined to surfactants of the same alkyl chain length (1). The data for n-decylphosphate ($Z_0 = -1$) and -sulfate, clearly show that even small changes in chemical structure have a pronounced effect on polymer-micelle interaction even within the class of monoanionic surfactants. The difference in the (unhydrated) headgroup volume between $-OPO_3H^-$ and $-OSO_3^-$ is relatively small as demonstrated by the limiting partial molar volume of HSO_4^- (35.67 $cm^3.mol^{-1}$) and $H_2PO_4^-$ (29.1 $cm^3.mol^{-1}$) (28). However, the effect of the headgroup on polymer-micelle interactions is considerable and, in contrast to expectation, the interaction is stronger for the micelles formed from the n-alkylsulfates. For example, we note that PEO 10k and 20k exert a relatively large and nearly equal stabilizing effect on SDeS micelles. By contrast, the stability of sodium n-decylphosphate micelles ($Z_0 = -1$) is hardly affected by PEO 10k whereas the stabilizing effect of PEO 20k is 50 % smaller than that for the sulfate. We have considered the possibility that the different behavior of n-alkylphosphates and -sulfates is caused by hydrogen-bonding interactions between phosphate headgroups, disfavoring penetration of polymer segments beyond the micellar surface.

Table I. Gibbs Energy of Micelle Stabilization by Polymers[a] for Va̶ı̶o̶
Surfactants

Surfactant	$-Z_0$[b]	$\Delta G_{mic\text{-}pol} - \Delta G_{mic}$, kJ.mol^{-1} [c]			
		PPO 1k	PEO 10k	PEO 20k	PVME 27k
n-C$_8$H$_{17}$OPO$_3$HNa	1.0	-0.7	0	0	
n-C$_{10}$H$_{21}$OPO$_3$H$_x$Na$_y$	1.0	-1.0	-0.05	-0.3	
n-C$_{10}$H$_{21}$OPO$_3$H$_x$Na$_y$	1.1	-1.5		-0.2	-1.0
n-C$_{10}$H$_{21}$OPO$_3$H$_x$Na$_y$	1.5	-1.1	-0.06	-0.3	-0.6
n-C$_{10}$H$_{21}$OPO$_3$H$_x$Na$_y$	2.0	-1.0	0	-0.1	-0.3
n-C$_{10}$H$_{21}$OSO$_3$Na	1.0		-0.6	-0.7	
n-C$_{12}$H$_{25}$OSO$_3$Na	1.0	-1.8	-1.0	-1.0	

Di-sodium n-decylphosphate micelles with Z_0 = -2 are expected to interact more strongly with PEO because at this structural charge interheadgroup hydrogen bonding will be reduced. In addition, the enhanced interheadgroup repulsions will cause a decrease in aggregation number (*23*) and an increased hydrophobic core-water contact. Nagarajan (*29*) suggested that the latter factor is expected to enhance polymer binding. However, in contrast to expectation, the interaction with polymer at Z_0 = -2 is *weaker* than that at lower Z_0 values. Particularly for PEO 10k and 20k, but also for PVME, this trend is clear. The unexpected decrease in interaction tendency may relate to the strong hydration of the phosphate headgroup at Z_0 = -2, hampering the presence of the polymer in the headgroup region. Another explanation hinges on the influence of the ions on the hydration sheaths of polymers, proposed by Witte (*30*) to explain the differences between anionic and cationic surfactants in their interaction with polymers. Anions exert a more pronounced influence on the hydration of, for instance, PEO than cations (*31*). The differences in $\Delta G_{mic\text{-}pol} - \Delta G_{mic}$ between alkylsulfate and -phosphate (Z_0=-1) and between alkylphospate at Z_0=-1 and Z_0=-2 might be the result of slight differences in the influence of the headgroups on the polymer hydration.

Spherical and Rod-like Micelles of n-Alkylmalonates. In principle, the 2-alkyl-malonate surfactants present an opportunity to study another surfactant for which the charge can be varied from -1 to -2 by deprotonation and compare the results with those for the alkylphosphates.

$$n\text{-}C_nH_{2n+1} \; CH(COOH)(COO^-) \quad \underset{}{\overset{K_a}{\rightleftharpoons}} \quad n\text{-}C_nH_{2n+1} \; CH(COO^-)_2$$

The pK_a value of the (second) dissociation step is 5.7 for malonic acid in water. This pK_a value of the 2-alkylmalonic acid will be slightly different and will depend on the aggregation state of the molecule.

The amount of published research on 2-alkylmalonates is extremely limited. It was anticipated that both the monoanionic and the dianionic surfactants form micelles. However, micellization has hitherto only been reported for the dianion (32-34). To the best of our knowledge, there are no literature data on micellization of the mono-anionic molecule. This lack of information probably stems from the high Krafft temperatures of the common alkalimetal monovalent salts (76°C for n-$CH_{12}H_{25}CH(CO_2H)CO_2K$). The Krafft temperature of the mono-NMe_4^+ salts lies below 20 °C. This is probably due to the reduced cation-anion Coulomb interactions in the solid, and to stronger binding of the slightly hydrophobic NMe_4^+ ion to the micelles resulting in a lower cmc.

Cmc values in the absence and presence of PEO 10k, PEO 20k and PVME are listed in Table II for several mono- and disalts of 2-alkylmalonic acids (15). The mono-NMe_4^+ salt of 2-dodecylmalonic acid forms viscoelastic solutions in water, indicative of the presence of flexible, rod-like aggregates. The viscoelasticity can be observed visually as the recoil of air bubbles when a swirling motion of the solution is abruptly stopped. This viscoelasticity is observed in the concentration range from the cmc (ca. 1.4 mM) up to approximately 13 mM. At present we have no explanation for the disappearance of the viscoelasticity at higher concentrations. The formation of rodlike micelles is confirmed by ^1H NMR spectral data. In the same concentration range in which viscoelasticity is observed, the proton resonances of the alkyl chain are severely broadened (15). This is characteristic for rodlike micelles and the effect has been attributed to increased T_2 values (35).

It is interesting to note the structural resemblance, shown in Figure 2, between mono-NMe_4^+ 2-dodecylmalonate and cetyltrimethylammonium salicylate (CTASal), the arche-type of a surfactant that forms rodlike micelles. Clearly, the relative orientation of the OH^- and CO_2^- groups is of decisive importance. This is also indicated by the fact that m- or p-hydroxybenzoate salts of cetyltrimethylammonium surfactants do not form rodlike micelles (36).

The cmc of mono-NMe_4^+ 2-dodecylmalonate is increased in the presence of PVME. This suggests a stabilization of the surfactant monomer by the polymer. Cloud point measurements support this conclusion (15). The most important observation, however, is the complete disappearance of the viscoelasticity of the aqueous micellar solutions of mono-NMe_4^+ 2-dodecylmalonate in the presence of PVME or PEO. This is supported by the observation that the ^1H NMR line broadening of the alkyl chain methylene protons has disappeared after addition of these polymers (15). Evidently,

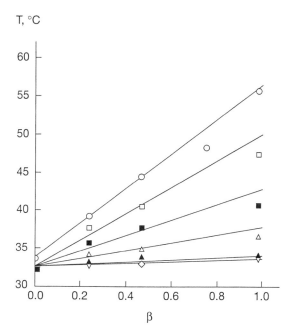

Figure 1. The cloud point of PVME in the presence of DDAO at various concentrations of DDAO, as a function of the degree of protonation (β) (a) 1.25 mM, ($\Diamond$); (b) 2.5 mM, ($\blacktriangle$); (c) 5 mM, (Δ); (d) 10 mM, ($\blacksquare$); (e) 15 mM, ($\square$); (f) 20 mM, (o).

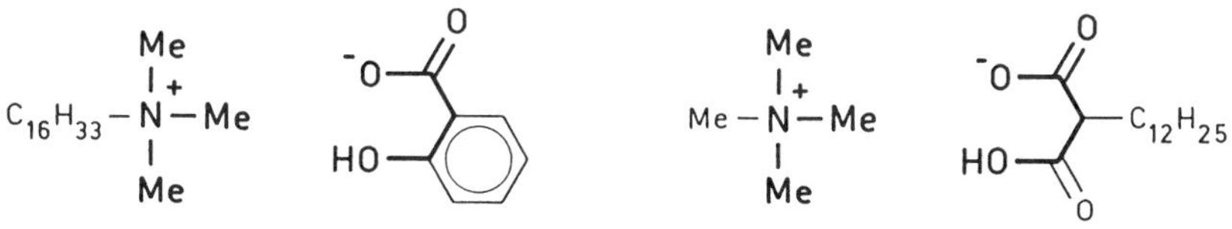

Figure 2. Structural resemblance between CTASal and mono-NMe$_4$ dodecyl-malonate.

Table II. Cmc Values (mM) for 2-Alkylmalonate Micelles in Water and in the
Presence of Polymers[a]

Surfactant	H_2O	PEO 10k	PEO 20k	PVME 27k
n-$C_{10}H_{21}CH(CO_2H)CO_2NMe_4$	4.9 ± 0.5^b	c	c	c
n-$C_{12}H_{25}CH(CO_2H)CO_2NMe_4$	1.43 ± 0.05^b	1.5 ± 0.1^b		1.8 ± 0.1^b
n-$C_{12}H_{25}CH(CO_2K)_2$`	55 ± 2^d	38.2 ± 2^d	41 ± 2^d	
	47 ± 2^e			
	48^f			
n-$C_{12}H_{25}CH(CO_2NMe_4)_2$	48.1 ± 1^b	40.1 ± 1^b	25.2 ± 2^b	13.8 ± 1^b
n-$C_{13}H_{27}CO_2NMe_4$	4.8 ± 0.1^b	4.3 ± 0.1^b		1.9 ± 0.2^b

SOURCE: Reprinted with permission from ref. 15. Copyright 1991.
[a] Polymer concentration: 0.5 g.dL^{-1}, temperature: 25 °C. [b] Pinacyanol chloride absorption method. [c] No accurate cmc values could be obtained. [d] Conductometric method. [e] pH method. [f] From reference 32.

spherical polymer-bound micelles are formed in favor of rodlike micelles, probably because the surface-to-volume ratio is higher for the spherical micelles. A comparatively large surface area of the micelles will result in a reduction of both headgroup-headgroup and headgroup-adsorbed polymer steric repulsions. The extra core-water contact of the spherical micelles is stabilized through binding of the polymer. Interestingly, Nagarajan (29) has theoretically predicted that rodlike micelles of an anionic and a nonionic surfactant will become ellipsoidal upon interaction with polymer.

The di-Na and di-K salts of 2-alkylmalonic acids are known to form spheroidal micelles (32,33). The behavior of the di-NMe$_4^+$ salts is quite comparable; the cmc values, shown in Table II, are similar. In sharp contrast to n-decylphosphate at $Z_0= -2$, the cmc values for the dianion of 2-dodecylmalonate are considerably reduced in the presence of PEO and PVME, which points to polymer-micelle interaction. For di-NMe$_4^+$ 2-dodecylmalonate/ PEO $\Delta G_{mic\text{-}pol} - \Delta G_{mic}$ ranges from -0.5 kJ.mol^{-1} to -1.6 kJ.mol^{-1} depending on the molecular weight of PEO. For comparison, $\Delta G_{mic\text{-}pol} - \Delta G_{mic}$ for SDS/PEO is -1.0 kJ.mol^{-1} both for PEO 10k (37) and PEO 20k. The alkyl chain length, which is known to affect polymer-micelle interaction significantly (1), is comparable for both surfactants since there is only one additional methine moiety in di-NMe$_4^+$ 2-dodecylmalonate. Thus, we conclude that micelles formed from di-NMe$_4^+$ 2-dodecylmalonate interact as strongly with PEO as SDS micelles do, despite the di-anionic and bulky malonate headgroup. This result is unexpected in view of theories (29,38) that hinge on the idea that a bulky headgroup hampers the binding of a polymer at the micellar surface. However, the surface of micelles formed from di-

NMe_4^+ 2-dodecylmalonate has structurally much in common with poly(carboxylates) like poly(methacrylic acid) (PMAA) and poly(acrylic acid) (PAA), which are known (*39*) to interact strongly with PEO. These interactions presumably involve (cooperative) hydrogen-bonding, but hydrophobic interactions may also play a role since PMAA has a much greater complexation tendency than PAA (*39*). The carboxylate groups in micelles formed from dianionic 2-dodecylmalonate will be protonated to some extent since electrostatic headgroup repulsions and the low local polarity will tend to increase their pK_a.

It may well be that no general rules can be formulated regarding the effect of surfactant charge on the strength of polymer-micelle interaction. In the case of the not fully charged surfactants, there will be competition between interheadgroup hydrogen-bonding and hydrogen-bonding interactions between the headgroups and the polymer. Furthermore, hydrophobic interactions and hydration shell overlap effects will play a role and the overall gain in Gibbs energy upon polymer-micelle complexation will be a compromise between a variety of not necessarily coupled interaction forces.

The Breakdown of Rod-like Micelles of CTAX

Cetyltrimethylammonium salicylate forms rodlike micelles even in dilute (ca. 10^{-4} M) solutions (*40,41*). We studied solutions containing 25 mM CTAB and 15 mM NaSal. These solutions exhibit non-Newtonian behavior. That is, the apparent viscosities vary dramatically with changing shear rate. The shear rate dependence of the viscosity of the CTAB/NaSal solutions is not at all affected by PVP, ethanol, or t-butanol (Figures 3 and 4) and only slightly by PEO (*11,12*). Three regions may be discerned in a plot of apparent viscosity versus shear rate. At very low shear rates, Newtonian behavior is displayed. Our data do not include shear rates low enough to observe this region. In the second region, the internal structure of the solution is altered by the shear forces. In the case of rodlike micelles, this causes a drop in apparent viscosity, due to aligning and disruption of the rods. A closer inspection of our data in the second region, where structural changes occur, shows that the shear stress is constant (0.8 Pa, cylindrical geometry) up to a shear rate of 84 s^{-1}. Rheopexy and thixotropy can be observed in the same shear rate range. Actually, the CTAB/NaSal system spoils a beginning rheologist since the more attractive and special aspects of rheology are encountered without recourse to further examination of the field. In the third region, at high shear rates, the structural changes are completed and Newtonian behaviour is again observed. For the present system, this transition occurs around a shear rate of 100 s^{-1}.

PEO exerts a slight influence on the apparent viscosity (Figure 4a) and also on the first normal stress differences measured for the CTAB/NaSal system (*11,12*). This may be due to (i) interference of the polymer chains with the intermicellar ordering and flow of the rods, or (ii) to a modest interaction of PEO with CTAX monomers or aggregates either influencing the structural relaxation time or the structures themselves. There appears to be no reason, however, why interference should occur for PEO but not for PVP, whereas there are indications that PEO has a very small but detectable effect on CTAB aggregation, which PVP does not (*7,42,43*). Thus, the latter explanation is more likely.

The apparent viscosities of CTAB/NaSal in the presence of PVME or PPO, and

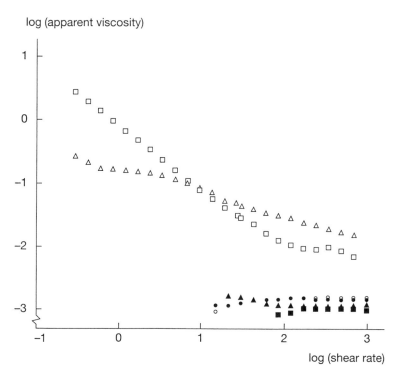

Figure 3. Double logarithmic plot of apparent viscosity vs. shear rate of the following aqueous solutions of CTAB (25 mM); no additives, (■); + PEO 20k (0.5 g.dL^{-1}), (▲); + PVME (0.5 g.dL^{-1}), (•); + NaSal (15 mM), (□); + NaSal (15 mM) and PEO 20k (0.5 g.dL^{-1}), (△); and + NaSal (15 mM) and PVME (0.5 g.dL^{-1}), (o). The data were measured with cylindrical geometry.

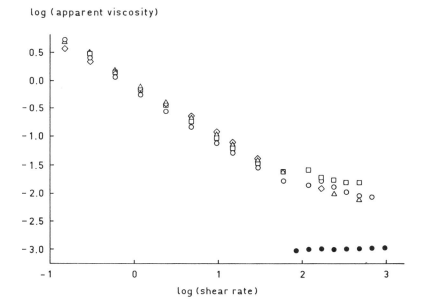

Figure 4. Double logarithmic plot of apparent viscosity vs. shear rate for the following aqueous solutions of CTAB (25 mM)/NaSal (15 mM): no additives, (o); + PVP (0.5 g.dL^{-1}), (□); + ethanol (0.5 g.dL^{-1}), (Δ); + t-butanol (0.5 g.dL^{-1}), (◊); + PPO (0.5 g.dL^{-1}), (•). The data were measured with cylindrical geometry.

of CTAB solutions without NaSal, are orders of magnitude lower and are independent of shear rate, indicative of Newtonian behavior. The change in rheological characteristics induced by PVME is so tremendous that it cannot be overlooked. The polymer-induced transition from a non-Newtonian to a Newtonian fluid is attributed to preferential binding of spherical rather than rodlike micelles onto the polymers, just as in case of mono-NMe$_4^+$ dodecylmalonate. Recently, the CTAB/NASal/PVME system has been investigated by NMR and the results confirmed our conclusions (44). The formation of smaller, spherical, polymer-bound micelles is consistent with the reduction in aggregation number of CTAB micelles from 70 to 30-40 (depending on the CTAB concentration) upon binding to PVME or PPO (polymer concentration 0.5 g.dL^{-1}). The cmc of CTAB (0.95 mM, ref. 36) is also reduced in the presence of PPO (cmc 0.37 mM) and PVME (cmc 0.46 mM). PEO and PVP have no effect on either the cmc or the aggregation number of CTAB micelles, which indicates the absence of (substantial) polymer-micelle interaction for these combinations.

The sphere-to-rod transition of CTATosylate is shifted to higher concentrations in the presence of PVME to an extent proportional to the polymer concentration (12). Apparently, the polymer is first saturated with spherical micelles before the activity of the surfactant rises and rod-like micelle formation starts. This finding is also in harmony with the above observations.

Thus, the slightly hydrophobic polymers that interact with CTAX micelles, bind preferentially to smaller, spherical micelles for which the surface to volume ratio is higher. The hydrophilic polymers PEO and PVP do not bind to CTAX micelles and, therefore, do not exert sizable effects on the morphology of these aggregates.

Literature Cited

1   Goddard, E.D. *Colloids Surf.* **1986**, *19*, 255.
2   Saito, S.J. *Biochemistry* **1957**, *154*, 21.
3   Breuer, M.M.; Robb, I.D. *Chem. Ind. (London)* **1972**, *13*, 530.
4   Lange, H. *Colloid Polym. Sci.* **1971**, *243*, 101.
5   Arai, H.; Murata, M.; Shinoda, K. *J. Colloid Interface Sci.* **1971**, *37*, 223.
6   Tokiwa, F.; Tsujii, K. *Bull. Chem. Soc. Jpn.* **1973**, *46*, 2684.
7   Schwuger, M.J. *J. Colloid Interface Sci.* **1973**, *43*, 491.
8   Cabane, B. *J. Phys. Chem.* **1977**, *81*, 1639.
9   Brackman, J.C. *Ph.D. Thesis*, University of Groningen, **1990**.
10 Brackman, J.C.; van Os, N.M.; Engberts, J.B.F.N. *Langmuir* **1988**, *4*, 1266.
11 Brackman, J.C.; Engberts, J.B.F.N. *J. Am. Chem. Soc.* **1990**, *112*, 872.
12 Brackman, J.C.; Engberts, J.B.F.N. *Langmuir* **1991**, *7*, 2097.
13 Brackman, J.C.; Engberts, J.B.F.N. *Langmuir* **1992**, *8*, 424.
14 Brackman, J.C.; Engberts, J.B.F.N. *J. Colloid Interface Sci.* **1989**, *132*, 250.
15 Brackman, J.C.; Engberts, J.B.F.N. *Langmuir* **1991**, *7*, 46.
16 Brackman, J.C. *Langmuir* **1991**, *7*, 469.
17 Brackman, J.C.; Engberts, J.B.F.N. *Chem. Soc. Rev.* **1993**, *85*.
18 Arnold, K.; Pratsch, L.; Gawrisch, K. *Biochem. Biophys. Acta* **1983**, *728*, 121.
19 Winnik, F.M. *Langmuir* **1990**, *6*, 522.
20 Herrmann, K.W. *J. Phys. Chem.* **1962**, *66*, 295.
21 Herrmann, K.W. *J. Phys. Chem.* **1964**, *68*, 1540.

22 Rathman, J.F.; Scamehorn J.F. *Langmuir* **1987**, *3*, 372.
23 Chevalier, Y.; Belloni, L.; Hayter, J.B.; Zemb, T. *J. Phys.* **1985**, *46*, 749.
24 Bunton, C.A.; Diaz, S.; Romsted, L.S.; Valenzuela, O. *J. Org. Chem.* **1976**, *41*, 3037.
25 Arakawa, J.; Pethica, B.A. *J. Colloid Interface Sci.* **1980**, *75*, 441.
26 Schulz, P.C. *Colloids Surf.* **1989**, *34*, 69.
27 Witte, F.M.; Engberts, J.B.F.N. *J. Org. Chem.* **1987**, *52*, 4767.
28 Marcus, Y. *Ion Solvation*; Wiley: New York, 1985, p 78.
29 Nagarajan, R. *J. Chem. Phys.* **1989**, *90*, 1980.
30 Witte, F.M.; Engberts, J.B.F.N. *Colloids Surf.* **1989**, *36*, 417.
31 Ataman, M. *Colloid Polym. Sci.* **1987**, *265*, 19.
32 Shinoda, K. *J. Phys. Chem.* **1955**, *59*, 432.
33 Vikingstad, E.; Saetersdal, H. *J. Colloid Interface Sci.* **1980**, *77*, 407.
34 Kvamme, O.; Kolle, G.; Backlund, S.; Hoiland, H. *Acta Chem. Scand.* **1983**, *A37*, 393.
35 Olsson, U.; Södermann, O.; Guéring, P. *J. Phys. Chem.* **1986**, *90*, 5223.
36 Gravsholt, S. *J. Colloid Interface Sci.* **1976**, *57*, 575.
37 Witte, F.M. *Ph.D. Thesis*, University of Groningen, **1988**.
38 Ruckenstein, E.; Huber, G.; Hoffmann, H. *Langmuir* **1987**, *3*, 382.
39 Saito, S.: Sakamoto, T. *Colloids Surf.* **1987**, *23*, 99.
40 Hoffmann, H.; Ebert, G. *Angew. Chem., Int. Ed.* **1988**, *27*, 902.
41 Gravsholt, S. In *Rheology*; Astarita, G.; Marucci, G.; Nicolais, L., Eds.; Plenum Press: New York., **1981**, Vol. 3, p 629.
42 Shirahama, K.; Oh-ishi, M.; Takisawa, N. *Colloids Surf.* **1989**, *40*, 261.
43 Hoffmann, H.; Huber, G. *Colloid Surf.* **1989**, *40*, 181.
44 Morris, K.F.; Johnson, C.S.Jr.; Wong, T.C *J. Phys. Chem.* **1994**, *98*, 603.

RECEIVED July 21, 1994

APPLICATIONS

Chapter 25

Viscoelastic Surfactants

Rheology Control Without Polymers or Particulates

Gene D. Rose and Arthur S. Teot

Central Research and Development, Dow Chemical Company,
1712 Building, Midland, MI 48674

Surfactants have been identified that impart controllable and useful viscous and elastic properties to a wide variety of aqueous liquids. These liquids include deionized water, aqueous solutions of ethylene glycol, and aqueous solutions of up to 75 wt% inorganic electrolytes. Both the surfactant structure and the composition of the liquid into which the surfactant is formulated influence the magnitude of the viscous and elastic response obtained. The rheological properties exhibited by these viscoelastic surfactant solutions are similar to those obtained with solutions of polymers or dispersions of particles without the disadvantages of irreversible mechanical shear degradation or insoluble particulates. These features have provided useful rheological control for a variety of applications.

This paper will provide a brief review of some of the interesting industrial applications of viscoelastic surfactant solutions. Its objective is to illustrate the utility of viscoelastic surfactant formulations in industrial applications that have widely different rheological requirements. Specific examples of industrial applications investigated in our laboratory will be used to show that the viscoelastic surfactant formulations can provide all of the desired viscous or elastic properties that could be obtained by the proper choice of a polymer or particulate (such as clays) as the rheology control agent. In addition, the viscoelastic surfactant formulations provide all of the benefits of normal surfactants plus rheological advantages, which result from the associative structure that produces their rheological properties.

   A wide variety of association structures can be obtained by the dissolution of even the simplest surfactant, such as dodecyltrimethylammonium chloride, in water. (*1*) The simplest of these structures is a roughly spherical micelle formed by the aggregation of 50 to 100 surfactant molecules as the surfactant concentration is increased above the critical micelle concentration (CMC), which for the dodecyl-trimethylammonium chloride occurs at about 2×10^{-2} M. (*2*) Other structures such as cylindrical micelles, isotropic cubic phases, anisotropic hexagonal phases, and anisotropic lamellar liquid crystals follow as the concentration of the surfactant is increased further. (*1*) For dodecyltrimethylammonium chloride the rheologically more interesting of these association structures only occur at high surfactant concentrations, e.g., at greater than 40 wt%. (*1*) Increasing the alkyl chain length of

0097–6156/94/0578–0352$08.00/0

the surfactant or changing the anion from chloride to bromide causes these rheologically more interesting association structures to occur at lower concentrations, but with cetyltrimethylammonium chloride the micelle size hardly increases even at surfactant concentrations up to 1 M (about 30 wt%) and with cetyltrimethylammonium bromide rodlike micelles are still not seen until a surfactant concentration of about 0.2 to 0.3 M (7 to 11 wt%). (3) However, through the right choice of surfactant structure or solution conditions surfactants form cylindrical micelles at surfactant concentrations as low as 100 ppm. For example, in contrast to the normal CMC (transition to spherical micelles) for cetylpyridinium chloride, which is reported (2) to be 9 x 10^{-4} M (about 250 ppm), Hoffmann, et al. have shown (4) that by changing the anion from chloride to salicylate a transition from spherical micelles to cylindrical micelles occurs at 4.4 x 10^{-4} M (about 100 ppm).

This short introduction to the association structures of surfactants illustrates that the cylindrical micellar association structures can be obtained at concentrations that would allow them to be economically attractive as rheology control agents in industrial applications. More detailed information on the relation between their viscoelasticity and micellar association properties is presented in the first four review chapters of this book authored by (1) H. Hoffmann, (2) M. E. Cates, (3) F. Lequeux and S. J. Candau, and (4) H. Rehage, respectively; therefore, it will not be discussed here. These micellar association structures and rheological properties are not restricted to cationic surfactants. They can be obtained by all types of surfactants. This is illustrated below by a short review of the use of viscoelastic surfactants to provide drag reduction (a decrease in the pressure drop) in turbulent pipeline flow. This review will also illustrate some key properties of viscoelastic surfactant systems that are generally applicable to a wide variety of applications.

Drag reduction has been historically the most often investigated rheological control application for viscoelastic surfactants. This work has been discussed earlier by others (5,6) and also provided the starting point for work on viscoelastic surfactants in our laboratory. (7,8,9,10) Two of the earliest viscoelastic surfactant systems to be investigated were the soaps, such as sodium oleate, illustrated by the work of Savins (11) and the cetyltrimethylammonium bromide complexes with α- or β-naphthol, illustrated by the work of White. (12) Both of these systems provided effective drag reduction that was stable to irreversible mechanical degradation. The stability to irreversible mechanical degradation is a particularly useful property of all viscoelastic surfactants.

However, both the soaps and the cetyltrimethylammonium bromide complexes with α- or β-naphthol viscoelastic surfactants have limitations. The soaps require large amounts of electrolyte (often >2 wt%) to be effective (11) and would be sensitive to divalent cations (5), whereas the naphthol complexes are oxidatively unstable. (5,12) Zakin and Lui (13) described nonionic viscoelastic surfactant systems which overcame these limitations. However, nonionic surfactants only provide effective drag reduction activity near their cloud point. (5,13) Thus, they have a narrow temperature range over which they produce effective drag reduction activity. Further comparisons of these systems can be found in an earlier review by Shenoy. (5)

In contrast to the above deficiencies, we found the following benefits could be obtained using certain cationic surfactant systems. First, chemically stable systems were identified. This is illustrated by the cetyltrimethylammonium salicylate system, which provided stable drag reduction for 2-1/2 months in a closed loop heating system (7) and stability in thickened bleach formulations, described later. However, it should be noted that for long term stability at high temperatures care should be taken to exclude oxygen. (14) Second, broad temperature ranges were seen for effective drag reduction. Excess organic counter ions added to such a system (such as salicylate) were shown to extend both the upper temperature and upper shear stress limits of this drag reduction activity. (7,8,9,10) For example, drag reduction was found over a temperature range of greater than 80°C for a 0.2 wt% erucyltri-methylammonium salicylate plus 0.2 wt% excess sodium salicylate system. (10)

Similar benefits are obtained with cationic surfactants in many other applications, some of which will be described below. Also, we found the quaternary ammonium surfactants, such as those illustrated in Figure 1, to be more easily adapted than anionic or nonionic surfactants to provide the desired rheological properties for various other applications. Because of these benefits for general industrial applications, only cationic surfactant systems are described in this paper. However, it should be emphasized that other types of surfactants and a great many combinations of surfactants can be used to impart useful rheological properties to applications as long as the environment to which they are subjected does not destroy their viscoelasticity. Examples of some of these are (a) bivalent metal alkyl sulfates and ether sulfates (15), (b) partially neutralized alkenylsuccinates (16), (c) anionic/cationic mixtures (17), (d) anionic/zwitterionic mixtures (18), and (e) alkylamines and amine oxides (19).

Both the solution environment of the surfactant and the surfactant structure strongly influence the type of micellar association structure formed by a surfactant, and, therefore, its rheological properties. Some of the factors affecting the viscoelastic properties of the aqueous formulations of the surfactants include the balance between the hydrophobic tail and hydrophilic head group; the counter ion for the surfactant ion; electrolytes, cosolvents, and other additives; the concentration of the surfactant; and the temperature.

The structures of the cationic viscoelastic surfactants that will be used to illustrate the effects of the above factors are given in Figure 1. The bis(hydroxyethyl)methyl head group, illustrated in Figure 1, has been particularly useful in several applications. With sufficiently high electrolyte level an aromatic anion is not required to develop useful viscoelasticity in aqueous solutions of these cationic surfactants. (3,20)

Experimental

Test Procedures. The drag reduction data were obtained at room temperature (~23°C) using a glass capillary lab scale apparatus. A 1.0 meter length, 2.65 mm diameter glass capillary was connected to a fluid reservoir to which pressure is applied by a regulated air supply. While the fluid flows through the capillary the pressure is noted and the flow rate is monitored by a recording balance. The flow properties of the test fluids were compared to each other using the Fanning friction factor, F, and the Reynolds number, Re, defined by the equations:

$$F = \Delta P \, D / (2 \, L \, \rho \, V^2)$$

$$Re = D \, V \, \rho / \eta$$

where: ΔP = pressure applied to the reservoir
D = diameter of the capillary
L = length of the capillary
ρ = density of the solvent
V = velocity of the test fluid
η = viscosity of the solvent

The solvent viscosities and densities were used for these calculations. This choice was made because of the non-Newtonian viscosity of the surfactant solutions. This leads to a displacement of the Fanning friction factor versus Reynolds number flow curves in the laminar flow region. However, the use of the solvent Reynolds number for the viscoelastic surfactant formulations provides a direct comparison of the level of pressure drop (drag reduction) relative to the solvent at the same flow rate (Reynolds number).

The oscillatory rheological measurements were made with a Rheometrics Model RFR-7800 Fluids Rheometer (Rheometrics, Inc., Piscataway, N.J.) equipped with a 10/100 g-cm full scale torque transducer, a 100 g full scale normal force transducer, and their RECAP computer analysis package. It is a computer controlled laboratory rheometer capable of measuring the time, temperature, and shear rate (or frequency) dependent viscoelastic properties of low to medium viscosity fluids. All of the programs for running the system and data reduction were obtained from Rheometrics, Inc. The complex viscosity (η^*), the storage modulus (G'), and loss modulus (G") were measured as a function of frequency at 20% strain and 25°C using the cone and plate fixture (25 mm radius, 0.04 radian cone angle). The equations for the rheological data reduction will not be discussed in this report. Further information can be obtained from standard texts. (*21,22*) Viscosity as a function of shear rate data were obtained with a Haake Rotovisco Model RV3 rotational viscometer equipped with a 50/500 g-cm dual range measuring head using the NV or MVI tools and the tempering unit connected to a Haake Model FE constant temperature circulator (Haake, Fisons Instruments, Paramus, NJ). Calculations were made automatically by an Apple II microcomputer interfaced to the viscometer using the equations for data reduction supplied by Haake. In both the dynamic and steady experiments, the samples were allowed to equilibrate about 5 minutes after loading before the tests were initiated. The Brookfield viscosity measurements were made at room temperature (~23°C) using a Model LVT with the U.L. cup and bob measuring system (Brookfield Engineering Laboratories, Inc. Stoughton, MA).

The adsorption experiments were carried out by preparing a master batch of 0.275 wt% devolatilized Ethoquad O/12 (see structure below) in 14.2 lb/gal brine. To six 20 g samples of the master batch were added various loadings of Rev Dust (described below). After warming to 65°C, mixing, and a 2 hour equilibration, the Rev Dust was removed by centrifugation and the supernatant was analyzed for the amount of remaining surfactant.

The Beckman Total Carbon Analyzer was used to measure the surfactant concentration in all of the adsorption measurements and in the reversible thickening experiments. The pH was monitored during the reversible thickening experiments with a Corning Model 125 pH meter.

Materials and Sample Preparation. Unless noted otherwise the test solutions were prepared from commercial surfactants obtained from Akzo Chemie America, Chicago, IL and Humko Chemical Division, Witco Chemical Corp., Memphis, TN. Those surfactants containing a salicylate counterion were prepared by combining the surfactant with an equivalent of sodium salicylate (NaSal) from Aceto Chemical Co., Inc., Flushing, NY plus deionized water and optional electrolyte to give the desired concentrations. Note that this preparation also forms one equivalent of sodium chloride (about 0.03 wt% for a 0.2 wt% surfactant solution). Except for the bleach formulations, the ceytltrimethylammonium salicylate (CTAS) was prepared using Arquad 16-50 from Akzo, the erucyltrimethylammonium salicylate (ETAS) was prepared using Kemamine Q2983-C from Humko, and the oleylbis(hydroxyethyl)-methylammonium chloride was Ethoquad O/12 from Akzo. For the bleach formulations the surfactant used was hexadecyltrimethylammonium bromide from

Aldrich Chemical Company. The p-toluene sulfonate salt was prepared by combining this surfactant with sodium p-toluene sulfonate from Hart Chemical.

The fluids into which the surfactants were placed are as follows: For the drag reduction, the ethylene glycol formulations were prepared from Dowtherm SR-1 heat transfer medium, which is a specially inhibited ethylene glycol-based fluid sold by The Dow Chemical Company. The various brines were prepared by using aqueous solutions containing the appropriate concentrations of sodium chloride, potassium chloride, calcium chloride, calcium bromide, and zinc bromide. The compositions of the brines discussed in this paper are listed in Table I.

Table I. Composition of Brines

Density, lb/gal	Electrolyte Composition*, wt%				
	KCl	NaCl	CaCl$_2$	CaBr$_2$	ZnBr$_2$
8.5	5.0	-	-	-	-
8.6	-	4.0	1.0	-	-
9.2	-	-	12	-	-
11.6	-	-	38	-	-
14.2	-	-	-	53	-
15.0	-	-	17.2	43.7	-
15.4	-	-	15.6	41.3	5.4
17.5	-	-	-	27	41.3
19.2	-	-	-	19	56

* Remainder of brine formulation is water.

In some experiments Rev Dust is added to the brines to simulate fines produced during drilling. The Rev Dust is a small particle (2 μm) calcium montmorillonite clay (Milwite, Inc. Houston). Fisher Scientific Company purified grade sodium hypochlorite was used to prepare the bleach solutions. The hypochlorite activity of the starting bleach was monitored by frequent titration to be certain of its activity. When it was not being used it was wrapped in aluminum foil to prevent exposure to the light. The procedures used to characterize the bleach formulations were: (1) Titration (iodide-iodine with thiosulfate and a stabilized starch indicator) for hypochlorite concentration. (2) Measurement of pH using a Corning Model 125 pH meter. (3) Measurement of the viscosity as described above.

Results and Discussion

Over a broad range of viscosities of interest for providing rheological control, the viscosity of viscoelastic surfactant solutions monotonically increases with increasing concentration of the surfactant in a manner similar to that seen with polymers. At sufficiently high concentrations, for example about 0.1 M (3 to 4 wt%) of alkyltrimethylammonium salicylate in water, the solutions become gel-like in appearance. These properties have also been reported by others (23) For aqueous solutions. We have found similar properties can also be obtained with aqueous surfactant solutions containing water miscible organic compounds. For example, Figure 2 shows the viscosity data for an aqueous cetyltrimethylammonium salicylate (CTAS) surfactant solution containing 20 wt% ethylene glycol. With the longer chain erucyltrimethylammonium salicylate (ETAS) viscoelastic properties were obtained in aqueous solutions containing ethylene glycol concentrations up to 50 wt%. Figure 3 shows that the elastic properties imparted to the 50 wt% ethylene

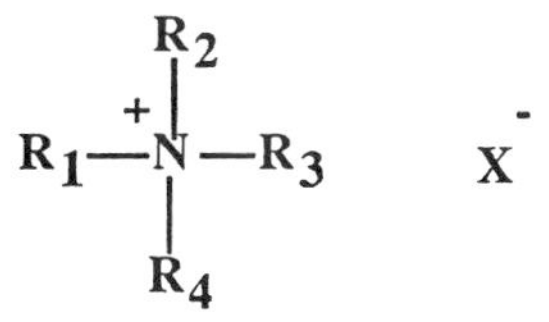

$$R_1-\overset{\overset{\displaystyle R_2}{|}}{\underset{\underset{\displaystyle R_4}{|}}{\overset{+}{N}}}-R_3 \qquad X^-$$

where:

$R_1 \geq C_{14}$ alkyl

R_2, R_3, R_4 = CH_3, CH_2CH_2OH, pyridinium

X^- = Aromatic carboxylate or sulfate, Cl^-, Br^-

Figure 1. Viscoelastic Surfactant Structure.

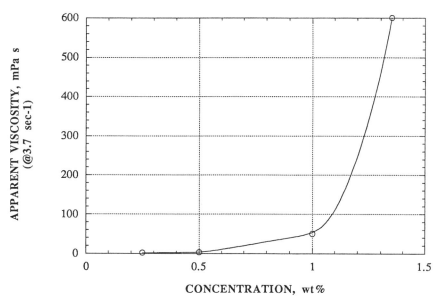

Figure 2. Viscosity of an Aqueous Solution Containing 20 wt% Ethylene Glycol as a Function of Cetyltrimethylammonium Salicylate Concentration.

glycol in water solution by the ETAS provides drag reduction for solvent Reynolds numbers greater than approximately 2000, which is the critical Reynolds number reported (24) for the transition from laminar to turbulent flow for a Newtonian fluid. Also, note the improved drag reduction performance with the excess sodium salicylate (NaSal).

The rheological response of viscoelastic surfactant solutions to temperature and shear rate may be similar to that exhibited by polymer solutions. For example, the viscosity of an aqueous solution containing 0.2 wt% CTAS plus 0.2 wt% excess NaSal decreased as the temperature or shear rate during the measurement increased. (10) Similar data have been provided by Hoffmann, et al. (25) for viscoelastic surfactant solutions without excess sodium salicylate. However, quite different behavior can often be seen as the examples in Figures 4 and 6 will illustrate.

Figure 4 shows the viscosity versus shear rate and temperature for 0.5 wt% oleylbis(hydroxyethyl)methylammonium chloride in 15 lb/gal brine. First, note that the viscosity of this viscoelastic surfactant solution is almost independent of both temperature and shear rate. This slide also illustrates the ability of surfactants to thicken a concentrated brine typical of those used in oil well service fluids (61 wt% total electrolyte with 17.2 wt% $CaCl_2$ and 43.7 wt% $CaBr_2$). Finally, note that no aromatic counterion is required to obtain this thickening behavior.

The thickening behavior provided by a surfactant in brines is highly dependent on the brine concentration. Different alkyl chain lengths or head group functionalities are required to maximize the thickening action in different brines. The maxima in the viscosity for a given surfactant as a function of density illustrated in Figure 5 is typical of many other combinations of surfactants and electrolyte concentrations (here plotted as fluid density).

Figure 6 shows the viscosity versus shear rate and temperature for 2.5 wt% erucyltrimethylammonium salicylate in water. Its rheological behavior is different from that of the solution in Figure 4 in two ways: First, note the much more rapid decrease in viscosity with increasing shear rate. Second, note that the viscosity at the higher temperature (85°C) is uniformly greater at all shear rates than that at the lower temperature (25°C). A final interesting feature illustrated by this figure is the very high viscosity of this viscoelastic surfactant solution at the low shear rates. This high viscosity would cause this solution to behave as if it had a yield value, and, therefore, it would be effective for supporting solids or stabilizing foams (26) for at least a limited period of time.

The examples above show that the rheological behavior in an industrial viscoelastic surfactant formulation will depend on the composition of the liquid into which the surfactant is formulated and the temperature it is subjected to. However, we have found the following general rules to be useful when formulating viscoelastic surfactants for various temperatures or brine concentrations. First, increasing the alkyl chain length of the surfactant will typically allow one to maintain viscoelastic properties at higher temperatures, greater brine concentrations, or greater ethylene glycol concentrations. Second, excess counterion concentration (e.g., addition of more than an equivalent amount of NaSal with an alkyltrimethyl-ammonium surfactant) will allow one to maintain viscoelastic properties at higher temperatures or greater shear stresses. For example, Figure 5 illustrates the improved thickening of concentrated brines provide by the erucyl hydrophobe compared to that of the oleyl hydrophobe and drag reduction results published earlier (10) have shown that a 0.2 wt% ETAS plus 0.2 wt% NaSal solution will maintain its activity to greater than 120°C compared to only about 70°C for a 0.2 wt% CTAS plus 0.2 wt% NaSal solution, and only about 50°C for a 0.2 wt% CTAS solution.

As described above, the elastic properties of viscoelastic surfactants have been extensively studied for their drag reduction properties. However, they can also impart antimisting properties to formulations. This is illustrated in Figure 7 using an aqueous solution of a surfactant formulation that was shown to provide effective

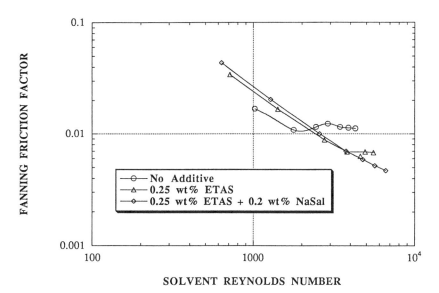

Figure 3. Drag Reduction Activity in an Aqueous Solution Containing 50 wt% Ethylene Glycol Using Erucyltrimethylammonium Salicylate.

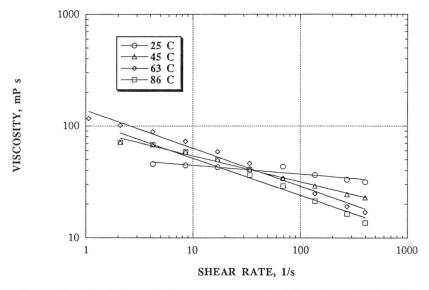

Figure 4. The Effect of Temperature on the Viscosity of 0.5 wt% Oleylbis(hydroxyethyl)methylammonium Chloride in 15 lb/gal Brine. (See Table I for the composition of the brine.)

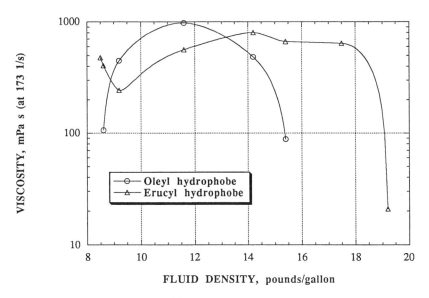

Figure 5. Viscosity of Alkylbis(hydroxyethyl)methylammonium Chloride
Surfactants in Brines. (See Table I for the composition of the brines.)

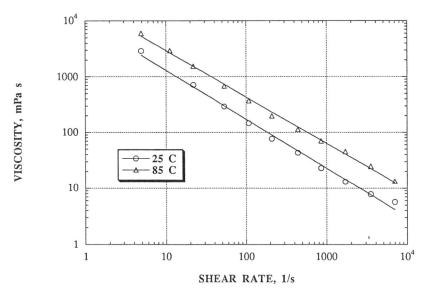

Figure 6. Viscosity versus Shear Rate and Temperature for 2.5 wt%
Erucyltrimethylammonium Salicylate in Water.

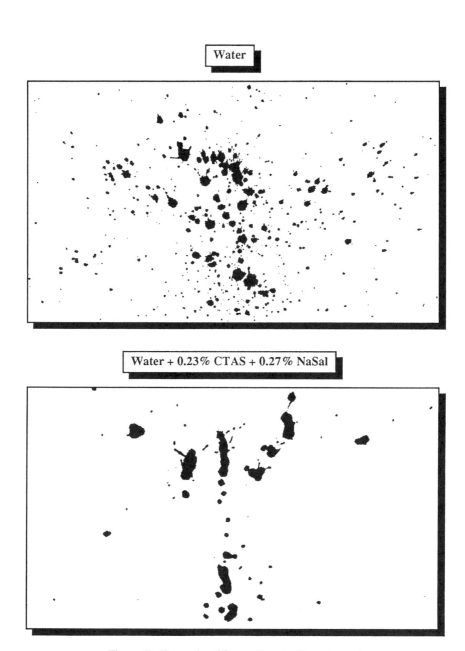

Figure 7. Example of Spray Droplet Size Control.

drag reduction to water. (7) This slide shows the spray droplet pattern generated from "splattering" a small sample of a colored liquid through a 10 mesh screen and recording the pattern of spray thus generated on an adsorbent sheet located behind the screen. Note the dramatic difference between the two spray patterns. The unmodified fluid has many more very small droplets than the fluid containing the viscoelastic surfactant formulation. This property would be useful in formulations where drifting of small droplets would be highly undesirable. Examples of these are the application of herbicide formulations or of hard surface cleaning formulations that contain bleach.

The viscoelastic properties of an aqueous bleach formulation suitable for use as a hard surface cleaner are illustrated in Figure 8. It contains 4 wt% sodium hypochlorite plus 0.9 wt% hexadecyltrimethylammonium p-toluene sulfonate. This aqueous solution contains a lower electrolyte concentration than the brines described in Figures 4 and 6 and thus still requires an aromatic counterion for the surfactant cation to provide viscoelasticity. However, the electrolyte, 4% sodium hypochlorite, is also a strong bleaching agent. This requires taking into consideration the chemical stability of the surfactant components as well as the rheological properties they impart. For example, the typical salicylate counterion must be changed to p-toluene sulfonate, and the alkyl groups must be saturated to provide the desired chemical stability in this hypochlorite formulation. Both the viscous and elastic properties imparted to the hypochlorite solution by this surfactant are useful properties for bleach containing formulations. (27) The low viscosity of the solution at high shear rate allows it to be readily pumped out of a hand sprayer, and the high viscosity at low shear rate prevents the solution from rapidly flowing down vertical surfaces onto which the bleach is sprayed. Finally, the low frequency at which the storage modulus (G') becomes greater than the loss modulus (G") shows that this solution will have elastic properties for processes taking even relatively long times. This elastic property imparts an anti-misting characteristic to the bleach formulation.

The viscoelastic surfactant approach to controlling mist can be applied in several ways. With the bleach system the hexadecyltrimethylammonium p-toluene sulfonate thickens an aqueous solution of bleach, which can be thought of as a water soluble active ingredient. Another example of an active ingredient which is compatible with a viscoelastic surfactant solution is provided by the addition of 2,4-dichlorophenoxyacetic acid (a herbicide) to an aqueous solution containing 0.23 wt% cetyltrimethylammonium salicylate (CTAS) plus 0.27 wt% sodium salicylate (NaSal). However, an active ingredient can interact with a surfactant cation to impart viscoelasticity to the surfactant solution. This is the case for water containing equimolar amounts of picloram (a herbicide) with cetyltrimethylammonium hydroxide to give a 0.5 wt% solution.

In some applications it would be useful to impart viscoelasticity to emulsions or suspensions. However, hydrocarbons (e.g., toluene) will "break" the viscoelasticity of a typical solution such as the 0.23 wt% CTAS plus 0.27 wt% NaSal described above. The process of "breaking" the viscoelastic properties results from solubilization of the hydrocarbon into the cylindrical micelles and transforming them into spherical micelles. This transformation process has been described by Hoffmann and Ebert. (28) A cationic surfactant containing a perfluorinated hydrophobic group (about C_8) can be used to provide viscoelastic emulsions of hydrocarbons. (29) For example, long lasting viscoelasticity can be imparted to an emulsion of 2 wt% toluene by combining 0.5 wt% Zonyl FSC fluorocarbon surfactant (DuPont Company, Wilmington, DE) plus an equivalent of sodium salicylate.

Although the breaking of a surfactant's viscoelasticity by a hydrocarbon can be undesirable, it can also be put to good use. An example of this is presented in Figure 9 which shows a schematic of a reversible thickening process for the

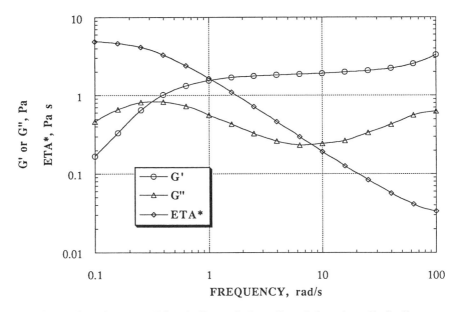

Figure 8.   Aqueous Bleach Formulation Containing 4 wt% Sodium Hypochlorite plus 0.9 wt% Hexadecyltrimethylammonium p-Toluene Sulfonate.

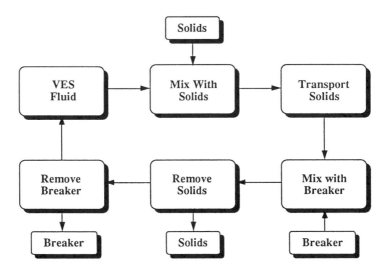

Figure 9. Reversible Thickening Process.

transport of solids. The process begins in the upper left hand corner of the figure with the addition of sufficient viscoelastic surfactant to the continuous phase to provide the desirable rheological response. Next, solids are added, suspended and transported to a point at which they are to be separated from the carrier fluid. At this point, a breaker is added which rapidly "breaks" the surfactant solution's viscoelastic properties and allows for easy solids separation, for example, by filtration. A typical breaker would be a low boiling hydrocarbon, e.g. toluene or 1,1,2-trichloroethene. The breaker is then removed, for example, by volatilization or absorption. Removal of the breaker allows the viscoelastic response to reform and the cycle can be repeated. Thus, a reversible thickening process can be provided with viscoelastic surfactants. (30)

One potentially important application of the reversible thickening process outlined in Figure 9 is a clear brine, well drilling fluid in which brine replaces the usual particulate weighting agent, e.g. barite. The major advantage of a clear brine drilling fluid is a faster drilling rate. (31,32,33) Since clear brines are expensive and solids build-up during drilling negates the faster drilling rate advantage, an effective solids removal procedure is necessary. A practical drilling fluid must also have the following properties: (1) low viscosity at high shear rates to cool and lubricate the bit (2) a high viscosity at low shear rates to suspend and carry the solid cuttings to the surface and (3) stability of the rheology control agent to mechanical degradation because the drilling fluid is recirculated. Thus, if the clear brine viscoelastic surfactant solutions provide the right rheology and are truly reversible, they would be an attractive process for drilling.

To demonstrate that a viscoelastic surfactant clear brine fluid in the absence of solids is truly reversible, a lab scale test was carried out employing a 14.2 lb per gal brine thickened with 1 wt% oleylbis(hydroxyethyl)methylammonium chloride. After measuring the initial viscosity, pH and surfactant concentration, 0.3 wt% 1,1,2-trichloroethene is added to break the viscosity. The breaker is then removed by distillation to restore the viscosity and water is added to adjust for the loss during stripping. This cycle is repeated five times. Figure 10 shows the results of this lab test. The initial rheology is of the required highly shear thinning type and the viscosity versus shear rate profile is the same after 5 cycles. Thus, it is reversible.

The rheological behavior of the viscoelastic surfactant drilling fluid in the presence of solids is an important question. The drilling operation will generate solids of different particle sizes. Since the large and medium size particles should be readily separated after addition of breaker, it is the fines that could be troublesome. Figure 11 illustrates the rheological data obtained with the same procedures as those in Figure 10 but with addition of about 5 lb/barrel (~0.9 wt%) of Rev Dust. It is described as a small particle (~2 µm) altered Ca-montmorillonate clay. The 0.9 wt% is based on the assumption of a solids loading during drilling of 20% total clay with 30% having $\leq$ 2 µm particle size. The Rev Dust was present during all rheological measurements and obviously does not adversely influence the reversibility at this concentration of Rev Dust.

Although the Rev Dust particles do not adversely affect the rheological response, since they are a small particle size clay they would be expected to adsorb the cationic surfactant. This indeed is the case, as is illustrated in Figure 12. These data were obtained by monitoring the concentration of remaining surfactant in a brine containing initially 0.275 wt% oleylbis(hydroxyethyl)methylammonium chloride after equilibration with increasing levels of the Rev Dust. As expected, the amount of surfactant remaining in the solution decreases for samples containing more Rev Dust. These data yield a value of about 35 mM/100 g for saturation of the Rev Dust by the oleylbis(hydroxyethyl)methylammonium chloride. This value is about equal to that of the cation exchange capacity of 32 meq/100 g reported for the Rev Dust by the supplier (34), and is well within the range of cation exchange values expected for montmorillonite. (35)

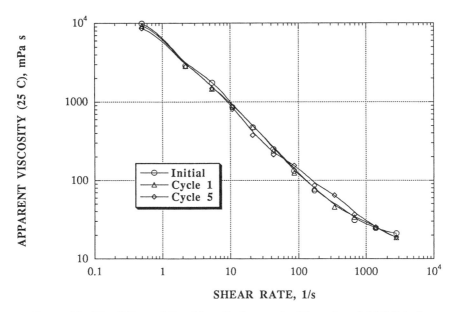

Figure 10. The Effect of Breaking Cycles on the Viscosity of 14.2 lb/gal Brine Thickened with 1% Oleylbis(hydroxyethyl)methylammonium Chloride.

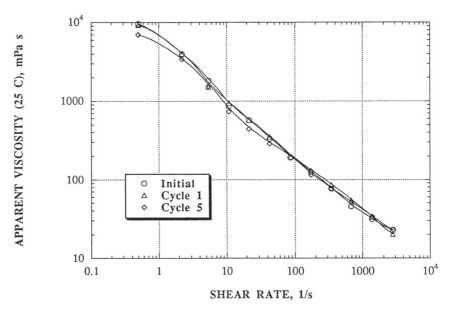

Figure 11. The Effect of Breaking Cycles on the Viscosity of 14.2 lb/gal Calcium Bromide Brine Thickened with 0.9% Rev Dust + 1% Oleylbis(hydroxyethyl)methylammonium Chloride.

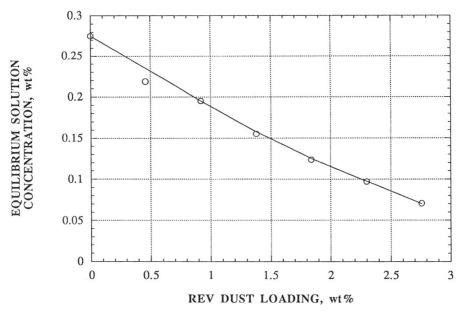

Figure 12. Oleylbis(hydroxyethyl)methylammonium Chloride Adsorption
onto Rev Dust in a 14.2 lb/gal Calcium Bromide Brine.

In extrapolating these results to drilling wherein small particle solid surfaces are continually generated by the bit, it is possible that adsorption could become a problem. The seriousness of the problem would depend on the accessible surface area of the rock being drilled. For example, at the other extreme from the adsorption on clays, adsorption of dodecyl- or cetyltrimethylammonium cations onto silica is reported to be about 5 μmoles/m^2 of surface area. (*36*) From this value the expected adsorption onto 2 μm diameter spherical silicon particles would be about 0.005 mM/g. Since adsorption on the debris from the rock formations encountered in drilling will lie between these two extremes, adsorption is not expected to be a prohibitive factor for the cost of this process. On the positive side, one other point illustrated by Figure 12 is the utility of a high surface clay to remove cationic surfactant from a formulation after it is no longer useful.

Thus, viscoelastic surfactants provide a shear stable, reversible thickening process for the removal of the solids generated from drilling. The same properties that make them useful in this process would also be useful for other solid transport processes in which solids separations is required. For long distance slurry transport of solids in a pipeline, the elastic properties of these solutions would provide the additional benefit of drag reduction. (*37*)

The cationic surfactants used to prepare the viscoelastic formulations described in this paper can impart additional beneficial features to the formulations into which they are included. Cationic surfactants are known to provide cleaning (*38,39*), to disperse solids (*38,40*), to prevent microbial growth (*38*), and to prevent corrosion. (*38,41*) The salicylate counterion is an effective chelant for metal ions (*42*) and p-toluene sulfonic acid is a well-known hydrotrope. (*43*) These properties could provide useful additional features for certain rheological applications.

Summary

Through the proper choice of surfactants, useful rheological properties have been demonstrated for a wide variety of aqueous formulations. These include increased viscosity for control of flow, elasticity for drag reduction and antimisting, and high viscosity at low shear rates to provide a pseudo yield value to stabilize dispersed phases. In addition, the viscoelastic surfactants offer unique advantages. These include no irreversible mechanical degradation, formulation versatility, and the ability to formulate them in concentrated electrolytes. These properties make them useful in a variety of applications including heat transfer fluids, hard surface cleaners, oil well service fluids, and solids transport.

Acknowledgments

The authors would like to thank P. A. Doty, K. L. Foster, S. Kropf, L. Manley-Ecker, K. G. Seymour, and D. J. Straka for their contributions to the work included in this chapter.

References

1. Laughlin, R. G. In *Cationic Surfactants Physical Chemistry*; Rubingh, D. N., Holland, P. M., Eds.; Surfactant Science Series, Vol. 37; Marcel Dekker: New York, NY, 1991; Chapter 1, pp 1-40.
2. Mukerjee, P.; Mysels, K. J. *Critical Micelle Concentrations of Aqueous Surfactant Systems*; U. S. Department of Commerce, National Bureau of Standards, Publication NSRDS-NBS 36; U. S. Government Printing Office: Washington, DC, 1970.

3. Zana, R. In *Cationic Surfactants Physical Chemistry*; Rubingh, D. N., Holland, P. M., Eds.; Surfactant Science Series, Vol. 37; Marcel Dekker: New York, NY, 1991; Chapter 2, pp 41-85.
4. Hoffmann, H.; Platz, G.; Rehage, H.; Schorr, W.; Ulbricht, W. *Ber. Bunsenges. Phys. Chem.* **1981**, *85*, 255-256.
5. Shenoy, A. V. *Colloid Polym. Sci.* **1984**, *262*, 319-337.
6. Ohlendorf, D.; Interthal, W.; Hoffmann, H. *Rheol. Acta* **1986**, *25*, pp 468-486.
7. Rose, G. D.; Foster, K. L.; Slocum, V. L.; Lenhart, J. G. In *Drag Reduction, Papers presented at the Third International Conference on Drag Reduction* 2-5 July 1984; Sellin, R. H. J., Moses, R. T., Eds.; , University of Bristol: Bristol, 1984; pp D.6-1 to D.6-7.
8. Rose, G. D. U. S. Patent 4,534,875, 1985.
9. Teot, A. S.; Rose, G. D.; Stevens, G. A. U .S. Patent 4,615,825, 1986.
10. Rose, G. D.; Foster, K. L. *J. Non-Newtonian Fluid Mech.*, **1989**, *31*, pp 59-85.
11. Savins, J. G. *Rheol. Acta* **1967**, *6*, 323-330.
12. White, A. In *Viscous Drag Reduction*; Wells, C. S., Ed.; Plenum: New York, NY, 1969; pp 297-311.
13. Zakin, J. L.; Lui, H.-L. *Chem. Eng. Commun.* **1983**. *23*, pp 77-88.
14. Schmidt, K.; Durst, F.; Brunn, P. O. In *Drag Reduction in Fluid Flows*, Sellin, R. H. J., Moses, R. T., Eds.; Ellis Horwood: Chichester, West Sussex, England., 1989; pp 205-213.
15. Hato, M.; Tahara, M.; Suda, Y. J. Colloid Interface Sci. **1979**, *72*, 458-464.
16. Tsujii, K.; Saito, N.; Takeuchi, T. J. Colloid Interface Sci. **1984**, *99*, 553-560.
17. Barker, C. A.; Saul, D.; Tiddy, G. J. T.; Wheeler, B. A.; Willis, E. J. Chem. Soc., Faraday Trans. 1 **1974**, *70*, 154-162.
18. Saul, D.; Tiddy, G. J. T.; Wheeler, B. A.; Wheeler, P. A.; Willis, E. J. Chem. Soc., Faraday Trans. 1 **1974**, *70*, 163-170.
19. Rice, H. L.; Becker, K. W.; Szabo, F. L. U. S. Patent 3,373,107, 1968.
20. Scheraga, H. A.; Backus, J. K. *J. Am. Chem. Soc.* **1951**, *73*, 5108-5112.
21. Bird, R. B.; Armstrong, R. C.; Hassager, O. *Dynamics of Polymeric Li quids*, Vol. 1; John Wiley and Sons: New York, NY, 1977.
22. Ferry, J. D. *Viscoelastic Properties of Polymers*, 3rd ed., John Wiley and Sons: New York, NY; 1980.
23. Hoffmann, H.; Rehage, H.; Reizlein, K.; Thurn, H. In *Macro- and Microemulsions Theory and Applications*; Shah, D. O., Ed; ACS Symposium Series 272, American Chemical Society: Washington, D. C., 1985, Chapter 4, pp. 41-66.
24. Boucher, D. F.; Alves, G. E. In *Chemical Engineer's Handbook;* Perry, R. H., Chilton, C. H., Eds; Fifth Edition; McGraw-Hill: New York, NY 1973, p 5-4.
25. Hoffmann, H.; Lobl, H.; Rehage, H.; Wunderlich, I. *Tenside Detergents* **1985**, *22* , 290-298.
26. Bonekamp, J. E.; Rose, G. D.; Schmidt, D. L.; Teot, A. S. U. S. Patent 5,258,137, 1993.
27. Rose, G. D.; Foster, K. L.; Teot, A. S. U. S. Patent 4,800,036, 1989.
28. Hoffmann, H.; Ebert, E. Angew. Chem. Int. Ed. Engl. **1988**, *27*, 902-912.
29. Rose, G. D.; Teot, A. S.; Seymour, K. G. U. S. Patent 4,880,565, 1989.
30. Rose, G. D.; Teot, A. S.; Doty, P. A. U. S. Patent 4,735,731, 1988.
31. Conners, J. H.; Bruton, J. R. Soc. Pet. Eng. Paper 8223, 54th Annual Fall Technical Conference, Las Vegas, NV, September 23-26, 1979.

32. Doty, P. A. Soc. Pet. Eng. Paper 13441, SPE/IADC 1985 Drilling Conference, New Orleans, Louisiana, March 6-8, 1985.
33. Clark, R. K.; Nahm, J. J. In *Kirk-Othmer Encyclopedia of Chemical Technology*, 3rd ed., Vol. 17,; Grayson, M., Ed., John Wiley & Sons: New York, NY, 1982; pp 143-167.
34. Milwite, Inc. P.O. Box 15038, Houston, TX.
35. Grim, R. E. *Clay Mineralogy*, 2nd ed., McGraw Hill: New York, NY; 1968, pp 188-235.
36. Keller, W. D. In *Kirk-Othmer Encyclopedia of Chemical Technology*, 3rd ed, Vol. 6; Grayson, M., Ed.; John Wiley & Sons: New York, NY, 1979; pp 190-206.
37. Poreh, M.; Zakin, J. L.; Brosh, A.; Warshavsky, M. *J. Hydraulics Division, Proceedings of the ASCE* **1970**, *96*, HY4, pp 903-909.
38. Myers, D. *Surfactant Science and Technology*; VCH Publishers: New York, NY, 1988; pp 62-67.
39. Rubingh, D. N. In *Cationic Surfactants Physical Chemistry*; Rubingh, D. N., Holland, P. M., Eds.; Surfactant Science Series, Vol. 37; Marcel Dekker: New York, NY, 1991; Chapter 10, pp 469-507.
40. Fuerstennau, D. W.; Herrera-Urbina, R. In *Cationic Surfactants Physical Chemistry*; Rubingh, D. N., Holland, P. M., Eds.; Surfactant Science Series, Vol. 37; Marcel Dekker: New York, NY, 1991; Chapter 8, pp 407-447.
41. "Applications of Armak Quaternary Ammonium Salts", Bulletin 81-8, Armak Company, Akzo Chemicals, Inc., Chicago, IL, 1981.
42. Harris, C. M.; Livingstone, S. I. In *Chelating Agents and Metal Chelates*; Dwyer, F. P., Mellor, D. P., Eds.; Academic Press: New York, NY, 1964; Chapter 3, p 106.
43. Davidson, A.; Milwidsky, B. M. *Synthetic Detergents*, Sixth ed; John Wiley & Sons: New York, NY, 1978; p. 79.

RECEIVED July 8, 1994

Chapter 26

Effect of Counterion Structure on Flow Birefringence and Drag Reduction Behavior of Quaternary Ammonium Salt Cationic Surfactants

Bryan C. Smith[1], Lu-Chien Chou[2], Bin Lu[3], and Jacques L. Zakin[3]

[1]Erling Riis Research Laboratory, International Paper, P.O. Box 2787, Mobile, AL 36652
[2]Baker Performance Chemicals, 9101 West Twenty-first Street, Sand Springs, OK 74063
[3]Department of Chemical Engineering, Ohio State University, Columbus, OH 43210

Quaternary ammonium cationic surfactants with appropriate counterions are friction reducers and can reduce pumping energy requirements, increase water throughput or reduce pump size or pipeline diameter in closed loop district heating and cooling systems. Drag reduction occurs in the temperature range in which rod-like micelles are present.

Variations in counterion chemical structures have major effects on the micelle structure and flow birefringence and drag reduction behavior of quaternary ammonium chloride surfactants. Isomers of ortho-, meta- and para-hydroxybenzoate, ortho-, meta- and para-chlorobenzoate and 1-hydroxy-2, 2-hydroxy-1- and 3-hydroxy-2-naphthoate as counterions were investigated by flow birefringence and drag reduction measurements. Those systems showing birefringence also showed drag reduction behavior and their upper temperature limits of effectiveness coincided. The results also correlated with FT-NMR data reported earlier. Only those counterions whose structure permits orienting their hydrophobic and hydrophilic portions in their preferred environments can stabilize the sphere to rod transition and induce flow birefringence and drag reduction.

Over 40 years ago, Agoston et al. (*1*), Mysels (*9*) and Toms (*17*) reported that the addition of aluminum disoap and polymer additives to hydrocarbons could reduce their turbulent friction energy losses in pipe flow significantly. While the phenomenon has been studied extensively since then, particularly for polymer additives, the details of its mechanism are not yet known. Efficient polymer additives have been discovered and they are used in crude and petroleum product pipelines. A number of excellent reviews on these additives exist (*8, 10, 13, 14, 18*). Applications have been limited, however, because of the sensitivity of the most effective high polymers (very high molecular weight) to mechanical degradation in regions of high shear such as in pumps. Thus, polymer additives can only be used by injecting them downstream of pumps and can not be used in recirculating systems.

In order to make more effective use of energy resources and to reduce environmental pollution by combustion products, expanded use of district heating and cooling systems is being encouraged in many countries. In district heating systems, cogeneration or waste heat is utilized to provide hot water (in the range of 80 to 130°C) for circulation to houses, industries, etc, within a district. The hot water is pumped through a 10 to 20 km loop, exchanging heat with the various installations along the way. It is then reheated and recirculated. The use of friction reducing additives in these systems could reduce pumping energy costs, reduce capital investments or increase capacity and, by utilizing energy from a large central station in place of many smaller inefficient units, reduce air pollution.

Cationic, nonionic and anionic surfactant additives have also been shown to be effective drag reducers. These surfactant additives have been shown to be "repairable" if they undergo mechanical shear degradation or if they exceed a critical upper temperature limit. Thus, they are more useful for recirculation systems even though the required additive concentrations for surfactants are higher than for polymers.

Quaternary ammonium surfactants with counterions have been shown to be particularly useful for district heating systems (*12*). Variations in the structure of the quaternary ammonium salt and of the counterion have major effects on micellar structure and on the effectiveness of the surfactant as a drag reducer and on the temperature range of effectiveness. In this paper we will describe experimental results showing how variations in the molecular structures of counterions with the cetyl trimethyl ammonium chloride cation (CTAC, $C_{16}H_{33}N(CH_3)_3Cl$) affect these drag reduction characteristics and relate them to flow birefringence.

Experimental

Drag Reducing Measurements.   Drag reduction experiments at temperatures up to 130°C were carried out in a recirculation system containing a 58-cm long, 0.617 cm diameter stainless steel test section. Low temperature studies were performed in a similar loop with a 122-cm long test section of the same diameter and containing a chilling unit permitting tests to be run down to 2°C. Entrance and exit corrections were applied. While these tube-lengths are probably not long enough for accurate engineering measurements, they were compared with each other in the range of 30-50°C and good agreement was obtained (*4*). Pressure drop measurements in these units are more than adequate for screening tests and for comparing the relative effectiveness of drag reducing additives.

Flow Birefringence Measurements.   Polarized light passing through an isotropic material experiences a single, scalar refractive index. In a birefringent material, this light experiences a tensorial refractive index. Consider a dilute suspension of rigid, ellipsoidal particles suspended in a Newtonian fluid. At rest, the particles have a random orientation, the solution shows no anisotropies, and light passing through it experiences a single refractive index. When simple flow is initiated, however, the particles will be oriented by the forces of the flow and the solution will exhibit polarization anisotropy. Birefringence is determined by the real part of the complex refractive index.

Early birefringence studies were limited almost exclusively to steady state experiments. With the advent of the phase-modulated flow birefringence technique, however, a much broader range of experimental systems could be investigated. The greatest advantage of this technique is the ability to simultaneously determine the flow induced birefringence and the orientation (extinction) angle of the particles (5). This type of measurement was used in all experiments.

The phase-modulated flow birefringence apparatus has been well described in the literature (5, 6). In the experiments performed here, monochromatic, partially polarized red light from a He-Ne laser was used. The light was linearly polarized when passed through an incident polarizer. It then passed through a photoelastic modulator which induced a time varying retardance in the phase of the light. The beam then traveled through the solution in a Couette cell with outer cylinder rotation and a one millimeter gap size. Shear rates could be varied from 0 to 700 1/s. The cell was jacketed for temperature control. The light then passed through a second polarizer and into a PIN photodiode detector connected to computer acquisition. The optical components were aligned using the method set forth by Galante (6).

The birefringence of the sample and the average extinction angle of the particles were simultaneously determined from the data (6).

Cationic Surfactant and Counterions. The cationic surfactant used was Arquad 16-50, a commercial product of Akzo Chemical which contains about 50% of active cationic surfactant in a solvent of water and isopropyl alcohol. The cationic is primarily (over 75%) cetyl trimethyl ammonium chloride with the remainder made-up of approximately equal amounts of tetradecyl and octadecyl trimethyl ammonium chloride The counterions were ortho-, meta- and para-hydroxy benzoate and chloro benzoate, and 1-2, 2-1 and 3-2 hydroxy naphthoate. All the counterions, except the sodium salt of 2-hydroxy benzoate (NaSal), were products of Aldrich Chemical Company, Inc. and were 99+% pure. NaSal is a product of MCB Manufacturing Chemicals, Inc. (GR purity grade).

Results and Discussion

The introduction of certain substituent groups onto the phenyl ring of benzoate has been shown to dramatically alter the effectiveness of the counterions for drag reduction and flow birefringence (4, 16). The effect depends on both the nature of the substituents and their position on the ring. Three geometric isomers of hydroxybenzoate, chlorobenzoate and hydroxynaphthoate counterions were tested with CTAC. The results are summarized in Table I.

Salicylate (2-hydroxybenzoate) has long been known to impart interesting rheological properties to cationic surfactant systems (7, 19). The CTAC/NaSal (5mM/12.5mM) system showed good drag reduction from 10 °C through 70 °C (Figure 1), and significant flow birefringence from 30 °C (the lowest temperature measured) through 60 °C and weak flow birefringence at 70 °C (Figure 2). The

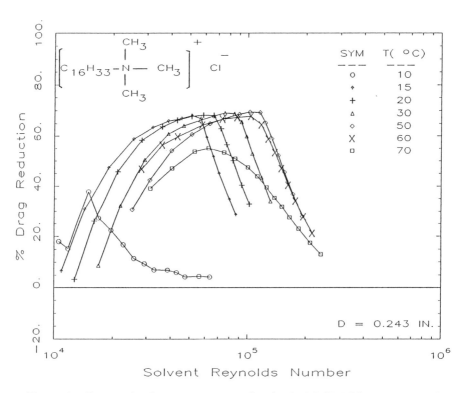

Figure 1. Drag reduction measurements for the 5mM CTAC/12.5mM NaSal system.

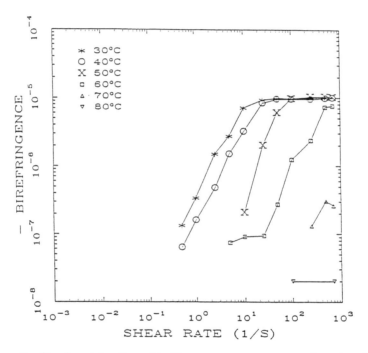

Figure 2. Steady state flow birefringence measurements for the 5mM CTAC/12.5mM NaSal system.

Table I. Comparison of Flow Birefringence and Drag Reduction Measurements for CTAC ($C_{16}H_{33}N(CH_3)_3Cl$) (5mM[a]) with Different Counterions

Counterion	Temperature Range with Significant Flow Birefringence (^0C)	Temperature Range for Effective Drag Reduction (^0C)
2-OH-benzoate (12.5mM)	30[b]--60, 70[c]	10--70
3-OH-benzoate (12.5mM)	None	None
4-OH-benzoate (12.5mM)	None	None
2-Cl-benzoate (12.5mM)	None	None
3-Cl-benzoate (12.5mM)	30[b]--50, 60[c]	30--50
4-Cl-benzoate (12.5mM)	30[b]--50, 60[c], 70[c]	20--70
1-OH-2-naphthoate (3.2mM)	30[c], 50[c], 70--100[b]	80--110
2-OH-1-naphthoate (3.2mM)	30[b]--100[b]	30[b]--100
3-OH-2-naphthoate (3.2mM)	30[b]--100[b]	40[b]--90, 100[c]

a: For hydroxy-naphthoate counterions, the concentration of CTAC is 3.2mM.
b: Highest or lowest temperature measured.
c: 20%--50% drag reduction or low birefringence measured.

effective temperature ranges for flow birefringence and for friction reduction are in good agreement with each other. The orientation of hydroxybenzoate counterions in the solutions deduced from proton shift and line broadening in NMR measurements is shown in Figure 3 (*15, 16*). From Figure 3, it can be seen that the hydrophilic hydroxy group is located in the water phase, an orientation favored by this hydrophilic group. This seems to disperse the charge on the nearby ammonium group allowing long, rod-like micelles to form. Rod-like micelles have been reported to be required for friction reduction (*3*).

No drag reduction and no flow birefringence were measurable in the CTAC/3-hydroxybenzoate system. While some rod-like structure was deduced from the NMR data of Smith, et. al. (*15, 16*), the effects were much smaller than for the 2-hydroxy-benzoate system. This is also consistent with earlier NMR results of Rao et. al. (*11*) and Bachofer and Turbitt (*2*) on the same system.

For the CTAC/4-hydroxybenzoate system, no drag reduction and no flow birefringence were measurable as expected. In this solution, the hydrophilic hydroxy group is turned into the micelle, which is an unfavorable orientation and rod-like micelles do not form. The NMR results of Smith et. al. (*15, 16*) support these observations. That is, there was little line shifting or peak broadening in the NMR tests for this system.

No birefringence and no drag reduction were obtained for the 2-Cl-benzoate/CTAC system. The chloro group is hydrophobic and the orientation inferred from the NMR measurements of Smith et. al. (*15, 16*) indicates that it is in the water phase (Figure 4). This is an unfavorable orientation for it and the micelles are either not stable or spherical.

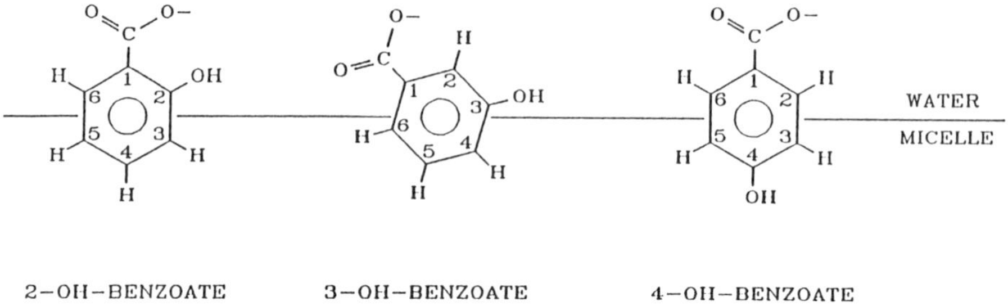

2—OH—BENZOATE   3—OH—BENZOATE   4—OH—BENZOATE

Figure 3. The orientation of hydroxybenzoate counterions at the micelle-water interface from NMR measurements. (Reproduced with permission from ref. 15. Copyright 1994 Journal of Rheology, American Institute of Physics.)

2—Cl—BENZOATE   3—Cl—BENZOATE   4—Cl—BENZOATE

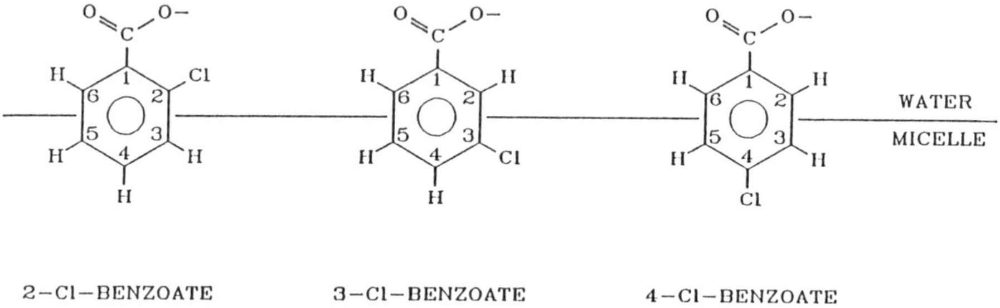

Figure 4. The orientation of chlorobenzoate counterions at the micelle-water interface from NMR measurements. (Reproduced with permission from ref. 15. Copyright 1994 Journal of Rheology, American Institute of Physics.)

For the 3-Cl-benzoate/CTAC system, the hydrophobic chloro group is oriented into the hydrocarbon phase at the interface (Figure 4). Significant flow birefringence and effective drag reduction were observed from 30 °C to 50 °C, and weak flow birefringence at 60 °C. The results from both measurements are in good agreement and indicate reasonable sized rod-like micelles, which was also inferred from the NMR measurements (*15, 16*).

The 4-Cl-benzoate/CTAC system has its hydrophobic chloro in its most favorable orientation (Figure 4) and showed the best drag reduction of all of the benzoate systems tested with significant effects from 20 through 70 °C. Significant flow birefringence was observed from 30 °C (the lowest temperature measured) through 50 °C, and weak flow birefringence at 60 °C and 70 °C. Again, the results from both measurements are in good agreement.

The critical wall shear stresses, above which drag reduction falls below 50%, for the 2-OH-benzoate/CTAC system (Table II) are greater than those for the 4-Cl-benzoate/CTAC at low temperatures (< 50°C). However, at 60 and 70 °C, the critical wall shear stresses of the 4-Cl-benzoate system are equal to or greater than those for the 2-OH-benzoate system. In the measurements of drag reduction for Kemamine Q2983C ($C_{22}H_{43}N(CH_3)_3Cl$ donated by Humko Corp.) with these two counterions, it was found that the critical wall shear stresses of 4-Cl-benzoate are higher than 2-OH-benzoate from 60 °C to 110 °C (*4*). Thus 4-Cl-benzoate is generally more effective than salicylate as a counterion in promoting drag reduction at high temperatures.

Table II. Critical Wall Shear Stress (Pa) of CTAC[a] with Benzoate Counterions

Counterions	T (°C)								
	10	15	20	30	40	50	60	70	80
2-OH-benzoate	M[b]	190	220	300	-	190	130	45	N[c]
4-Cl-benzoate	N[c]	-	156	145	149	-	130	92	N[c]

a: CTAC/counterion = 5mM/12.5mM.
b: M represents 20% to 50% drag reduction measured.
c: N represents less than 20% drag reduction measured.

The three isomers of hydroxy naphthoate were also tested as counterions. Significant flow birefringence was measured from 30 °C through 100 °C for 2-OH-1-naphthoate and 3-OH-2-naphthoate systems, and significant flow birefringence from 70 °C to 100 °C for 1-OH-2-naphthoate system and lower flow birefringence below 70 °C (Table I). The drag reduction studies showed effective drag reduction for the CTAC/2-hydroxy-1-naphthoate system between 30 and 100 °C, the CTAC/3-hydroxy-2-naphthoate system between 40 and 90 °C and the CTAC/1-hydroxy-2-naphthoate system between 80 and 110 °C. The CTAC/hydroxy naphthoate systems are not all soluble in water at room temperature. In particular, the 1-hydroxy-2-naphthoate system is only partially soluble and the 3-hydroxy-2-naphthoate, while apparently soluble, has a yellow color. The lower temperature limit for effective drag reduction is

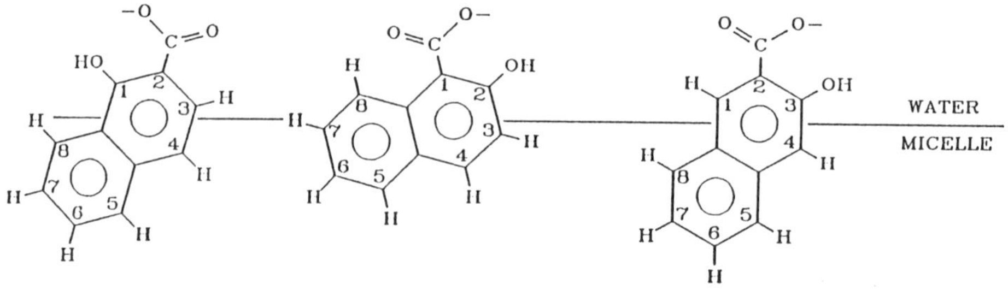

1−HYDROXY−   2−HYDROXY−   3−HYDROXY−
2−NAPHTHOATE   1−NAPHTHOATE   2−NAPHTHOATE

Figure 5. The orientation of hydroxy naphthoate counterions at the micelle-water interface from NMR measurements.

probably governed by solubility while the upper temperature limit may be enhanced by both deeper penetration of the counterion into the micelle and also by favorable conditions for hydrogen bonding. Figure 5 shows the orientation of the three naphthoate counterions inferred from NMR measurements (*15, 16*). It should be noted that the 1-2 carbon bond is shorter (1.36Å) than the 2-3 carbon bond (1.40Å) which would favor hydrogen bonding with the 1-hydroxy-2-naphthoate and the 2-hydroxy-1-naphthoate counterions compared with 3-hydroxy-2-naphthoate. The 1-hydroxy-2-naphthoate has both deep penetration into the micelle and the short carbon-carbon bond which may enhance high temperature effectiveness. The 2-hydroxy-1-naphthoate has less deep penetration but does have the short carbon-carbon bond. The 3-hydroxy-2-naphthoate also has deep penetration but a long carbon-carbon bond. Thus, the difference in the temperature limits for drag reduction may be a result of these effects.

It should also be noted that all of the systems which showed effective drag reduction and significant flow birefringence also showed viscoelastic recoil characteristics, while all the systems showing no drag reduction or flow birefringence are water-like.

Conclusions

1. Drag reduction effectiveness for alkyl trimethyl ammonium salt-counterion surfactant systems depends on the nature and the position of the substituent group on the benzoate or naphthoate and on temperature and wall shear stress.
2. Flow birefringence for alkyl trimethyl ammonium salt-counterion surfactant systems depends on the nature and the position of the substituent group on the benzoate or naphthoate and on temperature and wall shear rate.
3. There is a close correlation between the effective temperature range for drag reduction and for flow birefringence for these systems.
4. Only counterions whose structure permits orienting hydrophobic and hydrophilic groups into their preferred environments can stabilize the micellar interface and facilitate sphere-to-rod transition. Good drag reduction and strong flow birefringence were observed only in the systems whose counterions could do so.

Acknowledgments

Support from the Department of Energy (DE-F601-90CE-26605) under the supervision of Mr. Floyd Collins is gratefully acknowledged. The cationic surfactant samples were donated by Akzo Chemical.

Literature Cited

1. Agoston, G.A.; Hart, W.H.; Hottel, H.C.; Klemm, W.A.; Mysels, K.J.; Pomery, H.H.; Thompson, J.M. *Ind. Eng. Chem.* **1954**, *46*, 1017-1019.
2. Bachofer, S.J.; Turbitt, R.M. *J. Colloid Interface Sci.* **1990**, *135*, 325-334.
3. Bewersdorff, H.W.; Ohlendorf, D. *Colloid Polym. Sci.* **1988**, *266*, 941-953.
4. Chou L.C. Ph.D. Dissertation, The Ohio State University, Columbus, **1991**.
5. Frattini, P.L.; Fuller, G.G. *J. Rheol.* **1984**, *28*, 61-70.
6. Galante, S.R. Ph.D. Dissertation, Carnegie Mellon University, Pittsburgh, **1991**.
7. Gravsholt, S. *J. Colloid Interface Sci.* **1976**, *57*, 575-577.
8. Hoyt. J.W. *J. Basic Eng. Trans. ASME, Series D.* **1972**, *94*, 258-285.
9. Mysels, K.J. *U.S. Patent 2,492,173* December 27, **1949.**
10. Patterson, G.K.; Zakin, J.L.; Rodriguez, J.M. *Ind. Eng. Chem.* **1969**, *61*, 22-30.
11. Rao. U.R.K.; Manohar, C.; Valaulikar B.S.; Iyer, R.M. *J. Phys. Chem.* **1987,** *91*, 3286-3291.
12. Rose, G.D.; Foster, K.L.; Slocum, V.L.; Lenhart, J.G. *Proc. 3rd Int. Cong. on Drag Reduction,* Sellin, R.H.J.; Moses, R.T. eds., U. of Bristol, England, Paper D-6 **1984**.
13. Sellin, R.H.J.; Hoyt, J.W.; Pollert, J. *J. Hydr. Research* **1982**, *20*, 235-292.
14. Sellin, R.H.J.; Hoyt, J.W.; Scrivener, O. *J. Hydr. Research* **1982,** *20*, 29-68.
15. Smith, B.C.; Chou, L.C.; Zakin, J.L. presented at *6th Annual Society of Rheology Meeting.*, Santa Barbara, **1992**; *J. Rheol.*, **1994**, *38*, 73-83.
16. Smith, B.C. Ph.D. Dissertation, The Ohio State University, Columbus, **1992**.
17. Toms, B.A. In *Proc. 1st Int. Rheol. Cong.;* North Holland Publishing Co.; **1949;** p135.
18. Virk, P.S. *AIChE J.*, **1975**, *21*, 625-656.
19. Wan, L.S.C. *J. Pharm. Sci.* **1968,** *57*, 1903-1906.

RECEIVED July 8, 1994

Chapter 27

Interfacial Rheology of β-Casein Solutions

Diane J. Burgess[1] and N. Ozlen Sahin[2]

[1]School of Pharmacy, University of Connecticut, 372 Fairfield Road,
Storrs, CT 06269
[2]College of Pharmacy, University of Illinois at Chicago, Chicago, IL
60612

The effects of interfacial bulk concentration, pH, ionic strength,
temperature, aging and the addition of guanidine hydrochloride, urea
and Tween 80 on the interfacial rheology of β-casein solutions have
been investigated using an oscillating ring interfacial rheometer.
Interfacial tension data are also reported at pH 4.3 and 7.0 and at bulk
concentrations of 0.001 to 1% w/v. The Wilhelmy plate method was
used for the interfacial tension studies.

Interfacial rheology of surfactant solutions provides information on the structural-mechanical properties of the adsorbed surfactant films. Both elastic (solid-like) and viscous (liquid-like) activity can be measured. Interfacial film rigidity is an important factor in emulsion stability and has been related to the long term stability of emulsion formulations (1). A strong interfacial film will help resist coalescence following collision between emulsion droplets. Molecular segmental kinetics of film formation, molecular configurational changes and intermolecular interactions can be determined using interfacial rheology (2-5). Thus a comprehensive understanding of the formation and properties of the interfacial film can be obtained which may lead to improvements in emulsion formulation.

Proteins consist of both hydrophobic and hydrophilic amino acid residues and therefore behave like other surface active molecules. Their adsorption at interfaces can be explained by the hydrophobic effect (6). Configurational changes, reversible or irreversible, occur following protein adsorption. These intramolecular changes and intermolecular interactions between neighboring interfacial molecules may take time to complete. An interfacial oscillatory technique that allows dynamic measurement of aging films is used in this study. The instrument operates in the surface shear mode and, therefore, there is almost no disturbance of the interface during measurement and no consequent movement of molecules into and out of the interface. The measurement of interfacial properties of proteins using this instrument has been reported previously (4,5), where the effects of bulk concentration, pH (4), aging and

0097–6156/94/0578–0380$08.00/0
© 1994 American Chemical Society

temperature (*5*) on the interfacial properties of bovine serum albumin (BSA) and human immunoglobulin G (HI$_g$G) were investigated. The interfacial rheological and tension properties of the natural emulsion stabilizer β-casein are reported here. β-casein is the main emulsifying component in milk. It has a molecular weight of 23,900 Daltons and is composed of random coil clusters of hydrophobic and hydrophilic amino acid residues (*7*).

Materials and Methods

β-casein, from bovine milk, lyophilized and essentially salt free was obtained from Sigma Chemical Company, USA. Guanidine hydrochloride and urea were obtained from Sigma Chemical Company, USA. All other chemicals were of analytical grade and were obtained from Fisher Scientific, USA.

Single distilled, deionized water was filtered through a series of filters arranged in the following order: carbon; anion and cation exchange; and organic. Ultrapurified water was obtained by redistillation of the filtered single distilled, deionized water from acidic potassium permanganate solution, using an all-glass still. Ultrapurified water was stored prior to use in capped air tight bottles and was used within 24 hours. Ultrapurified water was used in the preparation of all aqueous solutions.

Preparation of solutions. β-casein solutions were prepared in phosphate buffer over the pH range 3.4 to 7.0. Isotonicity was adjusted using the sodium chloride equivalent method. The ionic strength was maintained at 100 mM for all solutions other than those where the ionic strength was varied (1 to 1000 mM). Concentrations in the range 1 to 4.0 % w/v were prepared at 25 and 37°C with minimal stirring to avoid foaming and subsequent denaturation. Solution preparation time was standardized at one hour to equilibrate the solutions at the final temperature and avoid error due to time dependent protein denaturation. All solutions were used immediately. Samples were siphoned from under the air/water interface using a glass pipette.

Interfacial Rheology Measurements. A Mark II Surface Rheometer was used (Surface Science Enterprises, UK). The instrument consists of four connecting systems: system I is a moving coil galvanometer; system II is a platinum Du Nouy ring, which is placed at the interface and is attached to a galvanometer; system III is the surface rheometer control unit which varies the driving frequency and monitors the amplitude of motion of the ring; system IV is an IBM PC which drives the instrument and analyzes the data. The equation of motion for the apparatus and the associated theory are explained by Sherriff and Warburton (*8*). The experimental technique consists of positioning a platinum du Nouy ring at the interface, either gas/liquid or liquid/liquid. The ring oscillates through a few degrees about a vertical axis. The amplitude of motion of the ring is measured by a proximity probe transducer, and automatic analysis of the signal generated gives both the dynamic surface rigidity modulus (surface elasticity, G'_s) [mN/m] and the dynamic surface viscosity (η'_s) [mNs/m] concurrently. G'_s and η'_s are defined as:

$$G'_s = g_f \cdot I \cdot 4\pi \left(f^2 - f_o^2 \right) \tag{1}$$

and

$$\eta'_s = g_f \cdot I \cdot NC (1/X - X_o) \tag{2}$$

where I is the moment of inertia, f is the sample interfacial resonance frequency, f_o is the reference interfacial resonance frequency, NC is the number of cycles of integration, X is the mean amplitude at the sample interface and X_o is the mean amplitude at the reference interface and g_f is the geometric factor. g_f is defined as:

$$g_f = 4\pi \left(R_1^2 R_2^2 \right) / \left[(R_1 + R_2)(R_2 - R_1) \right] \tag{3}$$

where R_1 is the radius of the ring and R_2 is the radius of the sample cell. The instrument is calibrated in air using a series of standard inertia bars of known weight. These bars are placed on the platinum Du Nouy ring which is then set to oscillate in air. Measurements are made with respect to a reference interface. The air/ultrapurified water interface is used as a reference for all experiments at the air/aqueous interface. Calibration against ultrapurified water is carried out prior to each measurement.

Samples were filled into a water-jacketed sample cell of internal diameter 3.853 cm. The temperature was maintained at the measurement temperature ± 0.1°C. The Du Nouy ring was lowered to sit at the interface. Solutions were used either freshly prepared or after aging at their final temperature. In all studies, with the exception of the continuous film formation studies, the interface was broken prior to each measurement by the action of pouring into the sample cell and by sweeping the surface with a clean glass rod. This removes any existing interfacial film allowing the induction of a new film. In the continuous film formation studies the interface was broken prior to the first measurement and the film was then allowed to form continuously in the sample cell and the Du Nouy ring was not removed to prevent any disturbance of the interface. All measurements were repeated three times and the mean values plotted.

Since it has been shown previously that adsorbed films of BSA and HI_gG exhibit more elastic (solid-like) interfacial activity than viscous (liquid-like) activity (5,6) interfacial elasticity values are reported here.

Dynamic Interfacial Tension Measurement. The apparatus consists of: a Cahn 2000 Recording Microbalance sensor and control unit (Cahn Instruments, Inc., Cerritos, California), a water jacketed sample chamber, a circulating water bath with digital temperature control, a thermocouple for the sample chamber, a platinum

Wilhelmy plate, an IBM PC compatible data acquisition computer interfaced with an analog-to-digital converter and a chart recorder. The temperature of the sample chamber was controlled to ± 0.1°C using the circulating water bath and monitored using a thermocouple sensitive to ± 0.1°C. The humidity over the sample chamber was maintained at 68 ± 2% during the experiments.

The procedure for dynamic tension measurement was as follows: (i) the upper and lower phases were equilibrated at the predetermined temperature in separate containers; (ii) the tension value was zeroed with the plate submerged in the upper oil phase, to account for buoyancy of the plate; (iii) the heavier aqueous phase was poured into the sample cell; (iv) the sample cell was raised slowly upward to the plate using a vibration-free motorized labstand connected to a variable voltage source until the entire plate was submerged in the aqueous phase; (v) the lighter oil phase was poured on the aqueous phase; (vi) the sample cell was lowered until the plate was completely above the interface, breaking away from the interface and surrounded by the upper oil phase; and (vii) the sample cell was raised slowly until the lower phase was attracted to the plate forming a meniscus around the plate. The last step reproducibly placed the bottom edge of the plate exactly at the interface, and hence, error due to buoyancy of the plate or inconsistency of placement was eliminated. Thirty minutes was allowed for temperature equilibration, step (i). Step (ii) took approximately 5 minutes and all the following steps took seconds to perform, at most twenty seconds. The recorded zero point was step (vii), which is not the actual zero time of the experiments (actual zero time is step (iii)). However, the elapsed time between steps (iii) and (vii) is small, of the order of seconds, and the experiments were continued for at least one hour.

The pulling tension of the interface was continuously monitored using the Cahn 2000 Microbalance and recorded by the computer. Interfacial tension was calculated from the pulling tension, the contact angle and the plate parameter using the following equation:

$$\gamma = \frac{p}{2(L+d)COS\ \theta} \tag{4}$$

where γ is the interfacial tension; p, the interfacial pulling force; L, the width of the plate; d, the thickness of the plate; and θ, the interfacial contact angle. The interfacial contact angles were measured by the dynamic contact angle method reported previously (9). Interfacial tension measurements were carried out in triplicate and the mean values plotted.

Results and Discussion

Interfacial Rheology. The effect of time on the interfacial elasticity of β-casein (pH 7.0, 100 mM, 25°C) was determined at different bulk concentrations of 1, 2 and 4% w/v. The interfacial elasticity increased rapidly initially and then more gradually (Figure 1). This trend is in agreement with previous studies on the interfacial rheology of BSA and HI$_g$G. The initial rapid phase was considered to be an adsorption phase, including both migration of the molecules from the bulk to the subsurface and adsorption of molecules from the subsurface into the surface. The more gradual increase phase is thought to be due to molecular rearrangement as well as lateral interactions between adsorbed molecules and multilayer adsorption. The presence of denatured (rearranged) protein molecules at the interface may promote further molecular adsorption. Surface adsorption studies on albumin reported by Damodaran and Song (*10*) showed a similar trend of rapid increase in interfacial adsorption at the air/aqueous interface over the first 30 minutes followed by a gradual increase over a 15 hour study period. Graham and Phillips (*11*) have reported similar effects for albumin adsorption. The initial rapid phase is not as pronounced for β-casein as that of BSA (*4*), taking approximately one hour compared to approximately 10 minutes for BSA.

Interfacial elasticity increased with increasing bulk concentration (Figure 1). At a bulk concentration of 1 % w/v β-casein no surface elasticity was detected over the 80 minute period of investigation, however elasticity was observed at bulk concentrations of 2 and 4 % w/v. The relationship between bulk concentration and interfacial elasticity was not linear. However, there is not necessarily a linear relationship between bulk and interfacial protein concentration. At low bulk concentrations the β-casein molecules may be spread out at the interface with little opportunity for intermolecular interaction. At higher bulk concentrations the β-casein molecules are likely to be relatively closely packed resulting in intermolecular entanglement and therefore the formation of elastic interfacial films. Networks of entangled molecular chains will form with time resulting in the formation of increasingly elastic films.

The effect of pH on interfacial elasticity of 2% w/v β-casein solutions (100 mM, 25°C) was investigated (Figure 2). The trend of a rapid adsorption phase followed by a slower rearrangement phase was observed at the different pH values. The absolute values were pH dependent with the lowest values occurring at pH 4.3, the isoelectric pH (pI) of β-casein. β-casein is expected to be in its most compact coiled configuration at its pI value and therefore the chance for intermolecular overlap and entanglement between extended interfacial molecular chains is minimized, minimizing elasticity. The interfacial elasticities of BSA and HI$_g$G were at a minimum at their pI values (*4*). At pH 4.3 the interfacial elasticity of β-casein solutions were zero over the 80 minute study period. Elasticity was detected at pH 3.4 and ph 6.2, but was not as high as that at pH 7.0. The extent of ionization and hence the extent of protein folding varies with pH. The more extended the protein molecules the greater the opportunity for lateral interactions between neighboring interfacial molecules leading to elastic interfacial films.

The effect of ionic strength on the interfacial elasticity of 2% w/v β-casein solutions (100 mM) was investigated at 25°C (Figure 3). Interfacial elasticity increased with increase in ionic strength over the range 1 to 1000 mM. The presence of ions causes unfolding of protein molecular chains with subsequent increase in interfacial intermolecular entanglement and elasticity. The microions can also compete with the polar amino acid residues of the protein molecules for water, increasing the lipophilicity of the interfacial film, decreasing the interfacial area and hence changing the curvature of the film (*12*). All of which will increase the interfacial elasticity.

The Interfacial elasticity of 2% w/v β-casein solutions (pH 7.0, 100 mM) were

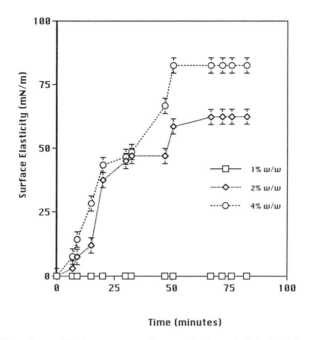

Figure 1. The effect of bulk concentration on the interfacial elasticity of β-casein (pH 7.0, 100 mM, 25°C).

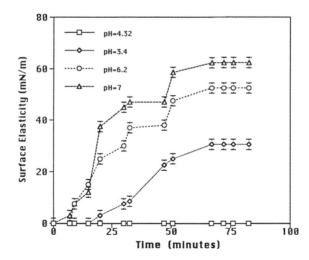

Figure 2. The effect of pH on the interfacial elasticity of β-casein solutions (2% w/v, 100 mM, 25°C).

investigated at 37°C and compared to the 25°C data (Figure 4). Interfacial elasticity decreased with increase in temperature. As temperature increases the molecules have more kinetic energy and hence the interfacial layer is more fluid and the probability of a molecule becoming adsorbed at the interface will decrease. This data is in agreement with bulk rheology, where increased fluidity is observed with increase in temperature. Continuous film formation studies were conducted on 2% w/v β-casein solutions (pH 7.0, 100 mM) at 25 and 37°C (Figure 5). Interfacial elasticity increased continuously over the 24 hour study period for samples held at both temperatures. The 25°C sample increased from 68 mN/m after one hour to 182 mN/m after 24 hours. A similar increase was observed for the 37°C sample which increased from approximately 45 mN/m at one hour to 122 mN/m at 24 hours. This increase in elasticity may be due to rearrangement (denaturation) of protein molecules, increase in lateral interactions between adjacent molecules and/or continued adsorption of molecules from the bulk solution.

The effect of aging of bulk molecules was investigated by measuring the interfacial elasticity of new interfacial films formed in 2% w/v β-casein solutions (pH 7.0, 100 mM) which had been stored at 25 and 37°C (Figures 6 and 7). Samples were tested after different aging periods. Samples were siphoned from under the interface of the aged solutions allowing the induction of a new interfacial film from bulk molecules. The initial adsorption phase kinetics were similar for each sample, however the plateau elasticity values were significantly different. The plateau values increased with sample age over the 72 hour study period, from 62 mN/m for the freshly prepared sample to 210 mN/m for the sample which had been aged for 72 hours (25°C). These data indicate that rearrangement (denaturation) of the bulk molecules must have taken place on aging and that the denatured molecules form more elastic interfacial films. Loss of structure may have produced a more expanded molecular shape, allowing increased lateral interactions between neighboring molecules.

The addition of guanidine hydrochloride to β-casein (2% w/v solutions) at pH 7.0, 100 mM and 25°C resulted in loss of interfacial elasticity (Figure 8). The higher the concentration of guanidine hydrochloride added the greater the loss in interfacial elasticity. Addition of 2% w/v guanidine hydrochloride resulted in no interfacial activity. Guanidine hydrochloride is a known protein denaturant (13), it interacts with the protein molecules and can result in aggregation and precipitation. The addition of urea, another known protein denaturant (14), to β-casein solutions (2% w/v at pH 7.0, 100 mM and 25°C) resulted in loss of interfacial elasticity and this effect was greater the higher the concentration of urea added (Figure 9). The addition of Tween 80 to 2% w/v β-casein solutions (pH 7.0, 100 mM) at 25°C also resulted in a loss of interfacial elasticity. The small surfactant molecules compete with the protein molecules for the interfacial space. The small surfactant molecules can diffuse to and be adsorbed at the interface faster than the larger protein molecules. The surfactant molecules are too small to have measurable interfacial elasticity.

Interfacial Tension. The interfacial tension of pH 7.0 and pH 4.3 β-casein solutions (100 mM, 25°C) decreased with time over a period of approximately 50 to 100 minutes and then reached equilibrium (Figures 10 and 11). The initial and equilibrium values decreased with increase in bulk concentration at both pH values. For all

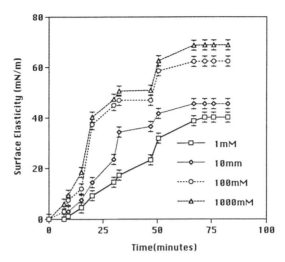

Figure 3. The effect of ionic strength on the interfacial elasticity of β-casein solutions (2% w/v, pH 7.0, 25°C).

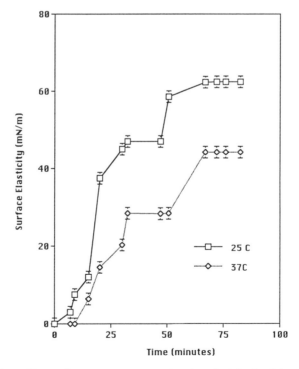

Figure 4. The effect of temperature on the interfacial elasticity of β-casein solutions (2% w/v, pH 7.0, 100 mM).

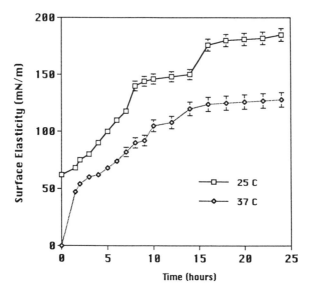

Figure 5. The effect of time on the interfacial elasticity of β-casein solutions (2% w/v, pH 7.0, 100 mM, 25°C and 37°C).

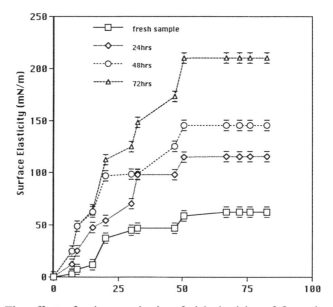

Figure 6. The effect of aging on the interfacial elasticity of β-casein solutions (2% w/v, pH 7.0, 100 mM, 25°C).

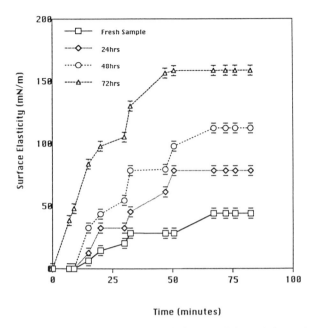

Time (minutes)

Figure 7. The effect of aging on the interfacial elasticity of β-casein solutions (2% w/v, pH 7.0, 100 mM, 37°C).

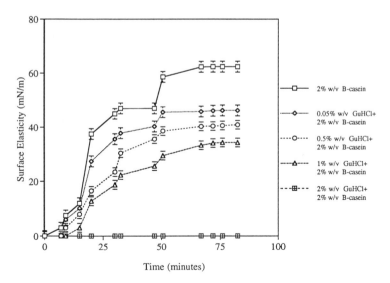

Figure 8. The effect of addition of Guanidine Hydrochloride on the interfacial elasticity of β-casein solutions (2% w/v, pH 7.0, 100 mM, 25°C).

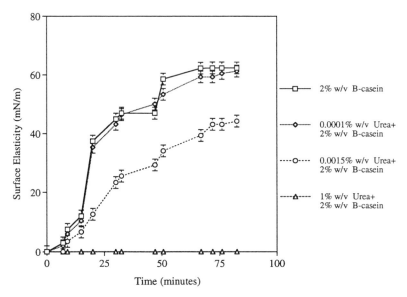

Figure 9. The effect of addition of Urea on the interfacial elasticity of β-casein solutions (2% w/v, pH 7.0, 100 mM, 25°C).

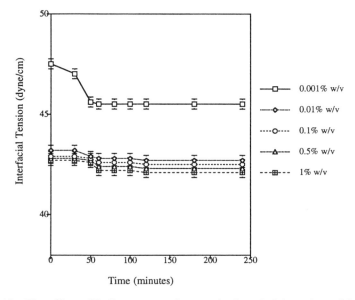

Figure 10. The effect of bulk concentration on the interfacial tension of β-casein (pH 7.0, 100 mM, 25°C).

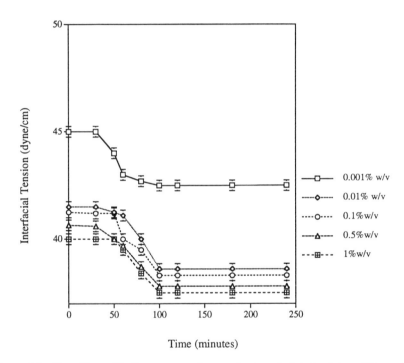

Figure 11. The effect of bulk concentration on the interfacial elasticity of β-casein (pH 4.3, 100 mM, 25°C).

concentrations studied the initial and equilibrium interfacial tension values were lower at pH 4.3 than at pH 7.0. This is the opposite trend to that observed for interfacial elasticity. Interfacial elasticity was at a minimum at the isoelectric pH whereas the reduction in interfacial tension is greater at the pI value (4.3) than at pH 7.0. Protein molecules are in their most compact and coiled configuration at their pI values and therefore more molecules are able to pack at the interface, maximizing the reduction in interfacial tension. The effect of pH on the interfacial tension of proteins has been reported by a number of authors.

Conclusions

The data reported here for β-casein are in general agreement with that reported previously for BSA and HI$_g$G. The interfacial elasticity values were time dependent for all systems studied and aging effects were observed over a 24 hour period. Interfacial rheology could also detect aging effects in bulk β-casein molecules. Denaturation of bulk molecules was evident as samples that had been aged for several days showed significantly higher interfacial elasticity values compared to freshly prepared samples. This technique appears to be useful for detecting configurational rearrangement (denaturation) effects in both interfacial and bulk molecules. These changes may be related to loss of biological activity of protein molecules.

The effects of pH and ionic strength on interfacial rheology were most interesting. Although interfacial adsorption of β-casein was at a maximum at the pI value, as was evident from the interfacial tension data the interfacial rheology values were lowest at this pH. The rheology data appears to be a result of decreased chain unfolding at this pH which minimizes lateral interactions between molecules. Similarly an increase in ionic strength which increases chain unfolding results in an increase in interfacial rheology. Competition between molecules for the interfacial space was evident on the addition of Tween 80 to β-casein solutions as no interfacial rheology was observed. The small surfactant molecules diffuse to and become adsorbed at the interface more rapidly than the larger protein molecules. The addition of guanidine hydrochloride and urea to β-casein solutions resulted in a loss of interfacial elasticity. Guanidine hydrochloride and urea bind to β-casein causing structural rearrangement, aggregation and precipitation.

Interfacial rheology is a measure of the strength of the interfacial film and can be related to the stability of emulsions prepared using protein as an emulsifier. High interfacial elasticity is likely to result in more stable emulsions.

Acknowledgments

The research was supported in part by a grant from Kraft general Foods. Ms. Sahin received a Fellowship for Postgraduate Study (Ph.D.) from the Turkish government.

Literature Cited

1.   Warburton, B.; In *Ions in Macromolecular and Biological Systems*, Everett, D. H. and Vincent, B. Eds.; Colston Symp. 29, Scientechnica: Bristol, U.K., 1981, pp 21.

2.   Kerr, H. R.; Warburton, B.; *Biorheology*, 1985, 16, 133.
3.   Moules, C.A.; Warburton, B.; In *Rheology of Food, Pharmaceuticals and Biological Materials with General Rheology*,   Carter, R.E. Ed.; Elsevier, Amsterdam, The Netherlands, 1990, pp 211.
4.   Burgess, D. J.; Longo, L.; Yoon, J. K.; *J. Parenteral Sci. and Technol.*, 1991, 45, 239.
5.   Burgess, D. J.; Yoon, J. K.; Sahin, N. O.; *J. Parenteral Sci. and Technol.*, 1992, 46, 150.
6.   Tanford, C.; *Formation of Micelles and Biological Membranes.*; 2nd Ed; Wiley; New York, N.Y., 1980.
7.   Charalambous, G.; Doxastakis, G.; In *Food Emulsifiers*, Doxastakis, G., Ed., Elsevier, amsterdam, The Netherlands. 1989, pp 1-50.
8.   Sherriff, M.; Warburton, B.; *Polymer*, 1974, 15,253.
9.   Burgess, D. J.; Yoon, J. K.; *Colloids and Surfaces: Biointerfaces*, 1993, 1, 283.
10. Damodaran, S.; Song, K. B.; *Biochimica et Biophysica Acta*, 1988, 954, 253.
11. Graham, D.E.; Phillips, M. C.; *J. Coll. Interface Sci.*, 1979, 70,403.
12. Kawashima, Y.; Hino, T.; Takenchi, H.; Niwa, T.; *Chem. Pharm. Bull.*, 1992,40,1240.
13. Lapanje, S.; In *Physicochemical Aspects of Protein Denaturation.* Lapanje, S., Ed., 1978, pp 89-101 and 272-281.

RECEIVED August 12, 1994

INDEXES

Author Index

Bates, F. S., 192
Berjano, M., 217
Blackburn, John C., 229
Brackman, Josephine C., 337
Bruinsma, Robijn F., 288
Burgess, Diane J., 380
Butler, P. D., 250
Candau, Sauveur J., 51
Cates, M. E., 32
Chen, Chih-Ming, 306
Chen, Liang Bin, 153
Chiruvolu, S., 86
Chou, Lu-Chien, 370
Danino, D., 105
Davis, H. T., 192
Diat, Olivier, 300
Doumaux, H. A., 192
Dubin, Paul L., 320
Engberts, Jan B. F. N., 337
Evans, D. F., 192
Gallegos, C., 217
Guerrero, A., 217
Hayter, J. B., 250
He, M., 192
Herb, Craig A., 153
Hill, R. M., 192
Hoffmann, H., 2
Hu, Y., 278
Idziak, Stefan H. J., 288
Imae, Toyoko, 140
Israelachvili, Jacob N., 288
Jamieson, A. M., 278
Kaler, Eric W., 120
Kaplun, A., 105
Khan, Saad A., 229
Kilpatrick, Peter K., 229
Koehler, Richard D., 120
Lequeux, François, 51

Li, Yingjie, 320
Liang, Keng S., 288
Lu, Bin, 370
Magid, L. J., 250
Muñoz, J., 217
Nallet, Frederic, 300
Naranjo, E., 86
Pearson, Dale S., 129
Phillies, G. D. J., 239
Rehage, H., 63
Ren, S. Z., 239
Richardson, C., 239
Rose, Gene D., 352
Rounds, Rhyta S., 260
Roux, Didier, 300
Safinya, Cyrus R., 288
Sahin, N. Ozlen, 380
Scriven, L. E., 192
Sheu, Eric Y., 167
Shields, Michael B., 167
Shikata, Toshiyuki, 129
Sirota, Eric B., 288
Smith, Bryan C., 370
Storm, David A., 167
Stott, J., 239
Streletzky, K., 239
Sun, Wei Mei, 153
Sushkin, N., 239
Talmon, Y., 105
Tayal, Akash, 229
Teot, Arthur S., 352
Wang, Shi-Quing, 278
Warr, Gregory G., 306
Wolff, Thomas, 181
Zakin, Jacques L., 370
Zana, R., 105
Zasadzinski, J. A., 86

Affiliation Index

Baker Performance Chemicals, 370
Becton Dickinson, 260
Case Western Reserve University, 239,278
Cavendish Laboratory, 32
Centre National de la Recherche Scientifique, 105,300
Dow Chemical Company, 352
Dow Corning Corporation, 192
Exxon Research and Engineering Company, 288
Helene Curtis, Inc., 153
Indiana University–Purdue University at Indianapolis, 320
International Paper, 370
Nagoya University, 140
North Carolina State University, 229
Oak Ridge National Laboratory, 250
Ohio State University, 370
Osaka University, 129

Technion-Israel Institute of Technology, 105
Technische Universität Dresden, 181
Texaco, Inc., 167
Universidad de Sevilla, 217
Université Louis Pasteur, 51
University of Bayreuth, 2
University of California–Los Angeles, 288
University of California–Santa Barbara, 86,129,288
University of Connecticut, 380
University of Delaware, 120
University of Essen, 63
University of Groningen, 337
University of Illinois at Chicago, 380
University of Minnesota, 192
University of South Dakota, 239
University of Sydney, 306
University of Tennessee, 250
Westvaco, 229
Worchester Polytechnic Institute, 239

Subject Index

A

Abelmoschus manihot root extract
  spinnability, 149
  viscoelasticity, 141
Alcohol chain-length effect on viscosity of sodium dodecyl sulfate micellar solutions
  alcohol partition coefficient, 169–172
  experimental procedure, 168–169
  long-chain alcohols, 178–179
  micellar growth vs. long-chain alcohol addition, 172–175
  short-chain alcohols, 175–178
Alcohol partition coefficient, theory, 169–172
Alcohols, micellar-state alteration, 167

Alkyl- and oleyldimethylamine oxides, aqueous NaCl solutions, spinnability, 148
n-Alkyl malonates, spherical and rodlike micelles, 342–345
n-Alkyl phosphates, spherical micelles, 339–341
Alkyldimethylamine oxides, rheological behavior, 13,14–16*f*
Amphiphilic systems, cryogenic temperature transmission electron microscopic investigations, 105–118
Anionic–cationic systems, critical packing parameter viscoelastic property relationship, 163,164*f*
Anionic–nonionic surfactant solutions, mixed, 153

Anionic surfactants
 dilute solutions, shear effects,
  265,266f,267,268f
 liquid crystals, shear effects, 270–274,276
Anionic surfactant–nonionic polymer
 complex, structure, 321–323
Anionic surfactant–polyanion complex,
 structure, 330
Aqueous NaCl solutions of alkyl- and
 oleyldimethylamine oxides,
 spinnability, 148
Aqueous sodium salicylate solutions of
 tetradecyl- and hexadecyltrimethylammonium salicylates, spinnability,
 145–147
Aqueous solutions
 cationic surfactants with aromatic
  counterions, spinnability, 140–141
 hexadecyl- and octadecyldimethylamine
  oxide, spinnability, 141
 tetradecyl- and hexadecyltrimethylammonium salicylates, spinnability,
  143–145
Aromatic solubilizates, photochemical
 transformations, 181

B

Bicontinuous cubic phases, ternary,
 steady-shear behavior, 306–316
Bipolar surfactants, description, 106
n,n-Bis(2-hydroxyethyl)dodecanamide–
 sodium dodecyl sulfate–water system,
 See Mixed surfactant solutions in
 normal micelle region

C

Carreau's model A parameters, sucrose
 laurate Newtonian aqueous systems,
 220,222–225
β-Casein, interfacial rheology, 381–392

Cationic–anionic systems, critical packing
 parameter–viscoelastic property
 relationship, 163,164f
Cationic micelles of dodecyldimethylamine
 oxide, water-soluble nonionic polymer–
 surfactant aggregate interactions,
 339,343f
Cationic surfactants
 dilute solutions, shear effects,
  261,263–266
 liquid crystals, shear effects,
  267,269f,270,271f
 quaternary ammonium salt, counterion
  structure effect on flow birefringence
  and drag-reduction behavior, 370–378
Cationic surfactant–nonionic polymer
 complex, structure, 324–325
Cetyltrimethylammonium bromide–
 sodium salicylate–water system
 micellar formation in flow, 278–286
 rheooptical behavior, 129–138
Cetyltrimethylammonium 3,5-dichloro-
 benzoate micelles, flexibility, 250–256
Cetyltrimethylammonium surfactants,
 production of viscoelastic solutions, 250
Cetyltrimethylammonium tosylate–sodium
 dodecylbenzenesulfonate rodlike
 micellar solutions, sheared, structure,
 See Structure of sheared cetyltrimethylammonium tosylate–sodium
 dodecylbenzenesulfonate rodlike
 micellar solutions
Chain-length distribution
 determination, 33
 wormlike micelles, 33
Characteristic length, nonionic
 trisiloxane surfactant solutions, 205
Charge density effect, viscoelastic
 surfactant solutions, 13,16f,17–20
Charged systems with excess salt,
 rheological behavior, 10–11,12f
Complex fluids
 distortion under flow, 288–289
 microstructure by electron microscopy,
 86–101

Complex fluids—*Continued*
  under flow and confinement advantages
    X-ray Couette shear cell, 297
    X-ray surface forces apparatus, 298
  experimental description, 289,290f
  importance, 289
  synchrotron X-ray scattering
    spectrometers, 289,290f,291
  X-ray Couette shear cell experiments,
    291–292,293f
  X-ray surface forces apparatus
    experiments, 292,294–297
Concentration, role in rheology of sucrose
  ester aqueous systems, 217–227
Concentration-driven microstructure
  progression
  large aggregates at midrange surfactant
    concentrations and above, 209–211
  small micelles at low surfactant
    concentrations, 209
Confinement, structure of complex fluids,
  288–298
Cosurfactants
  definition, 13
  rheological behavior, 13,14–16f
Counterion structure effect on flow
  birefringence and drag-reduction
  behavior of quaternary ammonium salt
  cationic surfactants
  applications, 371
  drag-reducing measurement procedure, 371
  experimental materials, 372
  flow birefringence measurement
    procedure, 371–372
  orientation effect, 375–378
  previous studies, 370–371
  substituent effect, 372–375
Critical molar ratio, definition, 329
Critical packing parameter
  determination in mixed surfactant
    solutions in normal micelle region,
    155–156
  vs. viscoelastic properties in mixed
    surfactant solutions in normal
    micelle region, 160,162f

Cross-link effect, entangled micellar
  solutions, 40–41
Cryogenic temperature transmission
  electron microscopy
  amphiphilic systems
    chemical formulas of amphiphiles,
      106,107f
    dimeric surfactants, 110–116
    distribution of distances between head
      groups, 106,108f
    experimental procedure, 106–110
    polyamphiphiles, 116–118
  complex fluid microstructure
    advantages, 88
    applications, 86–87
    contrast mechanisms, 94–101
    limitations, 88
    procedure, 88
    rapid freezing procedure, 88–92
    weak-phase object approximation,
      95–97,98–99f
  networks of entangled cylindrical
    micelles, 3–4
Crystal growth velocity, calculation, 91
Cubic phases
  characteristics, 306
  structural models, 306
  ternary bicontinuous, steady-shear
    behavior, 306–316
Cylindrical micelles
  description, 2
  equilibrium conformation in networks, 4

D

Detergency, requirements, 2
Didodecyldimethylammonium bromide
  cubic phase structure, 307–308
  phase diagrams, 307f
  steady-shear behavior, 308–316
Diffusion coefficients, viscoelastic
  surfactant solutions, 19,21f,22
Dilute aqueous micellar solutions, factors
  affecting rheological properties, 181

Dilute solutions, shear effects of surfactants, 261–268

Dimeric surfactants, cryogenic temperature transmission electron microscopy of amphiphilic systems, 110–116

Direct interaction effect, mutual and probe diffusion coefficients, 240–242

Distribution function, calculation, 73

Dodecyldimethylamine oxide, nonionic and cationic micelles, 339,343*f*

Drag reduction, use of viscoelastic surfactant solutions, 353

Drag-reduction behavior of quaternary ammonium salt cationic surfactants, counterion structure effect, 370–378

Dye assemblies, rheological properties, 63–83

Dye solutions, viscosities, 63

Dynamic modulus, entangled micellar solutions, 38–39

Dynamic structure, steady-shear behavior of ternary bicontinuous cubic phases, 314–315

Dynamical properties of wormlike micelles
Onsager regime, 60–61
salt concentration effect, 52–53
semidilute regime
scaling behavior to dilution, 56–59
stress relaxation function, 54–56
surfactant concentration effect, 52–53

E

Elastic force, definition, 304

Elasticity origin, wormlike micelles, 130

Electron microscopy
microstructure of complex fluids, 86–101
use to determine information about surfactant microstructure, 86

Elongational thickening, nonequilibrium micelle formation, 282–286

Entangled micellar solutions
cross-link effect, 40–41
dynamic modulus, 38–39

Entangled micellar solutions—*Continued*
linear viscoelastic spectra, 35–41
Maxwell time, 38
mean micellar length, 39–40
nonlinear viscoelasticity, 41–47
relaxation function, 35–38

Entanglement network, schematic sketch representation, 4,5*f*

Equilibrium size, calculation, 304

F

Fanning friction factor, definition, 354

First normal stress difference
calculation, 74–75
nonlinear viscoelasticity, 46

Flexibility of cetyltrimethylammonium 3,5-dichlorobenzoate micelles
apparatus, 251
bending rod plots, 251–255
experimental procedure, 251
future work, 257
quantitative estimates of persistence length, 255–256
techniques, 250–251

Flow, structure of complex fluids, 288–298

Flow birefringence behavior of quaternary ammonium salt cationic surfactants, counterion structure effect, 370–378

Flow curves, steady-shear behavior of ternary bicontinuous cubic phases, 314,315*f*

Flow property control in surfactant solutions via solubilizate photoreactions
future research, 190
light-scattering results, 186
micelle size effect, 186–187
microscopic solution structure effect, 187–188,189*f*
photorheological effects, 182–184
rheological results, 186–189
solubilizate effect, 187–188,189*f*,190
strategy, 184,186
water effect, 187

Freeze–fracture diagram, viscoelastic
    systems with yield value, 25,26f
Freeze–fracture replication of complex
    fluid microstructure
  advantages, 88
  applications, 86–87
  contrast mechanisms, 94–101
  etching, 92,94
  limitations, 88
  mass–thickness contrast, 97,100–101
  method, 92,93f
  procedure, 88
  rapid freezing procedure, 88–92
  replication, 94

G

Gemini surfactants, description, 106
Goodwin's parameters, sucrose laurate
    Newtonian aqueous systems, 219

H

Hexadecyl- and octadecyldimethylamine
    oxide, aqueous solutions,
    spinnability, 141
Hexadecyl- and tetradecyltrimethyl-
    ammonium salicylates
  aqueous sodium salicylate solutions,
    spinnability, 145–147
  spinnability, 143–145
High shear phase modeling, nonlinear
    viscoelasticity, 46–47
Hydrodynamic interaction effect, mutual
    and probe diffusion coefficients,
    240–242

I

Igepal 530–sodium dodecyl sulfate–water
    system, critical packing parameter–
    viscoelastic property relationship,
    162–163,164f
Incident electron wave function, 94–95

Intensity–intensity temporal correlation
    function, calculation, 240
Interfacial oscillatory technique,
    function, 380–381
Interfacial rheology
  β-casein solutions
    aging vs. interfacial elasticity,
        386,388–389f
    bulk concentration
      vs. interfacial elasticity, 384,385f
      vs. interfacial tension,
          386,390–391f,392
    experimental procedure, 381–383
    guanidine hydrochloride addition vs.
        interfacial elasticity, 386,389f
    ionic strength vs. interfacial
        elasticity, 384,387f
    pH
      vs. interfacial elasticity, 384,385f
      vs. interfacial tension,
          386,390–391f,392
    solution preparation procedure, 381
    temperature vs. interfacial elasticity,
        384,386,387f
    time vs. interfacial elasticity,
        383–386,388
    urea addition vs. interfacial
        elasticity, 386,390f
  surfactant solutions, information
      provided, 380
Interfacial tension, viscoelastic
    surfactant solutions, 22,23f
Ionic surfactant solutions, optical probe
    diffusion of surfactant solutions,
    243,245f,246
Ionic surfactants
  rheological behavior, 8,9f,10
  viscoelastic surfactant solutions, 8,9f,10
Isotropic solutions, microstructure
    progression patterns in surfactant–
    water binary systems, 212,213f

L

Light scattering, viscoelastic surfactant
    solutions, 19,21f

Linear charge density, definition, 326–327
Linear viscoelastic spectroscopy, entangled micellar solutions, 35–41
Liquid-crystalline phase identification, polycrystalline lyotrophic mesophases in cesium *n*-tetradecanoate–water system, 231–232,233*t*
Liquid crystals, shear effects on surfactants, 267–276
Loss modulus
  calculation, 70
  definition, 4,6
  polycrystalline lyotropic mesophase in cesium *n*-tetradecanoate–water system, 232–237
Lower consolute temperature, concentration effect, 192
Lyotropic lamellar phase
  composition, 301
  orientation diagram, 301
  orientation diagram vs. rheological properties, 300–301
  orientation states, 300
  shear effect, 300
Lyotropic surfactant mesophases
  applications, 229
  characteristics, 229–230

M

Magnitude of complex viscosity, calculation, 70
Mass–thickness contrast, freeze–fracture replication of complex fluid microstructure, 97,100–101
Maximum shear stress, relationship to shear modulus, 80
Maxwell material, description, 71–74
Maxwell time, entangled micellar solutions, 38
$(Me_3SiO)_2SiMe(CH_2)_3(OCH_2CH_2)_{12}OH$, microstructure and rheology, 192–193
Mean micellar length, entangled micellar solutions, 39–40

Metal thickness, calculation, 100–101
Micellar fusion and breakdown, kinetics model, 34–35
Micelles
  alcohol effect, 167
  shape prediction, 105–106
Micelle relaxation time determination, mixed surfactant solutions in normal micelle region, 155
Micelle size effect, flow property control in surfactant solutions via solubilizate photoreactions, 186–187
Microstructure
  complex fluids by electron microscopy
    contrast mechanisms in electron micrographs, 94–101
    etching procedure, 92,94
    experimental procedure, 86
    freeze–fracture technique, 92,93*f*
    mass–thickness contrast, 97,100–101
    rapid freezing procedure, 88–92
    replication procedure, 94
    weak-phase object approximation, 95–97,98–99*f*
  nonionic trisiloxane surfactant solutions, 192–213
  relationship to rheology of lyotropic lamellar phases, 300–304
Microstructure progression, concentration driven, *See* Concentration-driven microstructure progression
Microstructure progression patterns, surfactant–water binary systems of isotropic solutions, 212,213*f*
Mixed anionic–nonionic surfactant solutions, 153
Mixed surfactant solutions in normal micelle region
  critical packing parameter determination, 155–156
    vs. viscoelastic properties, 160,162*f*
  experimental procedure, 154–158
  generalization of critical packing parameter–viscoelastic property relationship to other systems, 162–165

Mixed surfactant solutions in normal
   micelle region–*Continued*
   micelle relaxation time, 155
   phase behavior, 158–159
   rheology, 159–160,161*f*
Mutual diffusion coefficient
   direct and hydrodynamic interaction
     effects, 240–242
   function, 240

N

NaCl solutions of alkyl- and
   oleyldimethylamine oxides, aqueous,
   spinnability, 148
Networks of entangled cylindrical
   micelles, cryogenic temperature
   transmission electron microscopy, 3–4
Neutron scattering, nonionic trisiloxane
   surfactant solutions, 202,203*f*
Newtonian flow systems, photorheological
   effects, 182–183
Non-Newtonian flow systems,
   photorheological effects, 184,185*f*
Nonequilibrium micelle formation in
   shear and elongational flow
   elongational thickening, 282–286
   experimental procedure, 279–280
   previous studies, 278–279
   shear thickening, 280–282,283*f*
   sodium salicylate concentration
     effect, 279
Nonionic micelles
   dodecyldimethylamine oxide, water-
     soluble nonionic polymer–surfactant
     aggregate interactions, 339,343*f*
   water-soluble nonionic polymer–surfactant
     aggregate interactions, 338–339
Nonionic polymer–anionic surfactant
   complex, structure, 321–323
Nonionic polymer–cationic surfactant
   complex, structure, 324–325
Nonionic polymer–nonionic surfactant
   complex, structure, 325

Nonionic polymer–surfactant aggregate
   interactions, water soluble, *See*
   Water-soluble nonionic polymer–
   surfactant aggregate interactions
Nonionic surfactant–nonionic polymer
   complex, structure, 325
Nonionic surfactant solutions, optical
   probe diffusion of surfactant
   solutions, 242–243,244*f*
Nonionic surfactants
   dilute solutions, shear effects, 267,268*f*
   liquid crystals, shear effects,
     274–275,276*f*
Nonionic trisiloxane surfactant solutions
   characteristic length, 205
   concentration-driven microstructure
     progression, 209–211
   experimental procedure, 194–196
   neutron scattering, 202,203*f*
   progression patterns, 212,213*f*
   pulse-field gradient spin-echo NMR
     spectroscopy, 206–209
   rheology, 196–202
   self-diffusion coefficients, 206–209
   small-angle X-ray scattering,
     202,204*f*,205–206
Nonlinear viscoelasticity
   first normal stress difference, 46
   high shear phase modeling, 46–47
   second moment of orientational
     distribution function, 42–43
   shear stress, 43–46
   stress tensor, 41–42
Nonsurfactant systems
   behavior studies, 6,8
   zero-shear viscosities, 6,8,9*f*
Normal stress effect, description, 75

O

Octadecyl- and hexadecyldimethylamine
   oxide, aqueous solutions,
   spinnability, 141

Oleyl- and alkyldimethylamine oxides, aqueous NaCl solutions, spinnability, 148

Onsager regime, dynamical properties of wormlike micelles, 60–61

Oppositely charged surfactant–polyelectrolyte complex, structure, 325–330

Optical anisotropy of flexible chains, calculation, 75

Optical probe diffusion of surfactant solutions
 experimental description, 239
 ionic surfactant solutions, 243,245f,246
 nonionic surfactant solutions, 242–243,244f
 principles, 239–242
 viscoelastic liquid effect, 246–247

Optical probe methods, applications, 239

P

Packing parameter
 applications, 106
 definition, 105

Persistence length
 calculation, 138,255
 influencing factors, 2
 quantitative estimation for cetyltrimethylammonium 3,5-dichlorobenzoate micelles, 255–256

Phase behavior, polycrystalline lyotrophic mesophases in cesium n-tetradecanoate–water system, 231,232f

Photochemical transformations, function, 181

Photoreactions of solubilizates, control of flow properties in surfactant solutions, 181–190

Photorheological effects
 function, 181
 Newtonian flow systems, 182–183
 non-Newtonian flow systems, 184,185f

Polyamphiphiles, cryogenic temperature transmission electron microscopy of amphiphilic systems, 116–118

Polyanion–anionic surfactant complex, structure, 330

Polycrystalline lyotrophic mesophases in cesium n-tetradecanoate–water system
 experimental procedure, 230–231
 liquid-crystalline phase identification, 231–232,233t
 phase behavior, 231,232f
 rheological measurements, 232–237

Polyelectrolyte–oppositely charged surfactant complex, structure, 325–330

Polymer–micelle association, description, 320

Polymer–surfactant complexes
 importance of understanding, 320
 nonionic polymer–anionic surfactant complex, 321–323
 nonionic polymer–cationic surfactant complex, 324–325
 nonionic polymer–nonionic surfactant complex, 325
 polyanion–anionic surfactant complex, 330
 polyelectrolyte–oppositely charged surfactant complex, 325–330
 previous studies, 320–321

Polysoaps, description, 106

Probe diffusion coefficient
 direct and hydrodynamic interaction effect, 240–242
 function, 240

Proteins, behavior measurement, 380–381

Pulse-field gradient spin-echo NMR spectroscopy, nonionic trisiloxane surfactant solutions, 206–209

Q

Quaternary ammonium salt cationic surfactants, counterion structure effect on flow birefringence and drag-reduction behavior, 370–378

R

Relaxation function, entangled micellar solutions, 35–38
Relaxation modulus, calculation, 66,70
Retraction, description, 42
Reynolds number, definition, 354
Rheological measurements, polycrystalline lyotrophic mesophases in cesium *n*-tetradecanoate–water system, 232–237
Rheological properties
  amphiphilic systems, 105–118
  dilute aqueous micellar solutions, influencing factors, 181
  rod-shaped micelles and dye assemblies influencing factors, 63–65
    linear rheological properties measurement, 66,70–74
      values, 76–78,79*f*
    nonlinear viscoelastic properties measurement, 74–76
      values, 78,80–83
    viscoelastic chromophore solutions, 65–66,67–69*f*
    viscoelastic surfactant solutions, 65,67*f*
Rheology
  β-casein solutions, interfacial, *See* Interfacial rheology of β-casein solutions
  nonionic trisiloxane surfactant solutions, 192–213
  polycrystalline lyotrophic mesophases in cesium *n*-tetradecanoate–water system, 229–237
  relationship to microstructure of lyotropic lamellar phases, 300–304
  sucrose ester aqueous systems
    entangled micellar systems, 220–225
    experimental procedure, 217–218
    lamellar liquid-crystalline phases, 220–225
    nonentangled micellar systems, 218–220,221*f*
    temperature effect, 225–227

Rheology control using viscoelastic surfactant solutions
  aqueous bleach formulation, 362,363*f*
  drag-reduction activity vs. surfactant concentration, 356,358,359*f*
  experimental materials, 355–356
  Rev Dust absorption of surfactant, 364,366*f*,367
  reversible thickening process via hydrocarbon breaking, 362,363*f*,364
  sample preparation, 355–356
  spray droplet size control, 358,361*f*,362
  test procedures, 354–355
  viscosity
    vs. breaking cycles, 364,365*f*
    vs. brine, 358,360*f*
    vs. surfactant concentration, 356,357*f*
    vs. temperature, 358,359*f*
Rheooptical behavior of wormlike micellar systems
  behavior
    under strong flow, 132,134
    under weak flow, 131–132,133*f*
  experimental procedure, 130–131
  persistence length, 138
  stress–optical coefficient, 135–136
  structural model, 136
  time dependence of birefringence and orientational angle, 136,137*f*,138
Rod-shaped micelles, rheological properties, 63–83
Rod-shaped particles, flow phenomena, 63–65
Rodlike micellar solutions, sheared cetyltrimethylammonium tosylate–sodium dodecylbenzenesulfonate, structure, *See* Structure of sheared cetyltrimethylammonium tosylate–sodium dodecylbenzenesulfonate rodlike micellar solutions
Rodlike micelles
  n-alkyl malonates, water-soluble nonionic polymer–surfactant aggregate interactions, 342–345

Rodlike micelles—*Continued*
  breakdown, water-soluble nonionic
    polymer–surfactant aggregate
    interactions, 345–348
  formation, 120

S

Salinity, viscoelastic system behavior
  effect, 22,25,26*f*
Salt concentration, role in dynamical
  properties of wormlike micelles, 52–53
Scaling behavior to dilution, dynamical
  properties of wormlike micelles, 56–59
Scattering
  limitations for microstructure
    determination, 87–88
  procedure, 87
Second moment of orientational
  distribution function, nonlinear
  viscoelasticity, 42–43
Self-diffusion coefficient, function, 240
Semidilute regime, dynamical properties
  of wormlike micelles, 54–59
Shampoo
  composition, 154
  importance of viscosity, 153–154
Shape of micelles, prediction, 105–106
Shear effects
  in surfactant solutions
    anionic surfactants
      in dilute solutions, 265,266*f*,267,268*f*
      in liquid crystals, 270–274,276
    cationic surfactants
      in dilute solutions, 261,263–266
      in liquid crystals, 267,269*f*,270,271*f*
    nonionic surfactants
      in dilute solutions, 267,268*f*
      in liquid crystals, 274–275,276*f*
    qualitative evaluation, 261,262*f*,264*f*
Shear modulus, definition, 3
Shear rate, step change in stress growth
  measurements, 260–261,262*f*
Shear rate relaxation at constant applied
  stress, definition, 316

Shear stress
  definition, 4,6
  nonlinear viscoelasticity, 43–46
Shear stress growth coefficient,
  calculation, 80,82–83*f*
Shear thickening, nonequilibrium micelle
  formation, 280–282,283*f*
Sinusoidal interaction potential, 313
Small-angle neutron scattering theory,
  structure of sheared cetyltrimethylammonium tosylate–sodium
  dodecylbenzenesulfonate rodlike
  micellar solutions, 121,122*f*,
  123–124
Small-angle X-ray scattering, nonionic
  trisiloxane surfactant solutions,
  202,204*f*,205–206
Sodium dodecylbenzenesulfonate–cetyltrimethylammonium tosylate rodlike
  micellar solutions, sheared, structure,
  See Structure of sheared cetyltrimethylammonium tosylate–sodium
  dodecylbenzenesulfonate rodlike
  micellar solutions
Sodium dodecyl sulfate–*n,n*-bis(2-hydroxy-
  ethyl)dodecanamide–water system, *See*
  Mixed surfactant solutions in normal
  micelle region
Sodium dodecyl sulfate–Igepal 530–water
  system, critical packing
  parameter–viscoelastic property
  relationship, 162–163,164*f*
Sodium dodecyl sulfate micellar solutions,
  alcohol chain-length effect on
  viscosity, 167–179
Sodium salicylate–cetyltrimethylammonium bromide–water system,
  micellar formation in flow,
  278–286
Sodium salicylate concentration effect,
  nonequilibrium micelle formation in
  shear and elongational flow, 279
Sodium salicylate solutions of tetradecyl-
  and hexadecyltrimethylammonium
  salicylates, aqueous, spinnability,
  145–147

Solubilizate effect, flow property control
in surfactant solutions via solubilizate
photoreactions, 187–188,189f,190
Specific viscosity, calculation, 218
Spherical micelles
n-alkyl malonates, water-soluble
nonionic polymer–surfactant aggregate
interactions, 342–345
n-alkyl phosphates, water-soluble
nonionic polymer–surfactant aggregate
interactions, 339–341
Spinnability
description, 140
viscoelastic surfactant solutions and
molecular assembly formation
aqueous NaCl solutions of alkyl- and
oleyldimethylamine oxides, 148
aqueous sodium salicylate solutions of
tetradecyl- and hexadecyltrimethylammonium salicylates, 145–147
aqueous solutions of tetradecyl- and
hexyltrimethylammonium salicylates,
143–145
comparison to polymers, 149
experimental description, 140
measurement, 141–143
types of viscoelasticity, 150–151
Spinnable liquids, properties, 140
Spurt effect, description, 44
Steady-shear behavior of ternary
bicontinuous cubic phases
compositions, 309
dynamic structure, 314–315
experimental procedure, 308–309
flow curves, 314,315f
stress relaxation, 315–316
yield stresses, 309–316
Steady-state value
first normal stress coefficient,
calculation, 75
flow birefringence, 74–75
Storage modulus
calculation, 70
definition, 4,6

Storage modulus—Continued
polycrystalline lyotropic mesophase in
cesium n-tetradecanoate–water
system, 232–237
Stress growth characteristics, examples,
261,262f
Stress optical coefficient, calculation,
75,135–136
Stress optical law, description, 74–75
Stress relaxation, steady-shear behavior
of ternary bicontinuous cubic phases,
315–316
Stress relaxation function, dynamical
properties of wormlike micelles,
54–56
Stress tensor, nonlinear viscoelasticity,
41–42
Structure
complex fluids under flow and
confinement, 288–298
sheared cetyltrimethylammonium
tosylate–sodium dodecylbenzenesulfonate rodlike micellar solutions
experimental procedure, 124
phase diagram of viscoelastic behavior,
120,122f
rotational diffusion coefficients vs.
shear rate, 136
scattering
vs. flow direction, 126,127f
vs. shear rate, 124,125f,126
small-angle scattering theory,
121,122f,123–124
Sublimation rate, calculation, 92,94
Sucrose ester aqueous systems, rheology,
217–227
Sucrose esters
applications, 217
formation of micellar and
liquid-crystalline phases, 217
Surfactants
rheological behavior, 260
shapes of micelles formed, 105
viscoelastic properties, 2

Surfactant aggregate–water-soluble
    polymer interactions, *See* Water-soluble
    nonionic polymer–surfactant aggregate
    interactions
Surfactant concentration, role in
    dynamical properties of wormlike
    micelles, 52–53
Surfactant microstructure, study
    techniques, 86–87
Surfactant obstruction factor, definition,
    206–207
Surfactant–polymer complexes, *See*
    Polymer–surfactant complexes
Surfactant solutions
  control of flow properties via
      solubilizate photoreactions, 181–190
  shear effects, 260–276
  viscoelastic, *See* Viscoelastic
      surfactant solutions
  viscosities, 63
  with globular micelles
    viscosity, 3
    volume fraction, 3
Surfactant–water binary systems of
    isotropic solutions, microstructure
    progression patterns, 212,213*f*

T

Temperature effect, rheology of sucrose
    ester aqueous systems, 225–227
Ternary bicontinuous cubic phases, steady
    shear behavior, 306–316
Tetradecyl- and hexadecyltrimethyl-
    ammonium salicylates
  aqueous solutions, spinnability, 143–145
  aqueous sodium solutions, spinnability,
      145–147
Theoretical modeling of viscoelastic
    phases
  linear viscoelastic spectra, 35–41
  nonlinear viscoelasticity, 41–47
  wormlike micelles, 32–35
Theoretical models, viscoelastic
    surfactant solutions, 25,27–29

Thixotropic loop measurements, qualitative
    evaluation of shear effects,
    261,262*f*,264*f*
Time-resolved fluorescence quenching of
    amphiphilic systems
  dimeric surfactants, 110–116
  experimental procedure, 109
  polyamphiphiles, 116–118
Triton X–100, optical probe diffusion of
    surfactant solutions, 242–243,244*f*

V

Viscoelastic liquid effect, optical probe
    diffusion of surfactant solutions,
    246–247
Viscoelastic properties, correlation with
    critical packing parameter for mixed
    surfactant solutions in normal micelle
    region, 153–165
Viscoelastic surfactant solutions
  alkyldimethylamine oxides and
      cosurfactants, 13,14–16*f*
  applications as rheology control agents,
      353–361
  association structures, 352–353
  charge density effect, 13,16*f*,17–20
  charged systems with excess salt,
      10–11,12*f*
  diffusion coefficients, 19,21*f*,22
  experimental objective, 352
  in normal micelle region,
      composition, 154
  interfacial tension, 22,23*f*
  ionic surfactants, 8,9*f*,10
  light scattering, 19,21*f*
  rheological behavior, 6,7*f*
  structures, 354,357*f*
  systems with yield value, 22–26
  theoretical models, 25,27–29
  use in drag reduction, 353
  viscosities, 19
  zwitterionic systems, 11,12*f*,14*f*

Viscoelastic systems
  surfactants used in preparation, 2–3
  with yield value
    freeze–fracture diagram, 25,26f
    rheological values, 22–26
Viscoelasticity, types, 150–151
Viscosity
  cetylpyridinium chloride vs. sodium
    salicylate concentration, 4,5f
  influencing factors, 3
  sodium dodecyl sulfate micellar
    solutions, alcohol chain-length
    effect, 167–179
  surfactant solutions with globular
    micelles, 3
Viscous force, definition, 304
Volume fraction, surfactant solutions with
  globular micelles, 3

W

Water effect, flow property control in
  surfactant solutions via solubilizate
  photoreactions, 187
Water-soluble nonionic polymer–
  surfactant aggregate interactions
  breakdown of rodlike micelles, 345–348
  experimental procedure, 338
  importance, 337
  nonionic and cationic micelles of
    dodecyldimethylamine oxide, 339,343f
  nonionic micelles, 338–339
  previous studies, 337–338
  spherical and rodlike micelles of
    n-alkyl malonates, 342–345
  spherical micelles of n-alkyl phosphates,
    339–341
  surfactant structure effect, 337
Weak-phase object approximation,
  cryogenic temperature transmission
  electron microscopy of complex
  fluid microstructure, 95–97,98–99f

Weissenberg effect, description, 75
Wormlike micellar systems, rheooptical
  behavior, 129–138
Wormlike micelles
  chain-length distribution, 33
  characteristics, 32–33
  dynamical properties, 51–61
  elasticity origin, 130
  reptation theory model application, 51–52
  reversibility of aggregation process,
    34–35
  surfactant vs. behavior, 129–130

X

X-ray Couette shear cell spectrometer
  advantages, 297
  apparatus, 289,290f,291
  experimental procedure, 291–292,293f
X-ray surface forces apparatus
  advantages, 298
  apparatus, 291
  experimental procedure, 292,294–297

Y

Yield stress
  definition, 313
  steady-shear behavior of ternary
    bicontinuous cubic phases, 309–316
Yield value, viscoelastic systems, 22–26

Z

Zero shear viscosity, definition, 3
Zwitterionic systems, rheological
  behavior, 11,12f,14f

Highlights from ACS Books

Good Laboratory Practice Standards: Applications for Field and Laboratory Studies
Edited by Willa Y. Garner, Maureen S. Barge, and James P. Ussary
ACS Professional Reference Book; 572 pp; clothbound ISBN 0–8412–2192–8

Silent Spring Revisited
Edited by Gino J. Marco, Robert M. Hollingworth, and William Durham
214 pp; clothbound ISBN 0–8412–0980–4; paperback ISBN 0–8412–0981–2

The Microkinetics of Heterogeneous Catalysis
By James A. Dumesic, Dale F. Rudd, Luis M. Aparicio, James E. Rekoske,
and Andrés A. Treviño
ACS Professional Reference Book; 316 pp; clothbound ISBN 0–8412–2214–2

Helping Your Child Learn Science
By Nancy Paulu with Margery Martin; Illustrated by Margaret Scott
58 pp; paperback ISBN 0–8412–2626–1

Handbook of Chemical Property Estimation Methods
By Warren J. Lyman, William F. Reehl, and David H. Rosenblatt
960 pp; clothbound ISBN 0–8412–1761–0

Understanding Chemical Patents: A Guide for the Inventor
By John T. Maynard and Howard M. Peters
184 pp; clothbound ISBN 0–8412–1997–4; paperback ISBN 0–8412–1998–2

Spectroscopy of Polymers
By Jack L. Koenig
ACS Professional Reference Book; 328 pp;
clothbound ISBN 0–8412–1904–4; paperback ISBN 0–8412–1924–9

Harnessing Biotechnology for the 21st Century
Edited by Michael R. Ladisch and Arindam Bose
Conference Proceedings Series; 612 pp;
clothbound ISBN 0–8412–2477–3

From Caveman to Chemist: Circumstances and Achievements
By Hugh W. Salzberg
300 pp; clothbound ISBN 0–8412–1786–6; paperback ISBN 0–8412–1787–4

The Green Flame: Surviving Government Secrecy
By Andrew Dequasie
300 pp; clothbound ISBN 0–8412–1857–9

For further information and a free catalog of ACS books, contact:
American Chemical Society
Distribution Office, Department 225
1155 16th Street, NW, Washington, DC 20036
Telephone 800–227–5558

Bestsellers from ACS Books

The ACS Style Guide: A Manual for Authors and Editors
Edited by Janet S. Dodd
264 pp; clothbound ISBN 0–8412–0917–0; paperback ISBN 0–8412–0943–X

Understanding Chemical Patents: A Guide for the Inventor
By John T. Maynard and Howard M. Peters
184 pp; clothbound ISBN 0–8412–1997–4; paperback ISBN 0–8412–1998–2

Chemical Activities (student and teacher editions)
By Christie L. Borgford and Lee R. Summerlin
330 pp; spiralbound ISBN 0–8412–1417–4; teacher ed. ISBN 0–8412–1416–6

Chemical Demonstrations: A Sourcebook for Teachers,
Volumes 1 and 2, Second Edition
Volume 1 by Lee R. Summerlin and James L. Ealy, Jr.;
Vol. 1, 198 pp; spiralbound ISBN 0–8412–1481–6;
Volume 2 by Lee R. Summerlin, Christie L. Borgford, and Julie B. Ealy
Vol. 2, 234 pp; spiralbound ISBN 0–8412–1535–9

Chemistry and Crime: From Sherlock Holmes to Today's Courtroom
Edited by Samuel M. Gerber
135 pp; clothbound ISBN 0–8412–0784–4; paperback ISBN 0–8412–0785–2

Writing the Laboratory Notebook
By Howard M. Kanare
145 pp; clothbound ISBN 0–8412–0906–5; paperback ISBN 0–8412–0933–2

Developing a Chemical Hygiene Plan
By Jay A. Young, Warren K. Kingsley, and George H. Wahl, Jr.
paperback ISBN 0–8412–1876–5

Introduction to Microwave Sample Preparation: Theory and Practice
Edited by H. M. Kingston and Lois B. Jassie
263 pp; clothbound ISBN 0–8412–1450–6

Principles of Environmental Sampling
Edited by Lawrence H. Keith
ACS Professional Reference Book; 458 pp;
clothbound ISBN 0–8412–1173–6; paperback ISBN 0–8412–1437–9

Biotechnology and Materials Science: Chemistry for the Future
Edited by Mary L. Good (Jacqueline K. Barton, Associate Editor)
135 pp; clothbound ISBN 0–8412–1472–7; paperback ISBN 0–8412–1473–5

For further information and a free catalog of ACS books, contact:
American Chemical Society
Distribution Office, Department 225
1155 16th Street, NW, Washington, DC 20036
Telephone 800–227–5558

RETURN TO: CHEMISTRY LIBRARY

100 Hildebrand Hall • 510-642-3753

LOAN PERIOD	1	2	1-MONTH USE	3
4		5		6

ALL BOOKS MAY BE RECALLED AFTER 7 DAYS.

Renewals may be requested by phone or, using GLADIS, type inv followed by your patron ID number.

DUE AS STAMPED BELOW.
